OXFORD MONOGRAPHS ON GEOLOGY AND
GEOPHYSICS NO. 16

OXFORD MONOGRAPHS ON GEOLOGY AND GEOPHYSICS

1. DeVerle P. Harris: *Mineral resources appraisal: mineral endowment, resources, and potential supply: concepts, methods, and cases*
2. J. J. Veevers (ed.): *Phanerozoic earth history of Australia*
3. Yang Zunyi, Wang Hongzhen, and Cheng Yuqi (eds): *The geology of China*
4. Lin-gun Liu and William A. Bassett: *Elements, oxides, and silicates: high-pressure phases with implications for the earth's interior*
5. Antoni Hoffman and Matthew H. Nitecki (eds): *Problematic fossil taxa*
6. S. Mahmood Naqvi and John J. W. Rogers: *Precambrian geology of India*
7. Chih-Pei Chang and T. N. Krishnamurti (eds): *Monsoon meteorology*
8. Zvi Ben-Avraham (ed.): *The evolutin of the Pacific Ocean margins*
9. Ian McDougall and T. Mark Harrison: *Geochronology and thermochronology by the* $^{40}Ar/^{39}Ar$ *method*
10. Walter C. Sweet: *The conodonta: morphology, taxonomy, paleoecology, and evolutionary history of a long-extinct animal phylum*
11. H. J. Melosh: *Impact cratering: a geologic process*
12. J. W. Cowie and M. D. Brasier (eds): *The Precambrian–Cambrian boundary*
13. C. S. Hutchison: *Geological evolution of southeast Asia*
14. Anthony J. Naldrett: *Magmatic sulfide deposits*
15. D. R. Prothero and R. M. Schoch (eds): *The evolution of persissodactyls*
16. Martin A. Menzies (ed.): *Continental mantle*

Continental Mantle

Edited by

MARTIN A. MENZIES
Department of Geology
Royal Holloway and Bedford New College
University of London

CLARENDON PRESS · OXFORD
1990

Oxford University Press, Walton Street, Oxford OX2 6DP
Oxford New York Toronto
Delhi Bombay Calcutta Madras Karachi
Petaling Jaya Singapore Hong Kong Tokyo
Nairobi Dar es Salaam Cape Town
Melbourne Auckland
and associated companies in
Berlin Ibadan

Oxford is a trade mark of Oxford University Press

Published in the United States
by Oxford University Press, New York

British Library Cataloguing in Publication Data
Continental mantle.
1. Earth. Mantle
I. Menzies, M. A. (Martin A.)
551.116
ISBN 0–19–854496–0

Library of Congress Cataloging in Publication Data
Continental mantle/edited by Martin A. Menzies.
(Oxford monographs on geology and geophysics; no. 16)
Includes bibliographical references.
1. Earth—Mantle. 2. Continents. I. Menzies, M. A. (Martin A.)
II. Series.
QE509.C677 1990 551.1'16—dc20 90–7223
ISBN 0–19–854496–0

Typeset by Cotswold Typesetting Limited, Gloucester, UK
Printed in Great Britain by
Courier International Ltd
Tiptree, Essex

Preface

At the turn of the century, seismic data constrained the upper mantle to being either peridotitic or eclogitic, and today geophysical features in the lower mantle can be linked to regions of Phanerozoic subduction and geochemical anomalies in volcanic rocks erupted at the Earth's surface (e.g. DUPAL). The lower mantle comprises zones of high and low seismic velocity that correspond respectively with areas of major downwelling around the Pacific basic and areas of major upwelling centred on the Pacific Ocean and the African continent. Continents tend to migrate away from these hot upwellings toward regions of cold downwelling thus ensuring their thermal stability over time. In addition, the development of subduction zones around the periphery of cratonic nucleii further protect the continents from thermal disturbances.

As seismic data suggest that hot spots are the surface expression of mantle upwelling, the presence of a 'crustal signature' in the Dupal hot spots perhaps points to a lower mantle reservoir fed by subduction. A major debate has arisen, however, around the correlation between seismic tomography, the geoid, and tectonic features and as to whether or not this indicates slab penetration through the upper mantle lower mantle boundary thus allowing a 'recycled component' to enter the lower mantle reservoir. Regardless of the eventual destiny of recycled lithosphere and the origin of plumes it is apparent that hot spot activity (intraplate processes) and slab subduction (subduction related processes) probably represent the two most important chemical fluxes on the Earth and that their influence is apparent in the evolution of the lithosphere.

Mantle upwelling and downwelling have been responsible for the chemical stratification of the subcrustal lithosphere and so to better understand the temporal evolution of the lithosphere one must consider evolution of the Earth as a whole. The lower mantle is presumed to be a pristine, relatively homogeneous and dense reservoir that may or may not have received significant amounts of recycled material from the upper mantle. The chemical character of the overlying upper mantle is hotly debated. Whilst from a geophysical standpoint one can argue for an enriched asthenosphere from a geochemical standpoint it is believed to be a chemically depleted, homogeneous, less dense reservoir whose origin involved partial melting/depletion in the first billion years of the Earth's existence. Melt extraction presumably contributed to the development of either oceanic or continental crust but possible recycling has meant that representatives of such early formed crust have not apparently survived at the Earth's surface.

The crust and mantle portions of the lithosphere are believed to have formed, respectively, from two chemically distinct components—a partial melt fraction highly enriched in incompatible elements, which formed the oxidized and hydrated crust and a partial melt residue markedly depleted in incompatible elements which formed the reduced and anhydrous lithospheric mantle that underplated the crust. It can be demonstrated that all types of mantle lithospheres were originally depleted protoliths and that post-Archaean lithospheric mantle evolved in a manner similar to modern oceanic mantle. In addition, post-stabilization contamination with recycled oceanic lithosphere, the asthenosphere, and possibly upwelling deep mantle has produced chemically stratified lithospheres.

Whilst in the Archaean proto-continental lithosphere may have been produced by thickening of Archaean 'oceanic' lithosphere or by underplating due to Archaean subduction, isotopic data preclude any link to modern oceanic mantle. This may indicate differing conditions in the Archaean particularly with regard to the temperature and chemical composition of the mantle. For example, the high temperatures of Archaean komatiitic magmas relative to the bulk of post-Archaean magmas substantiates the assertion that the Earth has cooled substantially in post-Archean times. If indeed higher temperatures existed in the Archaean then perhaps this explains why a considerable proportion of the Earth's crust was generated in the first few billion years of Earth's history. Much of this crust, however, was destroyed and that which survived was underplated by a lithospheric keel of refractory peridotite, the highly magnesian residua from the extraction of komatiites from Archaean mantle. This moderately reduced keel 'floated' on the upper mantle and, according to diamond inclusion work, reached thicknesses of 200 km in the first billion years of the Earth's existence. Indeed, it has been suggested that the chemical uniqueness of this keel was the key to the preservation of the overlying Archaean crust. The tendency of continents, however, to flee areas of hot mantle upwelling and to congregate on regions of cold mantle downwelling may have exposed the keel to intraplate and supra-subduction processes.

The presence of small volume melts in the asthenospheric substratum at the base of the Archaean keel has caused considerable modification of the continental lithosphere during upwelling of alkali carbonate melts now entrapped in diamonds. Furthermore, hydrous melting and metasomatism of the keel beneath the Archaean cratonic nucleii may have contributed respectively to the highly magnesian nature of the keel and its trace element enriched character. As a consequence of these post-stabilization processes the harzburgite protolith has been transformed into a repository for incompatible trace elements and the time-integrated inheritance of such elemental inhomogeneity is the present-day isotopic heterogeneity.

Asthenosphere–lithosphere contamination has been accompanied by post-Archaean plume–lithosphere contamination, again because of the propensity of continental plates to flee the site of mantle upwelling and to migrate to sites of mantle downwelling. The result is metasomatism of the lithosphere by upwelling mantle hot spots (e.g. Dupal). Much of this contamination by small volume melts from the asthenosphere and the lower mantle involves metasomatism either by interfacial energy-driven infiltration of carbonate and potassic melts or chemical transport of silicate melts and water/carbon dioxide rich fluids by crack propagation. Carbonate melts are potentially the most effective metasomatic agents because of their ability to infiltrate the peridotite matrix and to dissolve considerable amounts of silicate material. Much of the potassic metasomatism in the Archaean keel has occurred within the lower lithosphere (100–200 km) adjacent to the hot asthenospheric substratum. More sodic, carbon dioxide charged metasomatism is apparent closer to the crust–mantle boundary (50–75 km) perhaps the result of magma transfer to the surface. Although the Archaean lithospheric mantle has maintained a thickness of several hundred kilometers for some three billion years, its chemical integrity has been significantly eroded by post-stabilization processes.

At the time of the Archaean–Proterozoic boundary a change occurred in the character of the continental lithosphere. The post-Archaean crust was underplated by 'oceanic' lithosphere not the cratonic lithosphere so characteristic of sub-Archaean mantle. Accretion of the post-Archaean oceanic lithosphere appears to have been associated with upwelling of either deep or shallow mantle components and the movement of the continental plate toward regions of mantle downwelling may explain the later accretion of continental lithosphere with an identifiable subduction signature. The sub-Proterozoic mantle protolith is therefore markedly different from the Archaean mantle protolith with the implication that the Earth must have cooled down in the post-Archaean. Proterozoic (2.5–0.6 billion years) and Phanerozoic (<0.6 billion years) lithospheres are more clinopyroxene rich than their Archaean counterparts. They are essentially 'oceanic' in character with all the chemical and petrological attributes of mid-ocean ridge or active margin lithosphere. Post-Archaean lithospheres are chemically stratified due to transformation by subduction/recycling, plume–lithosphere, or asthenosphere–lithosphere interaction. The overall budget of incompatible elements, however, is less than in the case of Archaean lithosphere, and the resultant heterogeneity is not so well developed because post-Archaean lithosphere has been exposed to the hot asthenospheric substratum for half the time of Archaean lithosphere. Consequently isotopic heterogeneities in post-Archaean lithospheres tend to be significantly less than in Archaean lithospheres and generally match that of the modern ocean basins.

During the late Phanerozoic the major continental plates continued to migrate away from regions of hot upwelling toward areas of cold downwelling, where the American, Australian, and African plates are now located. Consequently, any lithospheres that interrupted major upwelling fluxes from the mantle became potential participants in continental flood volcanism. Archaean 'cratonic' lithosphere tended not to be activated by such processes, due probably to its chemical inertness. In addition, the physical dimensions of the keel may account for the fact that the majority of continental flood basalts were erupted through post-Archaean lithosphere.

Whilst it is generally agreed that the genesis of continental flood basalts has involved a depleted and enriched component the exact nature and origin of these components remains unresolved. Although traditionally these components are thought of as a MORB melt (depleted) contaminated with crust (enriched) such a simple model does not adequately explain the geochemical character of all flood basalts. Recently a flood basalt model has been proposed whereby mantle derived melts (depleted) 'mix' with melts from the lithospheric mantle (enriched) producing hybrids which in turn may encounter contamination with crust (enriched) en route to the surface. Despite post-Archaean lithospheric mantle being more fertile than Archaean lithospheric mantle (and potentially more capable of producing basaltic melts) the evidence for an extensive enriched reservoir beneath the post-Archaean crust is at best equivocal. Alternatively sub-lithospheric reservoirs may play an important role in the genesis of flood basalts. Whereas in some instances MORB mantle may provide a depleted component for flood basalts, in the case of the southern hemisphere flood basalt provinces associated with the break-up of Gondwanaland one might consider the upwelling Dupal hot spots as a viable source for the enriched isotopes and the subduction signature of so many of the southern hemisphere flood basalts.

The editor would like to thank all contributors for their immediate enthusiasm for the continental mantle and several colleagues are thanked for reviewing manuscripts, in particular, D. J. Blundell, F. R. Boyd, J. B. Dawson, D. McKenzie, and S. H. Richardson. The staff at the University of London libraries (RHBNC and Senate House) successfully located the most obscure references during production of this monograph. Christine, Keith, Kevin, and Sandra are thanked for being as responsive as ever to my demands for diagrams and photographs (for yesterday's deadline). Bruce Wilcock is thanked for initially planting the seed about this text and the staff at Oxford University Press are thanked for their assistance with its production.

Egham M.A.M.
1990

Contents

The plates for Chapter 6 fall between pp. 114 and 115

Contributors

Don L. Anderson

Seismological Laoratory, California Institute of Technology, Pasadena, California 91125, USA

Don R. Baker

Department of Geology, Rensselaer Polytechnic Institute, 101 Eight Street, Troy, New York 12181, USA

James M. Brenan

Department of Geology, Rensselaer Polytechnic Institute, 101 Eight Street, Troy, New York 12181, USA

Stephen E. Haggerty

Department of Geology, University of Massachusetts, Amherst, Massachusetts 01003, USA

Philip R. Kyle

Department of Geosciences, New Mexico Institute of Mining and Technology, Socorro, New Mexico, USA

T. Kurtis Kyser

Department of Geological Sciences, University of Saskatchewan, Saskatoon, Saskatchewan, S7N OWO, Canada

Martin A. Menzies

Department of Geology, University of London, Royal Holloway and Bedford New College, Egham, Surrey TW20 OEX, UK

Stephen H. Richardson

Department of Geochemistry, University of Cape Town, Rondebosch 7700, Cape Town, South Africa

E. Bruce Watson

Department of Geology, Rensselaer Polytechnic Institute, 101 Eight Street, Troy, New York 12181, USA

1

Geophysics of the continental mantle: an historical perspective

Don L. Anderson

1.1. INTRODUCTION

The continental lithosphere is a small fraction of the Earth and cannot be understood in isolation from the rest of the planet. It is at the top of the mantle and, therefore, is probably buoyant with respect to the bulk of the upper mantle in spite of the fact that it is also the coldest part of the mantle. The shield or cratonic lithosphere is ancient. Because of its long-term stability it is probably deficient in garnet, the densest abundant mantle mineral, and/or FeO relative to the rest of the upper mantle. It has high seismic velocities; this in addition to its anisotropy favours a high olivine, possibly forsterite-rich, content. High temperatures, high pyroxene contents, and high FeO contents all serve to decrease the seismic velocities. Since melts are enriched in CaO, Al_2O_3, and FeO compared to their source rocks the immediate implication of long-term buoyancy is that the cratonic lithosphere represents, at least in part, the refractory, buoyant residue, i.e. olivine and orthopyroxene, of mantle differentiation processes. In so far as it is related to the overlying continental crust, at least in age, it may be a product of the differentiation of the whole mantle rather than just the upper mantle. The argument here is that the continental crust is so enriched in the most incompatible elements that it would require efficient processing of the whole mantle in order to obtain the observed concentrations.

The term 'continental lithosphere' has often been used for that part of the mantle, underlying continental crust, which appears to have some generic relationship to the crust and which, presumably, accompanies the crust in the wanderings associated with continental drift. The term is sometimes restricted to shield or cratonic mantle. It is difficult to assign great thickness or permanence to any mantle keel under tectonically active continental regions. In fact, delamination is an attractive mechanism for causing uplift and high heat flow in mobile belts.

The term 'lithosphere' implies strength and the word is most accurately associated with that part of the crust and mantle that deforms elastically and reversibly to long sustained loads. This is in contrast to the 'asthenosphere' which flows under an applied load. Lithospheric thicknesses are determined from flexure profiles in regions of loading, unloading, or bending, as at deep sea trenches, mountain belts, or in regions of glacier unloading. Thicknesses so determined are generally less than 50 km for oceanic lithosphere and range up to 100+ km for continental lithosphere in stable cratons. Petrologists are often interested in the part of the mantle that is attached to the continental crust. It is not necessarily only the strength that matters in this context; it may be the long-term buoyancy and, therefore, the major element composition and mineralogy. The important parameters of the continental lithosphere are temperature, chemistry, mineralogy, stress level, and age. Lithosphere and asthenosphere may differ in several or all of these parameters (see also Menzies, Chapter 4, this volume).

In a homogeneous mantle the boundary between elastic and viscous behaviour depends on temperature, pressure, mineralogy, stress, and time. High stress and long time tend to thin the lithosphere as does the process of lithosphere overriding hot asthenosphere. For short-duration processes, such as are associated with post-glacial rebound, the lithosphere appears to be thick. If the load is larger, or longer duration, such as ocean island loading or bending at an island arc, the lithosphere appears thinner. The load effect is due to the non-Newtonian, or stress-dependent, rheology. The time effect is simply due to the fact that flow is a time-dependent process and appreciable flow might not be evident for some time. Lithosphere can be thinned if its base is heated or intruded. Since continental nuclei are ancient, the continental mantle root should flow away and thin with time if its existence were primarily due to rheology. If the continental lithosphere were simply a thermal boundary layer it would drag down the surface as it cooled. For a stationary continent, cooling by conduction, the thermal boundary layer thickens with time until it becomes unstable. Since continents move around over the mantle there may be a shear boundary layer between the base of the continental mantle, (or plate, keel, or root) and the top of the flowing mantle. In a homogeneous mantle the depth of the continental keel will depend on stress, temperature, and time. The alternative is that the thickness of the continental plate depends on the relative buoyancy or chemistry of the crust-plus-mantle system and the history of the system, and not primarily on temperature or rheology. The continental plate, the continental lithosphere, and the conductively cooled layer are different concepts and they need not correspond in thickness.

These and several other notions are often confused with, or used interchangeably with, lithosphere. In a convecting planet there are thermal boundary layers (TBL) at the surface and at any internal boundaries which separate regions that differ in intrinsic chemistry and density enough so that material cannot cross the boundary. Through these layers heat is transferred by conduction and this establishes a steep thermal gradient (10–20°C km^{-1}), much steeper than the nearly adiabatic gradient (0.3–0.5°C km^{-1}) associated with convection. The thickness of a conductive layer depends on thermal conductivity, age of the layer, or cooling time, and heat flow from below. Its long-term stability also depends on viscosity. Ancient continents, cooling by conduction alone and unperturbated by later thermal events, can be expected to have dense thermal boundary layers more than 200 km in thickness (see also Richardson, Chapter 3, this volume). At the bottom of the conductive layer, the temperature approaches the temperature of the deep mantle adiabat. The depth at which the material starts to flow may be less than the depth to which the cooling effect has penetrated but there should be a general increase of the thickness of the elastic or rheological lithosphere with an increase in the thickness of the thermal boundary layer, which, in simple cases, increases as the square root of time. The oceanic lithosphere becomes unstable with time because, as it cools, it becomes denser. The amount of density increase depends on the coefficient of thermal expansion and on temperature-dependent phase changes such as partial melting and the basalt–eclogite phase change. On the other hand, the cratonic lithosphere appears to be very stable for long periods of time. This is an argument for a chemically distinct layer beneath continents, one that remains buoyant even when cold (see also Haggerty, Chapter 5, this volume). The oceanic lithosphere is also not necessarily a pure TBL. The upper layers, at least, are chemically buoyant. The deeper portions may become denser with time because of temperature-induced phase changes in addition to thermal contraction. Younger parts of continents apparently do not have thick permanent lithospheres. Tectonic lithospheres appear to be vulnerable to melt penetration or delamination.

Stable shields exhibit very high mantle seismic velocities and deep, relatively modest, low-velocity zones (LVZ). The high-velocity layer is called the LID and is sometimes equated with the lithosphere. To avoid confusion it can be called the seismic lithosphere but there is no connotation as to strength except that low temperatures generally give both high seismic velocities and high viscosity. The high velocities and positive velocity gradients under shields generally extend to 130 to 165 km depth. Below this depth there is a rapid decrease of velocity, followed by a gradual increase. Between about 165 and 400 km the seismic velocities in shields, tectonic continental regions, and oceanic region all converge, the shield gradients being low or negative, possibly implying a high thermal gradient. Below some 400 km the fastest regions, on average, are beneath old oceans, subduction zones, and regions where oceanic lithosphere was overridden by continental dispersal in the past 100 Ma. The velocity gradient below shields, starting at a depth of about 200 km, is appropriate for a self-compresssed material following the 1400°C adiabat (Anderson and Bass 1984). Thus, the mantle at a depth of some 200 km, under the stable interiors of continents, apparently behaves as convecting mantle rather than as a thermal boundary layer or a rigid lithosphere and there is no evidence that the deeper mantle is fixed to the continental plate. If the mantle under shields, below about 150 km, were much hotter or if it contained more water it would be above the mantle solidus and the seismic velocities would be much lower. The probable reason for a shallower and more pronounced low-velocity zone under tectonic and oceanic regions, regions of high heat flow, is the presence of a melt phase due to slightly higher temperatures. A partially molten mantle is implied between depths of the order of 50–100 km and at least 300 km under oceans and tectonic regions. The amount of melting is unknown since the amount by which a melt phase decreases the seismic velocity depends on the shapes of the melt pockets (see also Watson *et al.*, Chapter 6, this volume). Thin-grain boundary melt films are more effective than equidimensional melt pockets such as spheres, or elongate shapes such as needles or tubes. All we know for sure is that the seismic velocities in the lowest-velocity regions of the upper mantle are lower than the high-frequency laboratory velocities of any combination of the abundant mantle minerals (olivine, pyroxene, and garnet) at reasonable temperatures. At high temperature, clinopyroxene and garnet are replaced by a low-rigidity melt phase. It should be repeated that the subshield mantle need be only slightly hotter in order to be above the solidus and, therefore, to have very low seismic velocities. There is little disagreement about the sign of the temperature difference between the mantles of ancient shields and other younger terrains. The velocity contrasts between the various kinds of mantles, to depths of about 390 km, are likely to reflect the lower temperatures and absence of a melt phase under stable continental interiors. The chemically distinct, and intrinsically buoyant, part of the stable continental mantle need extend no deeper than 150 km to be consistent with the geophysical and petrological data. High stress and high dislocation density can also contribute to a lowering of seismic velocities under subsolidus conditions.

Why is subshield mantle colder than suboceanic mantle? There are several reasons. First of all, the shield upper mantle has been cooling by conduction longer than oceanic mantle or the mantle under young continental crust. In fact, as the continent plus its mantle keel moves around it can cool the underlying mantle. When continents override oceanic lithosphere the cold subducted material replaces hot material and also cools the mantle into which it subducts. Continents tend to collect over cold downwelling parts of the mantle and to move away from hot upwelling parts of the mantle. The presence of cold mantle, to great depth, under the stable interiors of continents does not imply that continents have correspondingly thick roots or keels or tectospheres which move around with the continents. The actual roots of continents are probably characterized by those regions which have much higher seismic velocities than average, say 5 per cent, and which occur above the depths at which the shield shear velocities start to drop, and converge toward the velocities elsewhere. This restricts the thickness of the mantle roots under continents to be less than about 120 km, giving a total thickness of about 150 km for the continental crust–mantle system (Fig. 1.1).

The hot upwellings which show up in mantle tomographic studies under oceanic basins and other regions of extension can be traced to at least 400 km depth (Nataf *et al.* 1984). These may be tapping a chemically different, basalt-rich region of the mantle. If so, the shallow sublithospheric mantle under oceanic and tectonic regions may indeed differ from subshield mantle, where the upwelling material, at least today, is generally absent. At greater depth, if the basalt source region is global, the mantle may be chemically similar everywhere. However, the main difference in seismic velocity from place to place, above 400 km depth, must be due to variations in melt content and other temperature-induced phase changes and, possibly, changes in crystal orientation and dislocation density. Changes in composition have relatively little effect on seismic velocity. For example, both eclogite and peridotite can match the seismic velocities in the upper 400 km of the mantle (Anderson and Bass 1984, 1986; see also Menzies, Chapter 2, this volume). This is because clinopyroxene and garnet (the eclogite minerals) and olivine and orthopyroxene (the peridotite minerals) bracket the observed seismic velocities of the upper mantle except where a melt content is implied by extremely low seismic velocities (Table 1.2). In these cases a partially molten rock, either eclogite or peridotite, can satisfy the seismic velocities. On the other hand an olivine–eclogite (piclogite) composition is preferred for the transition region (Duffy and Anderson 1989).

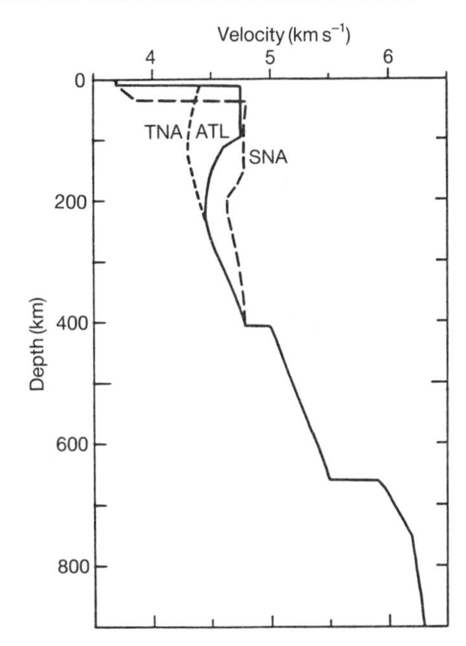

Fig. 1.1. Shear velocity (SH) profiles beneath shield areas of North America (SNA) tectonic areas of western North America (TNA), and the western Atlantic Ocean off of North America (ATL) (after Grand and Helmberger 1984).

1.2. EARLY VIEWS OF THE UPPER MANTLE

Early body wave results were plagued by problems associated with low-velocity (LVZ) zones and discontinuities which frustrate using travel times of first arriving phases. In fact, most of the discrepancy between models due to Jeffreys and Gutenberg was in the depth interval where we now know that velocity reversals and discontinuities occur, and also where we know that lateral variations, and, possibly, anisotropy, are the greatest. Surface waves provided an alternative method of studying the upper mantle but their full utilization required long-period instrumentation, computers, inversion techniques, and large earthquakes. These all came together in the early 1960s when long-period instrumentation was being installed around the world and when several very large earthquakes stimulated an interest in free oscillations of the Earth and the related extra long period surface waves, waves which sample deeply into the mantle. The early surface wave studies resolved some of the problems associated with the radial structure of the mantle and also gave the first reliable indications of lateral variations in the upper 400 km of the mantle. They also suggested the presence of a new complication, anisotropy.

Gutenberg (1959) believed that there was no definite indication of any first-order discontinuity in the mantle but that there were regions of rapid increase of velocity in the upper 1000 km. The transition region, a region of high velocity gradient, was assigned the depth range of 200 to 950 km by Gutenberg and 410 to 1000 km by Bullen (1953). Gutenberg, Jeffreys, and Lehmann repeatedly investigated the so-called '20°-discontinuity', the break in the travel time curve of seismic waves at an angular distance of about 20°. Gutenberg (1959) preferred to place the break at 15° where the amplitudes of seismic waves suddenly increase. The term '20°-discontinuity' has been frequently applied to the corresponding discontinuity in the Earth at a depth of several hundred or more kilometers. Gutenberg considered the '20° (15°) discontinuity' to be the outermost limit of the shadow zone caused by an LVZ centred at 100–150 km depth. Jeffreys attributed it to a marked increase in the rate of increase of velocity with depth (second-order discontinuity). He variously placed it at 481 km (1936), 413 km (1939), and, in 1952, at about 500 km. Still later he determined (Jeffreys 1958) that the travel times of P-waves 'imply that at the transition there must be a considerable discontinuity of velocity gradient but only a small one of velocity'. He assigned a depth of 200 km. In his velocity models he always arranged for there to be a break in slope of the travel time curve but only a minor cusp or triplication. Later work showed, however, that there were two breaks in the travel time curves, each associated with a triplication and later arrivals. Lehmann (1962) consistently found evidence for a first-order discontinuity at about 220 km depth. The significance of this boundary, particularly with regard to the base of the continental plate, is discussed in Anderson (1979). Larger mantle discontinuities were later found near depths of 400 km and 650 km (Anderson and Toksöz 1963; Niazi and Anderson 1965; Johnson 1967).

Trains of long-period surface wave repeatedly circle the Earth after great earthquakes. These waves provide world-wide averages of upper mantle structure and require no knowledge of the source spectrum or even location. Because of the vast distances involved in the transits of these waves, and the cancellation of source characteristics, the velocities of the waves at various periods (the dispersion relation) can be determined with high precision. Not only that but both the phase velocity and the group velocity (a derivative of the phase velocity) can be determined and this permits small inflections in the phase velocity curve to be determined. Ultimately, this high precision permitted subtleties of the radial velocity distribution to be determined and differences between different great circle paths to be discovered.

The first detailed analyses of the world-encircling waves were performed by Sato (1955) and Ewing and Press (1954). Dorman *et al.* (1960) computed the dispersion expected for a variety of models and concluded that low velocity was a necessary feature of the average upper mantle and that the structure proposed by Gutenberg was far superior to the Jeffreys–Bullen (JB) model, the standard at that time. This study led to the widespread acceptance of a global low-velocity channel in the upper mantle and to a discrediting of the monotonic JB model. Ironically, the long-period data used by Dorman *et al.* (1960) can equally be satisfied by an anisotropic upper mantle with no need for an LVZ (Anderson 1966; Dziewonski and Anderson 1981). Much shorter period data is required to resolve a shallow velocity reversal with depth (Anderson and Dziewonski 1982). A high-velocity lid is a feature of most current oceanic and continental models of the upper mantle, including shield regions. The LVZ is often equated with the asthenosphere but this is not necessarily valid. Likewise, the seismic LID is not necessarily the same as the lithosphere or elastic plate.

Most of the disagreements between early seismological studies, such as those of Gutenberg and Jeffreys, were in the upper 500 km of the mantle. Surface waves showed, in fact, that there were regional, or lateral, changes in velocity down to about this depth (Dorman *et al.* 1960; Anderson 1966). The situation as of 1963 was summarized by Nuttli (1963).

1. There are lateral or horizontal variations of P- and S-wave velocity in the upper mantle.
2. There is a world-encircling low-velocity channel for

shear waves in the upper mantle, but there are lateral variations in the velocity in this channel and possibly also in the depth and thickness of the channel.

3. There is no convincing evidence of the existence of a low-velocity channel for compressional waves, except in tectonically active regions such as Japan, Andes, California–Nevada, etc.

Anderson and Toksöz (1963) determined a high-precision Love wave dispersion curve to ultralong periods (>600 s). They showed that standard models did not satisfy this new data over the entire period range. In order to obtain a fit they needed to decrease the velocities between 220 and 360 km and between 500 and 700 km, resulting in major seismic discontinuities near 360 and 700 km. Smoother models without discontinuities or major inflections near these depths did not satisfy inflections in the phase velocity data nor the levels of the group velocity curves. The rapid increase in velocity between some 350 and 500 km depth required to satisfy the dispersion data was much greater than proposed by Jeffreys to explain the '20° discontinuity' and resulted in cusps or triplications in the travel time curves.

Accurate surface wave data are surprisingly sensitive to details of mantle structure provided that data is available over a broad period range (Anderson 1964). The use of higher modes in addition to fundamental modes increases the resolution. The results of high-precision surface wave studies as of 1964 were summarized as follows.

1. The low-velocity zone is a widespread phenomena although it may be locally absent.
2. The velocity starts to decrease at a depth of about 20 km below the base of the crust.
3. The LVZ, at least under oceans, extends to some 350 to 400 km depth.
4. There is an extremely rapid increase in velocity between some 350 and 500 km depth, a much greater increase than proposed by Jeffreys.
5. The LVZ is also a zone of high attenuation. The attentuation of shear waves in the LVZ is an order of magnitude higher than in the lower mantle.
6. Theoretical travel times of shear waves exhibit a shadow zone which ends at 15°, a minor travel time discontinuity at 18°, and a more pronounced discontinuity at 26.4°. (These observations are consistent with an upper mantle LVZ and two discontinuities.)
7. The structure which satisfied Love-wave data gives theoretical Rayleigh-wave velocities appreciably above the data. This discrepancy can be removed by a 10 per cent anisotropy in the LVZ.

The refinement of mantle models to satisfy the increasingly accurate surface wave dispersion data was greatly facilitated by an analytical method to compute structural partial derivative curves (Anderson 1964; Archambeau and Anderson 1964). Progress was rapid from this point onwards, including the determination of radial and lateral variations.

Model CIT11 (Anderson and Toksöz 1963) was a result of several iterations from earlier models of Kovach and Anderson (1962). CIT11 has an LVZ which is more pronounced than the corresponding feature of the Gutenberg model and a rapid change in properties near 400 km which is more pronounced than the corresponding feature of the Jeffereys model. A new feature is a major discontinuity near 700 km. This model also satisfied higher-mode data (Kovach and Anderson 1964).

The presence of two major seismic discontinuities in the upper mantle was confirmed by a direct determination of apparent velocities across a seismic array in western North America (Niazi and Anderson 1965). Two relatively abrupt changes in apparent velocity were detected at arc distances from seismic events of 17° and 24°. These are most readily interpreted as two second-order discontinuities in the region of the mantle above 1000 km. Making assumptions about the nature of the shallow mantle Niazi and Anderson proposed mean depths of 360 and 660 km for these discontinuities. These are now referred to as the '400 km' and '650 or 670 km' discontinuities.

The 400 km 'discontinuity' was previously thought to be the start of a broad transition region extending to about 1000 km depth and was attributed to a second-order transition from olivine to a cubic spinel structure. As seismic resolution improved it became clear that the transition region was actually a region of moderate seismic velocity gradient containing major discontinuities at 400 and 650 km (e.g. Johnson 1967). Thermochemical modelling showed that phase transitions in olivine, including the olivine–spinel and spinel–post-spinel transformations, provided a good first-order fit to the seismic profiles (Anderson 1967), although a hint of a chemical variation with depth was evident.

As seismic resolution continued to improve and elastic data on the spinel forms of olivine became available it became apparent that the olivine–spinel interpretation of the 400 km discontinuity was not completely satisfactory (Bass and Anderson 1984; Duffy and Anderson 1989). The main problem is that the velocity increase associated with α-olivine to β-spinel transition is much greater than the jump of velocity at 400 km. Olivine, of course, is not the only component in the mantle. The second most abundant mineral in upper mantle peridotites is orthopyroxene, (Mg, Fe)SiO_3. This also undergoes a phase change near 400 km, although probably spread out over at least 50 km, that involves a large jump in compressional velocity as pyroxene transforms to a garnet-like structure, majorite. Garnet and, possibly, clinopyroxene do not undergo pressure-induced phase changes until about 500 km and their presence will decrease the size of the velocity jump near 400 km. Olivine and orthopyroxene probably constitute less than 50 per cent of the mantle between 400 and 650 km (as the spinel and garnet/majorite phases).

1.3. UPPER MANTLE BENEATH CONTINENTS

Until about 30 years ago the main preoccupations of seismologists were in the determination of lateral variations in crustal structure and in the determination of the average structure of the mantle and core. Hardly anyone considered that the radial structure of the mantle might vary from place to place.

By 1962 crustal seismic refraction surveys had shown that the P velocity just below Moho ranges from about 7.5 km s^{-1} beneath the centre of active tectonic regions to 8.6 km s^{-1} beneath certain stable shields. In the United States the velocity varies from 8.2 km s^{-1} under the midcontinent to 7.8 km s^{-1} under some of the mountainous regions in the west (Herrin and Taggart 1962). The deeper structure is more difficult to determine but the general tendency of low velocity under tectonic regions and high velocity under shield regions was known to extend at least to 400 km depth (Anderson 1966). Lehmann (1964) used nuclear explosions to show that the P velocity structure of the upper mantle differs strikingly between the western tectonic regions and the eastern region of the United States.

The first detailed velocity structure beneath a shield was determined by Brune and Dorman (1963). Their model CANDSD, for the Canadian Shield, had upper mantle velocities higher than found anywhere else. Ibraham and Nuttli (1967), Kovach and Robinson (1969), Amderson and Julian (1969), Helmberger and Engen (1974), and Romanowicz (1979) reported several different upper mantle structures beneath the North American continent, suggesting the existence of lateral heterogeneity within a single continental area, which might reflect different evolutionary stages of a craton. Nakada and Hashizume (1983), using higher-mode surface waves, concluded that the upper 200 km of the Canadian Shield had distinctly high seismic velocities but that between 200 and 400 km the velocities were similar to other continental and global mantle models. The variations across the North American continent are primarily found above 200 km.

Anderson and Kovach (1964) studied the travel times of multiply reflected ScS waves and concluded that the shear velocities averaged over the crust and mantle were about the same in oceanic and continental regions. This indicated to them that the continents are more likely to have been derived from the underlying mantle than 'to have been superimposed on the mantle either from external sources or from adjacent regions of the mantle. The low-density, low-velocity crustal material was envisaged as a differentiate of mantle material which leaves a higher density, higher velocity residue. The upper mantle under oceans would therefore be less dense and have a lower velocity than the upper mantle under continents.' Actually, the removal of crustal material from the mantle can either increase or decrease the velocity and density, depending on the state of the crustal component while in the mantle (basaltic, eclogitic, or melt).

The great Alaskan earthquake of 28 March 1964 provided an excellent set of long-period surface wave data that made it possible to discuss deep lateral variations of the mantle. Toksöz and Anderson (1966) analysed the surface waves from this earthquake at three stations and combined them with data from four other great events to determine travel times along eight great circle paths. This was enough data to do a crude tomographic analysis. Average great-circle phase velocities are markedly affected by the character of the continental fraction of the path. Shield areas raise the average phase velocity; tectonic and mountainous areas have the opposite effect. The tectonic–shield distinction is as important as the more obvious continental–oceanic distinction. Inversion of this data suggested that tectonic regions had slower average shear velocities to depths of the order of 400 km. Clearly, there is more than one kind of continental mantle.

By combining data from different composite great-circle paths it was possible to estimate dispersion curves appropriate to a 'pure path'. The composite-path data was interpreted in terms of segments designated oceanic, shield, and mountain-tectonic. Toksöz and Anderson (1966) noted that great-circle phase velocity data was not ordered, as expected, with phase velocity increasing with the oceanic fraction of the path. Clearly, the percentage of ocean, although it is the major part of each great circle path, does not, by itself, control the ordering of the phase velocity curves. They concluded that all continents were not equal, that the distinction between stable shields and mountain-tectonic areas is as important and fundamental as the ocean–continent distinction, and that this distinction must extend to great depths in the mantle.

Toksöz *et al.* (1967) in a detailed discussion of lateral inhomogeneities in the mantle concluded that the upper mantle under shields is cold and represents regions where heat sources and acidic components have been differentiated into a crust. Temperatures under oceanic regions were concluded to be about 300°C hotter at a depth of 100 km (100°C hotter at 200 km), and the suboceanic mantle is less differentiated, containing 'more radioactive heat sources and associated lighter components to account for high temperatures and lower velocities'. Tectonic regions were concluded to be in an in-between stage where the differentiations may still be in process. These conclusions were based primarily on the lateral variation of seismic velocities.

It should be recalled that until about 1968 most Earth scientists regarded the Earth as a more-or-less static body. The plate tectonic revolution was just beginning and mantle convection was not a well developed, or widely believed, concept. We now tend to attribute temperature variations in the mantle to convection, with the hotter portions representing upwellings. The surface plates are, to some extent, cold thermal boundary layers but they also include buoyant crustal, and probably upper mantle, material. In some extreme versions of 'pure plate tectonics' the variation of elevation, bathymetry, gravity, and heat flow is entirely a consequence of cooling, with essentially no contribution from the deeper mantle. We now suspect that there are significant lateral variations in both the temperature and mineralogy (including melt content) of the plate and the asthenosphere. Lateral variations in physical properties can be quite large because of temperature-induced phase changes (Anderson 1987).

Knopoff (1972) used surface waves to delineate five different structural provinces with quite different seismic characteristics: shields, aseismic continental platforms, rifts, mountains, and ocean basins. Rifts essentially have no LID; LVZ material appears to rise directly to the base of the crust. Some continental aseismic regions have a thick LID and a large drop of velocity near 80 km into a pronounced LVZ. Shields have the highest mantle shear velocities and mountains have velocity–depth curves parallel to shields at about 0.1 km s^{-1} less. Oceans have a thin high-velocity LID and a very low-velocity LVZ. These general conclusions have been confirmed and extended by more recent work.

1.4. HIGH-RESOLUTION UPPER MANTLE MODELS

Don Helmberger and his group at Caltech have developed and refined waveform modelling techniques which provide more detail about mantle velocity variations than other techniques. They have derived compressional and shear-wave velocity profiles for many tectonic regions including shield, active tectonic, oceanic ridge, old ocean, and so on. Some of their results are shown in Figs 1.1 and 1.2.

Most of the lateral variation is in the upper 150 km and amounts to about 10 per cent. Velocities converge rapidly below this depth and there is little variation below about 300 km, at least in the regions studied so far. In general, the nature and magnitude of the lateral variation and the depth extent agree with surface wave results and global tomographic studies (Fig. 1.3).

The shield (SNA) and tectonic (TNA) profiles of North America bracket the profiles found in other

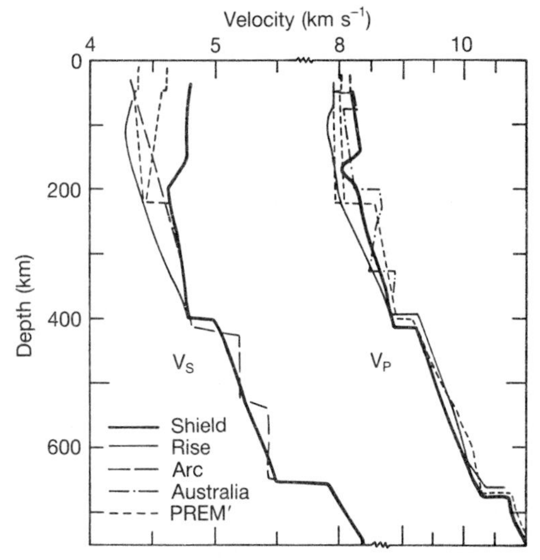

Fig. 1.2. Shear velocity and compressional velocity profiles in different tectonic regions. Although there are small (<3 per cent) differences below some 300 km the largest and most consistent differences between tectonic regions are shallower than 150 km. Shields generally have faster velocities than other regions in the depth interval between 100 and 200 km but subduction regions and old oceanic regions often have the highest velocities at greater depth in the upper mantle (Anderson and Bass 1986).

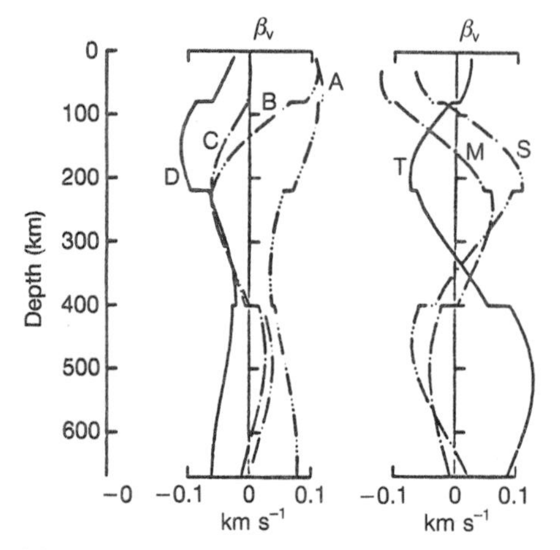

Fig. 1.3. Shear wave velocities (SV or β_v), relative to average values, versus depth for oceanic (A, oldest; D, youngest) and continental regions (S, shield; M, mountains; T, tectonically active). The high velocities in the transition region for A and T probably represent cold subducted material rather than rigid roots (after Nataf *et al.* 1986).

regions. Shields have the thickest and highest velocity lids. The velocity starts to decrease at a depth of 165 km under the Canadian shield and this is a good candidate for the thickness of the cratonic plate. Anderson and Bass (1984) inferred that the high-velocity seismic lid is cold and buoyant, that temperature increased rapidly below about 165 km, and, probably, that the composition (mineralogy) changed as well. A rapid increase of temperature means that the strength and viscosity rapidly decrease so that the seismic thickness may be similar to the rheological thickness. In fact, if the plate has high velocity, low density, and strong minerals compared to the minerals of the underlying mantle then the lithosphere–asthenosphere boundary may be controlled more by mineralogy than by temperature. It is only in these special cases that the various definitions of plate, lithosphere, and lid may coincide and have little to do with the thickness of the thermal boundary layer.

It is interesting that flexural studies of ancient cratonic areas are now approaching a value of 135 km (western Australia) for the effective thickness of the elastic plate (Don Forsyth, personal communication). It was previously shown (Regan and Anderson 1984) that, when anisotropy is properly taken into account, the seismic and flexural thicknesses of the oceanic lithosphere also are about the same. Other seismic models for shields have a velocity reversal setting in at depths as shallow as 140 km. In some tectonic regions the velocities are already low at the top of the mantle and decrease immediately with depth. In a profile in an older part of the Atlantic the high-velocity layer extends to a depth of about 100 km. It is evident from Fig. 1.1 that lithospheric doubling of the ATL profile would, in time, give a quite acceptable SNA profile.

1.5. THE SHALLOW MANTLE

The P-wave velocity at the top of the mantle is typically in the range 8.0–8.5 km s^{-1} in stable shield areas and 7.8–8.1 km s^{-1} in less ancient terrain. The pressure at the base of the average continental crust (10 kbar) increases velocities in silicate minerals, relative to their $P=0$ velocities by about 0.1 km s^{-1}. The temperature effect for $\Delta T \sim 600°C$ decreases velocities by about 0.3 km s^{-1}. Therefore, shield velocities, when corrected for pressure and temperature, range from about 8.2 to 8.7 km s^{-1} and younger terrains have upper mantle velocities of about 8.0 to 8.3 km s^{-1}, making a similar temperature correction in both regions. Velocity differences of 0.2 to 0.4 km s^{-1} imply temperature differences of 700 to 1400°C. This is probably greater than can be accommodated and a change in mineralogy or state, such as partial melting, may be involved. Tectonic mantle may have more pyroxene or a larger basaltic component or a higher melt content, at least at the top, than shield mantle. Deep ocean mantle velocities are in the range of shield velocities or even higher.

The compressional velocity in the immediate subcrustal part of the continental mantle is often used as an argument for a peridotite upper mantle, dominated by olivine. This is consistent with exposures of upper mantle and xenoliths but it is not a unique interpretation of the seismic data, as Table 1.1 shows. Eclogite, lherzolite, and harzburgite all have appropriate veloci-

Table 1.1. Elastic velocities of mantle rocks.

	ρ (g cm^{-3})	V_P (km s^{-1})	V_S (km s^{-1})
Harzburgite	3.30	8.40	4.90
Dunite	3.30	8.45	4.90
Garnet lherzolite	3.53	8.29	4.83
	3.47	8.19	4.72
	3.46	8.34	4.81
	3.31	8.30	4.87
Eclogite	3.46	8.61	4.77
	3.61	8.43	4.69
	3.52	8.29	4.49

Table 1.2. Elastic velocities of mantle minerals.

	ρ (g cm^{-3})	V_P (km s^{-1})	V_S (km s^{-1})
Olivine			
Fo	3.21	8.57	5.02
Fo_{93}	3.31	8.42	4.89
Orthopyroxene			
En	3.21	8.08	4.87
En_{80}	3.35	7.80	4.73
Clinopyroxene			
Di	3.29	7.84	4.51
Jd	3.32	8.76	5.03
Garnet			
Py	3.56	8.96	5.05
Al	4.32	8.42	4.68
Gr	3.60	9.31	5.43

ties. Eclogite pods in a peridotite matrix would increase the density and shear velocity over that of a barren peridotite but the compressional velocity is little affected. Table 1.2 shows that similar seismic velocities, particularly the compressional velocity, beneath the crust in oceanic and shield terrains do not necessarily imply similar mineralogy or composition. Olivine-rich barren or depleted harzburgite can have P-wave velocities similar to those of fertile peridotite or ultra-fertile eclogite, the reason being that either garnet or olivine can offset the slow velocities of clinopyroxene and orthopyroxene. Olivine, of course, is a refractory residual mineral and garnet is an easily fusable mineral in the shallow mantle.

The shear velocities, V_S, of the shallow mantle are less well determined than the P-wave velocities, V_P, but typical values are about 4.7 km s^{-1} for shield areas and 4.4–4.5 km s^{-1} for areas such as plateaux and extensional continental areas. Corrections for temperature and pressure would make these fall in the range 4.6–4.85 km s^{-1}, well within the range of upper mantle rocks and minerals (Tables 1.1 and 1.2). Shear velocities alone are not a good mineralogical or chemical discriminant but, if both V_P and V_S are measured accurately at the same place, their ratio may help distinguish between possible mineralogies. The seismic anisotropy is also a mineralogical discriminant since peridotites are much more anisotropic than eclogites. Evidence from xenoliths favour an olivine-rich shallow mantle and this is consistent with the seismic data. It should be pointed out, however, that the xenolithic evidence is neither widespread nor necessarily typical.

Velocities as high as 8.5–8.6 km s^{-1} are found at depths of 10–30 km below the crust mantle boundary in layered zones embedded in normal upper mantle of 8.0–8.2 km s^{-1}, in the continental lithosphere in France, Britain, Germany, and Scandinavia (Fuchs 1986). Azimuthal anisotropy, at least as deep as 100 km, is also found. The anisotropy and high-velocity layers may be due to changes in the orientation of olivine crystals. Silver and Chan (1988) inferred anisotropy in shield mantle to a depth of the order of 200 km. Lateral or radial variations in the degree or direction of crystal orientation may contribute to interpretations of seismic heterogeneity.

1.6. HEAT FLOW

Heat flow through continents is relatively low, compared to oceans, and is particularly low in shield areas. Only a small fraction of the total heat lost by the mantle is currently escaping through the continental crust to the surface. This may be surprising since the continental crust contains a large fraction of the radioactive heat-producing elements. Up to 40 per cent of the heat flow through shield areas can be accounted for by heat produced in the crust so not only is the total heat flow low but the mantle contribution is particularly low. The thermal gradient in the subshield mantle must be low. It is likely that the shield mantle is depleted in radioactive elements or that less heat is delivered to the base of old continental lithosphere, or plate, than elsewhere or that continental shields tend to seek out colder than average mantle to settle in. It should be pointed out that not all shields are adequately sampled and that heat flow, as well as other geophysical generalizations, is based on a possibly biased sample. However, there is a good correlation of surface wave velocities and heat flow (Nakanishi and Anderson 1983, 1984).

There are two contributions to sublithospheric heat flow—radioactive decay and cooling of the mantle and core. There are lateral variations in the rate of convective heat delivery to the shallow mantle and crust. In regions of mantle upwelling the lithosphere and thermal boundary layers are generally thin and a high conductive gradient is required to remove the heat delivered by convection. Thus, conductive heat flow is high near oceanic ridges, hot spots, and regions of continental extension. Heat flow is expected to be low over regions of mantle downwelling or in regions where the subcontinental mantle has already cooled substantially, so that the cooling component of heat flow is not large. It appears that mantle heat flow is either deflected away from cratons, or continental drift tends to move shield areas toward regions of low mantle heat flow.

The heat flow through continents is not necessarily a steady-state process or a radial process controlled only by the thermal conduction through the continental crust and upper mantle. Continents can readily move about on the Earth's surface and they should tend to move from hot to cold areas of the mantle. Mantle convection, of course, is directed from hot upwellings toward cold downwellings. Downwellings are also the sites of subduction, where cold oceanic lithosphere is carried down into the hot mantle, thereby displacing hot material and cooling the remainder. Subduction, in fact, is the most efficient way of cooling the mantle and we can be sure that the coldest parts of the upper mantle, apart from the surface boundary layer, are where subduction cooling has been most effective. Diagnostics of cold mantle are broad geoid lows and high seismic velocities. In fact, most of the continents lie in a narrow polar belt of long-wavelength geoid lows (Anderson 1989*b*), in spite of the fact that an isostatically compensated continent should give a mild geoid high because of the high centre of gravity of thick buoyant crust. It is likely that these geoid lows are generated beneath the continental lithosphere and represent cold mantle over which the continents have drifted and, perhaps, come to rest. Africa is the only continent which lies in a geoid high and it, of course, formed the centre of Pangaea.

The average heat flow though Archaean shields is about 41 mW m^{-2} (Chapman and Furlong 1977). The mean heat flow of continents is 57 mW m^{-2} (Sclater *et al.* 1980). Polyak and Smirnov (1968) noted that there was a general decrease of heat flow with tectonic age. Heat flow rises to about 61 mW m^{-2} in Late Palaeozoic and 71 mW m^{-2} in Cainozoic terrains. The age refers to the last tectonothermal mobilization of the terrain which can be the stabilization of the platform, the age of intrusion or extrusion, the youngest age of folding or deformation, or the age of metamorphism. Low heat flow is observed along continental margins situated above ongoing or recent subduction. High heat flow is observed in ancient terrains undergoing extension or rifting. The heat flux into the base of the continental lithosphere is about 25 mW m^{-2} (Sclater *et al.* 1980).

The stable parts of the continental lithosphere have apparently been thick and cold for a long period of time (Boyd and Gurney 1986; see also Richardson, Chapter 3, this volume). The thermal stability of the cratonic lithosphere, as deduced from geothermometry of mantle nodules, has been somewhat of a paradox. Archaean komatiites have high melting temperatures and this has been taken as evidence that the upper mantle has cooled appreciably in the past 10^9 years or so. Certainly, the decay rate of radioactive elements has declined with time and there is evidence that a large fraction of the Earth's heat flow, perhaps 50 per cent, is due to cooling rather than to radioactivity. On average, the outer part of the Earth should be colder, and have a lower thermal gradient, than in the past.

If the upper mantle has cooled of the order of 200–400°C in the past 10^9 years why are cratonic elevations and lithospheric temperatures stable over time? The reason could be that continents tend to spend most of their time over cold mantle. Not only do they move from hot toward cold mantle but they often subduct cold oceanic lithosphere under their leading edges. In principle, they can be cooled from both above and below.

Thermal cooling models were so successful in explaining oceanic bathymetry that it was suggested that continental plates were thicker simply because they cooled longer and that their elevation was controlled by the presence of a thick sialic crust (McKenzie 1969). The thermal boundary layer model of the evolution of the continental lithosphere was explored by Crough and Thompson (1976), Pollack and Chapman (1977), and Kono and Amano (1978).

An alternative model is that the continental crust and lithosphere were chemically differentiated from the asthenosphere or the upper mantle (Ringwood 1969; Jordan 1975; Oxburgh and Parmentier 1978). Garnet lherzolite, a popular source rock, produces basalt and an olivine–orthopyroxene-rich residue upon partial melting. The residue is less dense than the parent rock and could be a major constituent of the continental lithosphere. This type of model assumes a primitive, undifferentiated lower mantle (see also Menzies, Chapter 2, this volume).

Another alternative is that the crust and various parts of the upper mantle, including the continental lithosphere, are the results of chemical differentiation of the whole mantle, probably during the accretion of the Earth. The buoyant products of differentiation, including the incompatible elements, tend to concentrate in the outer part of the Earth. In this model the source of the continental crust cannot be so clearly identified with the immediately underlying mantle (Anderson 1989*b*).

The purely thermal models of the continental lithosphere ignore the seismic evidence regarding LID thicknesses, seismic velocities, and the presence or absence of low-velocity zones (LVZ) and partial melt in the mantle. They also assume that, below the lithosphere, the mantle is the same everywhere. These models predict a variation of elevation with time that is not observed.

1.7. DEEP CONTINENTAL ROOTS?

Satellite-derived gravity anomalies do not appear to be related systematically to the distribution of continents and oceans. This is partly to be expected since the major surface features are all in approximate isostatic equilibrium so that the surface elevation effect on gravity is approximately cancelled by the opposite effect of the root. In fact, a small positive residual gravity anomaly is expected because the elevation anomaly is closer to the observer than the root and because of the decreasing lateral extent of the compensating masses due to the change of circumference with depth.

There are several other possible effects involved in the lack of a continental signature in the global gravity field. The light continental crust may be counterbalanced by denser than average upper mantle, either lithospheric or sublithospheric. The continental lithosphere is colder than oceanic lithosphere but it also appears to be buoyant relative to adjacent mantle. The sublithospheric continental mantle (below 150–200 km) may be denser than oceanic mantle because it is colder and because of the absence of partial melting and other high-temperature 'phases'.

The oceanic crust–shallow mantle system (lithosphere) probably overlies hotter and lighter than average mantle. The very low seismic velocities under some tectonic and oceanic regions require partial melting, or some anelastic effect, to reduce the velocities below those of the constituent minerals. In a partially molten mantle the dense mineral garnet is reduced or eliminated, to be replaced by a light melt phase. Even at great depth where melts may become denser than the residual phases, olivine and orthopyroxene, they are unlikely to become denser than garnet or majorite, a high-pressure refractory garnet-like form of pyroxene. Partially molten rock is therefore lighter than the same rock below the solidus. Melts may become denser than the residual crystals at great depth (>200 km) because of the high FeO content and high compressibility of melts, but both are less dense than the unmelted rock. Dislocation relaxation and dehydration are also effective in decreasing seismic velocities.

The similarity of heat flow and gravity between continents and oceans has been used to argue that, on average, mass and radioactivity per unit area are equal under the continents and oceans (MacDonald 1963) and that vertical segregation has been the dominant feature of the process of continent formation. On average, it was proposed, material initially placed along a radial direction remains in the vicinity of that radius. MacDonald (1963) proposed that the material to depths of 400 to 700 km, under continents, provided the material for the crust and, therefore, continental mantle differed from oceanic mantle to this depth. Earthquake zones were explained as a concentration of thermal stresses at the continent–ocean boundary. MacDonald (1963) found it difficult to imagine any style of convection in the mantle or large-scale horizontal motions that would result in equality of heat flow and mass per unit over continents and oceans. We now know, of course, that there are, in fact, large variations in heat flow and that isostasy is effective in levelling out geoid anomalies. The propensity for continents to drift toward cold, dense mantle and to override cold oceanic lithosphere also contributes to an ironing out of heat flow and geoidal differences. Even so, it is true that moderate lateral variations in seismic velocity exist at all depths. These are most prominent above 200 km but substantial (~ 3 per cent) variations, somewhat correlated with surface features, extend to about 400 km. The correlation, however, decays rapidly below 200 km and most of the remaining correlation is caused by fast velocities over most of the upper mantle below areas of old oceanic crust, and low velocities under the youngest oceans (Nataf *et al.* 1986). In fact, if roots are defined as continuity of a given type of seismic anomaly (fast or slow), then old and young ocean would have the deepest roots (Fig. 1.3). Actually, however, it is broad mantle upwellings associated with young ocean, including ridges, and broad downwellings associated with oldest ocean, including trenches, that give this correlation. Other tectonic features, including shields, tectonic regions, and intermediate age ocean, tend to have velocity anomalies that reverse sign between the shallow mantle and 670 km depth.

Other characteristics which lead MacDonald to a static view of the mantle, with permanent deep continental roots, can also be explained by cold but buoyant continental lithosphere overriding dense oceanic lithosphere or cold downwelling mantle.

Jordan (1975) revived MacDonald's (1963) deep stable continental root idea by again proposing that cratonic mantle perhaps extended to the base of the upper mantle. He termed this deep shield keel the continental tectosphere. The same objections that were applied to MacDonald's hypothesis can be applied to the tectosphere hypothesis. There is abundant evidence that the continental root associated with shields extends to depths of the order of 150–250 km (see also

Richardson, Chapter 3, this volume) but litle support for the tectosphere conjecture which extends this continental root concept to depths in excess of 400 km.

Sipkin and Jordan (1975) used ScS waves to argue for deep continental structure perhaps extending throughout the entire upper mantle. ScS waves sample the entire mantle and it is only by assumption that the large regional variations in travel time are attributed to the upper mantle. In this regard it is of interest that Tanimoto's (1989) maps of the lowermost mantle generally have fast velocities in areas where Okal and Anderson (1975) and Sipkin and Jordan (1975) record fast ScS times. At least some of the ScS travel time variations are therefore acquired in the deep mantle. Most of the variation is caused by effects in the upper 200 km (Okal and Anderson 1975). There is also evidence that ScS waves are split by anisotropy, probably somewhere in the upper mantle, and this complicates previous interpretations of deep continental roots which assumed an isotropic mantle.

The thickness of the continental mantle root or keel has become controversial, unnecessarily so in my opinion. Seismic studies, even prior to the development of plate tectonics, showed that the high-velocity LID extended to depths as great as 150 km under continental shields, much greater than under tectonic and oceanic regions (e.g. Anderson 1966). The LVZ zone under shields was deep and less pronounced than elsewhere. When plate tectonics and a more open attitude toward continental drift came along it was natural to assume that the maximum thickness of the continental plate was about 150 km, somewhat thicker than the thickest oceanic plate. Since old oceanic plates become unstable and subduct while old continental plates exhibit remarkable stability there was a hint that the continental mantle may be intrinsically buoyant. The slightly higher than average subshield mantle velocities between 200 and 400 km were attributed to the absence of an appreciable melt phase beneath stable shields (Anderson and Sammis 1970; Anderson and Bass 1984) rather than to a compositionally distinct root that extended deeper than about 150 km. The shield LVZ would then correspond, at least approximately, to the asthenosphere.

The thickness and composition of the continental root, at least as determined by seismology, are complicated by two factors; anisotropy and anelasticity. The depth and magnitude of the low-velocity zone and, hence, the properties of the seismic LID depend on what assumptions are made in reducing the seismic data. For example, if it is assumed that the mantle is isotropic one tends to obtain very high-velocity LID velocities and very pronounced LVZs with low channel velocities from the inversion of surface wave data. Oscillatory solutions such as this are often the sign of an improperly parametrized problem. When anisotropy is allowed for, the models become simpler and the velocity extremes are damped out (Anderson 1966; Anderson and Dziewonski 1982).

Anisotropy shows up when both Love wave and Rayleigh wave data are used in the inversion or when there is adequate azimuthal control to detect an angular variation in velocity. There is now abundant evidence that the upper 200 km of the mantle is markedly anisotropic and that this is primarily due to the preferred alignment of olivine crystals. If only Love wave data is used to invert for a mantle velocity model, under the assumption of isotropy, the resulting model generally does not fit the Rayleigh wave data. Love waves involve shear waves with horizontal particle motion polarization (SH). In an anisotropic solid the propagation velocity depends on polarization as well as direction of propagation. Rayleigh waves involve shear waves with vertical shear wave polarization (SV) which, in general, travel with different velocities than SH waves. In a mantle composed primarily of olivine crystals with their a-axis oriented in the horizontal plane the SH velocity will be greater than the SV velocity. If this is not taken into account in seismic inversions a model which satisfies the Love wave data will predict Rayleigh wave velocities which are too fast. In fact, satisfactory seismic models of the upper mantle require anisotropy in the upper 200 km (Dziewonski and Anderson 1981; Regan and Anderson 1984). For isotropic models we have what is known as the Love wave–Rayleigh wave discrepancy.

Jordan (1975) noted a discrepancy between lateral variations in travel times of near vertical ScS waves (shear waves which reflect off the core) and Love waves. Since Love waves only sample the outer part of the mantle and ScS waves sample the entire mantle, it was proposed that large lateral variations of shear velocity must extend deep into the mantle. However, this may be just the Love wave–Rayleigh wave discrepancy in disguise. ScS waves involve nearly vertically travelling waves with horizontal polarization. Love waves involve horizontally travelling SH waves. In an anisotropic mantle these velocities are not the same. In fact, in a transversely isotropic medium with a vertical axis of symmetry the vertically travelling SH velocity is the same as the horizontal travelling SV velocity. One should, therefore, compare ScS waves with Rayleigh waves not Love waves, in order to avoid the trade offs between depths of heterogeneity and anisotropy. There are many other examples of heterogeneity trading off with anisotropy. Mantle minerals are very anisotropic and they tend to be oriented by flow or recrystallization.

The percentage variations of seismic velocity, due to changes in direction or polarization, are generally much greater than variations due to reasonable changes in chemistry or temperature. If anisotropy is ignored these variations will be mapped into deep heterogeneities such as deep continental roots or deeply penetrating slabs. The largest variations in seismic velocity are due to anisotropy, partial melting, and solid–solid phase changes. Temperature and composition are second-order effects, causing relatively modest changes in velocity.

In fact, Silver and Chan (1988) used the SKS phase to demonstrate that the upper 200 km of the Canadian Shield mantle is markedly anisotropic. The SKS wave travels as a shear wave in the mantle and a compressional wave in the core and exhibits an anisotropy induced shear wave splitting of nearly 2 seconds. Silver and Chan suggest that a 200 km thick layer beneath the Canadian Shield has been stable for over 2.5 billion years and this supports the idea that continents have chemically distinct roots to about 200 km which are mechanically strong because of their low temperature. The observed anisotropy removes another element of the interpretation of ScS and Love waves in terms of a deep tectospheric root beneath continents, an interpretation that assumed isotropy as well as a negligible contribution from deep mantle variations.

Later high-resolution seismic studies (Grand and Helmberger 1984) confirmed that shield velocities were particularly high to depths of 165 km and that shield and oceanic velocities rapidly converged below this depth.

The partial melt interpretation of the extremely low seismic velocities beneath oceans and tectonic regions also needs some comment. In some parts of the upper mantle the seismic velocities are lower than any plausible (isotropic) combination of olivine, orthopyroxene, clinopyroxene, and garnet at reasonable temperatures. This suggests some anelastic process such as grain boundary melting, which partially decouples the elastic grains, or dislocation relaxation. Anelastic phenomena tent to reduce seismic velocities from the so-called high-frequency unrelaxed limit to a low-frequency, or high-temperature, relaxed value. This introduces a physical dispersion into the velocities of seismic waves and causes the elastic velocities to vary with frequency and more strongly with temperature than measured in the laboratory at high frequency. Associated with the maximum rate of change of velocity, with either temperature or frequency, is a high attenuation or absorption. The low-velocity zone is also a zone of high attenuation, even under shields (Der *et al.* 1986). In my opinion the LVZ is the natural place to look for decoupling of the motion of the plate and the underlying mantle. The LVZ and the asthenosphere may be coincident, even though the time scales and stress levels used to define them are quite different. Unfortunately, petrologists and geochemists tend to attribute trace element and isotopic characteristics to the lithosphere, asthenosphere, and LVZ. There is no evidence, for example, that the LVZ is either depleted or uniform.

1.8. THE ELASTIC OR FLEXURAL THICKNESS OF THE CONTINENTAL LITHOSPHERE

The elastic, or flexural, thickness of the lithosphere can be determined from topographic and gravity profiles across features which are caused by loads on, within, or just below the elastic plate. Estimates of the effective elastic plate thickness in Africa vary from 64 to 90+ km in the stable cratonic areas to about 45 km under uplifted plateaux; severely faulted and volcanically active regions give thicknesses of 21–36 km (Ebinger *et al.* 1989). Generally hot mantle under Africa plus its near stationarity relative to the mantle during the past 25 Ma may have resulted in local thermal thinning of the lithosphere. Thinning by a factor of 2 is not unreasonable (Yuen and Fleitout 1985). The estimated elastic thickness of the continental lithosphere in mountain belts ranges from 80 to 130 km for the Appalachians, 80 to 100 km for the Himalayas, and 25 to 50 km for the Alps (Karner and Watts 1983). The Deccan Traps in India were emplaced on an elastic lithosphere of 100 km thickness (Watts and Cox 1989). Western Australia has an effective elastic thickness of 135 km (Forsyth, personal communication, 1989). In some other continental areas the effective elastic thickness is much smaller, possibly the result of an extensively fractured crust or thermally weakened or delaminated upper mantle.

The thickness of the elastic plate (flexural studies) in continental regions varies from 10 km to more than 100 km (McNutt *et al.* 1988). The 100+ km values for elastic plate thicknesses of ancient shield areas, if controlled by a temperature-only dependent rheology, suggest a thermal plate thickness (the thermal boundary layer) of at least 250 km which is about twice the thermal thickness of the oldest oceanic plate. Continents, of course, have had longer to cool than oceanic plates. A given isotherm deepens roughly as the square root of age for a purely conductive situation. In oceanic regions the base of the elastic plate corresponds roughly to the 600°C isotherm (McNutt and Menard 1982). Variations in thickness of the continental plate, however, seem to reflect stress levels and extent of bending rather than temperature. The largest elastic thicknesses that have been inferred to date from continental data are 105 ± 25 km for the Appalachians on the North American platform, 135 km for Western Australia, 90 ± 10 km for the Himalayas on the Indian shield, 85 ± 10 km for the Caucasus on the East European platform, and 75 ± 25 km for the Urals on the East European platform. These were all loaded more than 10^9 years ago. These are only slightly less than the thickness of the seismic LID under the Canadian shield.

1.9. THE MINERALOGY OF THE CONTINENTAL LITHOSPHERE

There are trade offs between composition and temperature when one is attempting to interpret seismic velocity. Furthermore, even if temperature is known the seismic velocities do not uniquely determine the chemistry. Further complications involve the frequency dependence of elastic moduli, meaning that relatively low-frequency seismic data must be compared with high-frequency laboratory data only with care.

Seismic velocities are better mineralogical discriminants than chemical discriminants. For example, olivine and garnet are relatively fast minerals compared to the pyroxenes but there is little variation of seismic velocity for reasonable changes in chemistry (e.g. FeO/MgO ratios). Of course, the stable mineral assemblage at a given P and T depends on the chemistry and, therefore, the mineralogy is an indirect measure of composition, if equilibrium prevails.

In addition to the two seismic velocities, V_P and V_S, which characterize an isotropic solid there are a few other seismic parameters which have been used in discussions of mantle mineralogy and composition. These include the V_P/V_S ratio, which is related to Poisson's ratio, the seismic parameter $\Phi = V_P^2 - (4/3)V_S^2$, the seismic anisotropy, and the seismic quality factors. For example, the V_P/V_S ratio increases with the garnet, clinopyroxene, or FeO contents. This allows eclogitic rocks to be distinguished from peridotitic rocks. Likewise, peridotites tend to be much more anisotropic than eclogites. Temperature is relatively unimportant when interpreting V_P/V_S.

Large lateral variations in seismic velocities and the effects of high-temperature gradients, anisotropy, and partial melting have made it difficult to infer the composition and mineralogy of the upper 400 km of the mantle. The important components of this region are olivine, orthopyroxene, clinopyroxene, and garnet and possible melt or vapour phases. The dominant rock types are presumably harzburgites, lherzolites, pyroxenites, and eclogites (see also Menzies, Chapter 2, this volume). The composition and mineralogy may change both vertically and laterally. The largest lateral variations occur above 200 km but lateral variations are still appreciable between 200 and 400 km. This variation may be due to differences in petrology, temperature, crystal orientation, or partial melt content. Temperature has a direct effect on seismic velocity and an indirect effect caused by temperature dependent phase changes.

Figure 1.4 shows compressional and shear velocity profiles for several types of tectonic provinces (Anderson and Bass 1984). Also shown are calculations for two hypothetical rock types, pyrolite and piclogite. These rock types both have variable compositions but pyrolite is an olivine–orthopyroxene rich rock and piclogite is richer in clinopyroxene and garnet and contains less olivine and orthopyroxene. Thus, piclogite is more 'fertile' and contains more of the easily fusable basaltic elements while pyrolite is a more refractory rock type. Piclogite is denser throughout much of the upper mantle. The actual compositions (in Wt %) used in the calculation are: pyrolite—olivine (57), opx (17), cpx (12), and gt (14); piclogite—ol (16), opx (6), cpx (33), gt (45). In a later study Duffy and Anderson (1989) derived higher olivine and cpx contents and lower garnet contents for the transition region. Note the similarity in computed velocities, an illustration of the non-uniqueness of seismic velocity interpretations. It is often stated that seismic velocities in the upper mantle are consistent with a peridotitic (pyrolite) upper mantle but they are equally consistent with olivine eclogite (piclogite). Xenolith data and upper mantle anisotropy, however, favour the peridotite interpretation, at least for the upper 200 km and, possibly for the upper 400 km. One interpretation of the smallness of the seismic velocity jump at 400 km involves a change in composition from peridotite to piclogite (Duffy and Anderson 1989). In regions under young oceans and in tectonic regions, the velocities are so low that partial melting or some anelastic process is required and no estimate of mineralogy can be made. It is quite possible that in these regions the shallow mantle is invaded by basalt-rich material which can be eclogitic at depth (e.g. a fertile piclogite transition region).

Also shown in Fig. 1.4 are estimates of the dry solidi of the mantle. Curves which fall below these lines are in the partial melt field and the calculated velocities are upper bounds. Note that in oceanic and tectonic regions the data fall below the partial melt lines to depths of about 300 km. The mantle in these regions may also contain a melt phase at greater depth but some

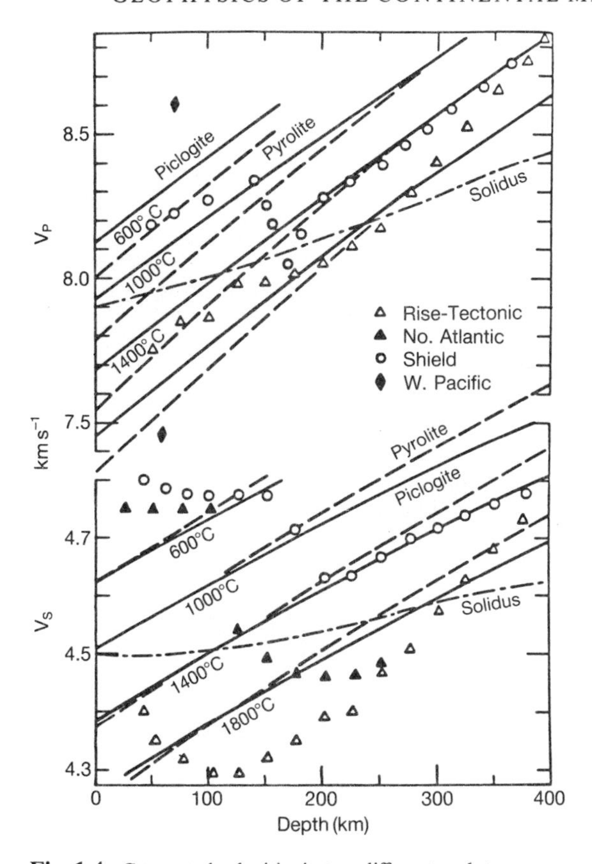

Fig. 1.4. Computed velocities in two different rock types; one olivine–orthopyroxene-rich (pyrolite), one clinopyroxene–garnet-rich with some olivine (piclogite). Note the very cold (400–800°C) temperatures that are implied for the upper 150 km of the shield model, and the high temperature gradient implied for 150 to 200 km depth. Curves represent upper bounds for velocity when they extend below the solidus curves (after Anderson and Bass 1984).

combination of the predominant mantle minerals can explain the velocities at subsolidus temperatures.

The seismic velocities for shield data are relatively high to depths of about 150 km and then drop rapidly. The rapid drop can be due to a high temperature gradient or a change in mineralogy or a change in anisotropy. In any case it could also represent the base of the plate. This is a more plausible location for the base of the plate than the maximum depth to which shield velocities consistently exceed oceanic velocities which is close to 390 km. Below 200 km the shield velocities fall near the 1400°C adiabat. Below this depth we are probably in normal convecting mantle. The velocities are still higher than in other regions but only a slight rise in temperature would cause partial melting and then subshield and subocean velocities would probably not differ much.

The V_P/V_S ratio for the shield data is very low in the upper 150 km. This is consistent with a high olivine content or a low FeO content or both. The seismic velocities of the shield lithosphere are best explained by a very olivine-rich, possibly low-FeO, peridotite with low temperatures and a low temperature gradient (but higher than adiabatic because the shield velocity points cross adiabats (Fig. 1.4). For a uniform harzburgite LID the inferred temperature gradient is only about 2–4°C km^{-1}, substantially less than the values of 8–10°C km^{-1} expected for a conductive geotherm. The subsequent rapid drop in velocity, if due to temperature alone, requires a gradient of at least 10 to 14°C km^{-1}. The large increase in V_P/V_S with depth suggests that a change in mineralogy is involved, possibly an increase in garnet and clinopyroxene relative to olivine and orthopyroxene. This implies a more fertile sublithospheric mantle. An increase in FeO content will also drive the velocities and velocity ratios in the right direction, and ensure that the lithosphere is chemically buoyant.

The inferred high olivine content for shield lithosphere and the increased fertility implied for depths greater than 150 km agree with evidence from kimberlite xenoliths. It is also consistent with long-term stability of the shield lithosphere. However, such an olivine-rich lithosphere is not necessarily appropriate for other regions. The lower oceanic lithosphere may be more eclogitic. The continental mantle is fairly anisotropic down to a depth of about 200 km (Nataf *et al.* 1986; Silver and Chan 1988), consistent with a high olivine content. The shield lithosphere is infertile.

The extremely low temperatures at the base of the shield lithosphere deserve some comment. Mechanisms for cooling at such great depth include overriding of cold oceanic lithosphere (flat subduction) or having the continent move over cold, downwelling mantle. These mechanisms can actually cause reversals in the geotherm and downward conduction of heat. In fact, one mechanism for generating continental lithosphere is the doubling of oceanic lithosphere. Unless the oceanic crust gets scraped off in the process there would be basalt–eclogite pockets in the resulting cold overthickened lithosphere. The cold harzburgite part of the oceanic lithospheric may also rise back into the shallow mantle as it warms up. The eclogites and harzburgites delivered to the surface by kimberlite pipes may, in fact, be samples of ancient lithosphere underplating the craton (see also Richardson, Chapter 3, and Menzies, Chapter 4, this volume).

1.10. ATTENUATION OF SEISMIC WAVES AND ELECTRICAL CONDUCTIVITY

The shield mantle is remarkably transparent to seismic waves compared to tectonic and oceanic areas. For example, the quality factor for 1 second shear waves, Q_S, is about 400 for the upper 100 km of the northern shield area of Eurasia (Der *et al.* 1986). There is a minimum of Q, or a maximum in attenuation, between about 100 and 200 km depth but the attenuation is still much less than in tectonic regions. Attenuation is much less under the craton of North America than in the west (Der *et al.* 1982). These results are consistent with a relatively cold upper mantle under shields and a high temperature gradient below about 100 km. Thus, the Q results give about the same lithospheric thickness as do shear velocities and flexural studies. This is only expected to be the case if the temperature rise is so great near 100 km that the various measures of weakening all set in at about the same depth, or if changes in mineralogy, dislocation density, or melting control the rheology.

Electrical conductivity studies give depths to a high conductivity layer of about 45 km in the Basin and Range, 80 km in the Colorado plateau, and 160 km in the Great Plains (Gough *et al.* 1983). The high conductivity layer is often attributed to partial melting although other mechanisms have been proposed, such as grain boundary carbon coatings. Regions of low mantle seismic velocity, high attenuation, and high heat flow generally have high mantle electrical conductivity (Gough 1989). In many regions of the world there is a rapid increase in conductivity inferred between depths of 100 and 300 km. In Fennoscandia, for example, there is a marked rise in conductivity between 155 and 185 km depth (Jones 1982), which is, by now, a familiar depth range.

Thus, we have a large number of studies which indicate a change in physical properties at depths between about 100 and 200 km under shield areas, and smaller depths under other continental regions, that are consistent with a rapid increase in temperature and/or a change in composition or mineralogy. None of these measures directly indicate the thickness of the plate but, in so far as strength and viscosity decrease rapidly with temperature, the layer of decoupling or extensive shearing should occur somewhere in the depth interval indicated by the rapid decrease in mechanical properties such as velocity, Q, flexural strength, and resistivity. I propose that the thickness of the plate is at approximately the depth of maximum rate of decrease of the shear velocity. This is at about 165 km for the Grand–Helmberger shield model.

1.11. SEISMIC TOMOGRAPHY

Tomography, or three-dimensional imaging, does not have the resolution of body wave profiling but it provides a view of the long-wavelength features of the Earth's mantle. There is a great deal of lateral and radial smearing because of the smoothing effect of present techniques applied to a relatively sparse data set and this must be kept in mind when interpreting the images. Figure 1.5 shows the average SH-velocity at depths ranging from 220 to 400 km. Note that the fastest velocities are centred near the oldest parts of the oceans and continents. The lowest velocities are associated with tectonic regions and the younger parts of the oceans (Fig. 1.6). The total range of velocities is about 4.5 per cent. Above 220 km the total range is 9 per cent. This agrees with the range found by detailed body wave profiling in areas such as the East Pacific rise and the Canadian shield. Above 220 km there are low velocities in south-east Asia, where there are subduction zones, back-arc basins, and extensional tectonics, and near other extensional areas, particularly triple junctions. It appears that the high velocities found in the upper part of the Canadian shield may be typical of other shield areas. In Fig. 1.6 the pattern is similar but even more reduced in amplitude. Note that south-east Asia is now a region of fast velocity, possibly the result of subduction of cold oceanic lithosphere. Fig. 1.7 shows the top of the lower mantle.

A slightly higher resolution tomographic image of part of the western and northern hemisphere is shown in Fig. 1.8 (175–250 km depth). Note the large difference in velocity between eastern and western North America. The more stable parts of North America are clearly evident as high-velocity regions. The areal extent and magnitude of the velocity anomaly rapidly diminishes with depth. Figure 1.9 shows the 325–400 km depth interval. Note the very low velocities associated with the oceanic spreading centres, evidence that they are rooted deep in the upper mantle. This is one reason why one should not talk about the 'depleted asthenosphere' or to attribute the MORB source to the shallow mantle. Below 400 km there is little correlation with surface features. Note the change in scale from previous figures.

From Fig. 1.9 one might argue that:

(1) continental nuclei have deep high-velocity roots; or

(2) young oceans and extensional tectonic domains have deep low-velocity 'roots', perhaps due to partial melting; or

(3) the high-velocity regions appear to be associated with present subduction or regions where previously

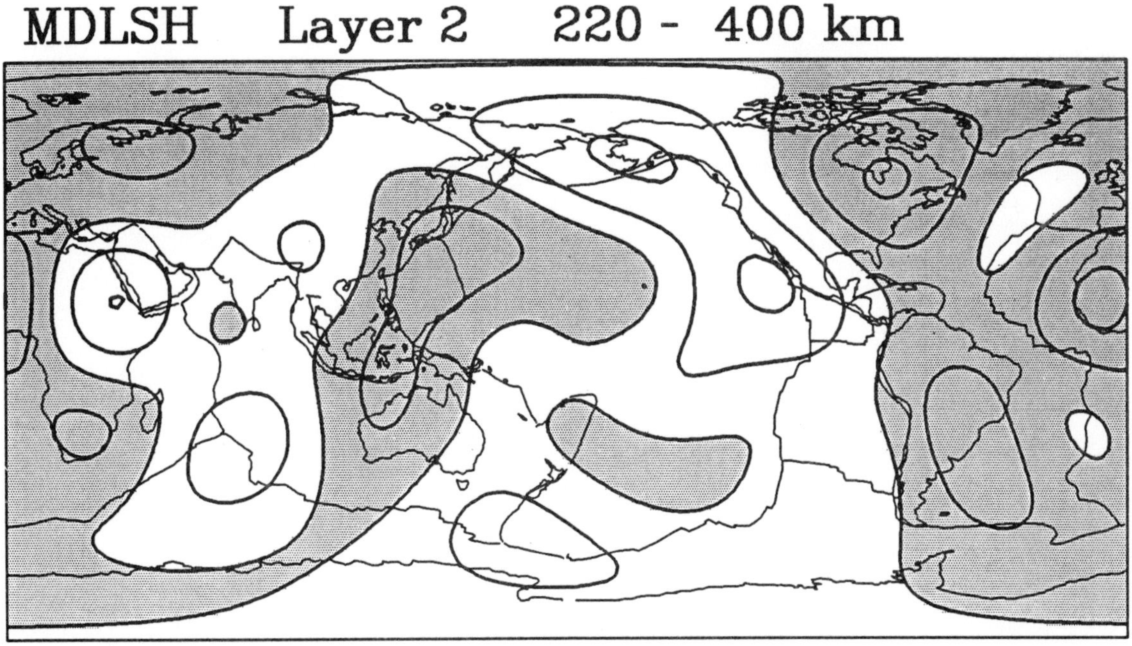

MDLSH Layer 2 220 - 400 km

Fig. 1.5. Tomographic map (SH velocity) for the 220–400 km depth range. Shaded is faster than average and contour lines 1 per cent apart. Note that the faster regions underlie the oldest oceans and the oldest continental areas. The total range in velocity is about one-half the variation in the upper 220 km. Note that most large tectonically active continental areas are slow (after Tanimoto submitted).

existing oceans have disappeared by continents overriding oceanic lithosphere.

In the second two alternatives the high velocities under continents represent overridden fast material or the absence of partial melting rather than a rigidly attached root.

In the mantle transition zone, the range of velocities is only 3.5 per cent. There is little correlation with surface tectonics but ridges are generally in slow regions or in less fast regions (Atlantic) (Fig. 1.6). One could argue that the Afar–Red Sea–East African Rift area, the East Pacific Rise, the Indian Ocean Rise, and the North Atlantic Rise are rooted in the transition region. Duffy and Anderson (1989) argue that the transition region is more fertile than the shallow mantle. There is also a large amount of fast, presumably cold, material in the vicinity of the western Pacific subduction zones.

1.12. LOWER MANTLE

Seismic tomography is starting to reveal the secrets of the lower mantle, in particular, its three-dimensional structure. Radially, the lower mantle appears to be homogeneous except near its top and bottom. By homogeneous we mean that the variation of physical properties with depth is about what one would calculate from the increase of pressure with depth. The lower mantle may have more FeO than the upper mantle and, therefore, be intrinsically denser (e.g. Anderson 1989*a*,*b*). The lateral variations are very small (of the order of 0.5 per cent), very much smaller than in the various regions of the upper mantle. Because seismic sources and receivers are near the surface of the Earth all seismic waves which we use to probe the lower mantle must first pass through the inhomogeneous upper mantle. Even long-period surface waves and free oscillations are sensitive, to some extent, to the properties of the shallow parts of the Earth. We are therefore looking for a small signal (the heterogeneity of the lower mantle) embedded in a high-noise background (the variance of seismic parameters due to the large lateral variability of paths through the crust and upper mantle).

Nevertheless, we now have maps of the lateral heterogeneity of the lower mantle (Figs 1.10 and 1.11). Several things are immediately clear. The pattern is relatively simple, meaning that much of the resolved

MDLSH Layer 3 400 - 670 km

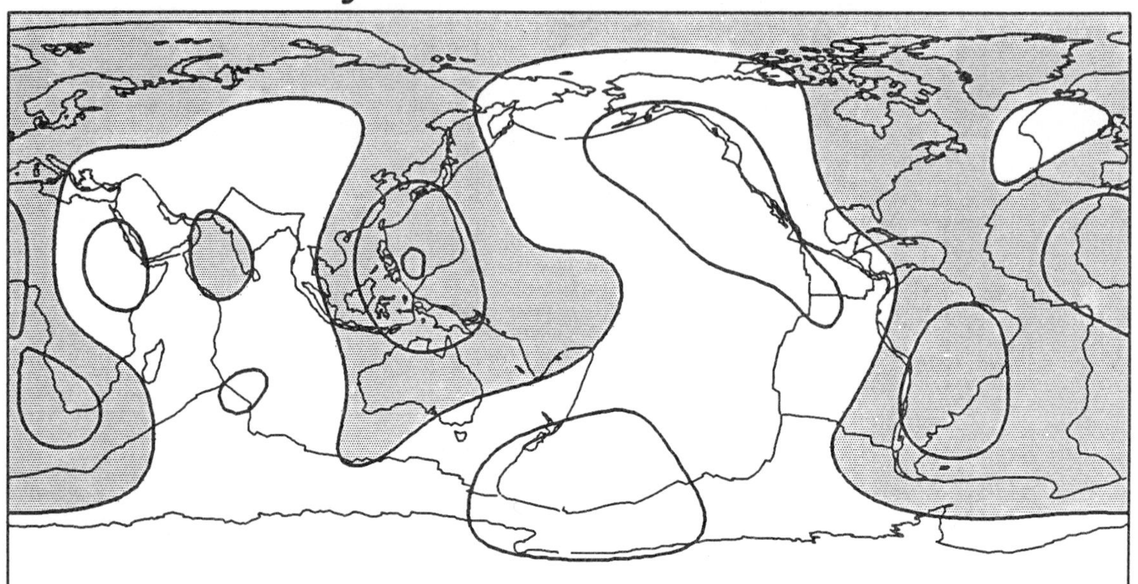

Fig. 1.6. Same as Fig. 1.5 except this gives average SH velocities in the transition region. Most young oceanic areas are slow or relatively slow as are large areas of continental extension. The fastest regions may represent areas of long sustained subduction (after Tanimoto submitted).

MDLSH Layer 4 670 - 1022 km

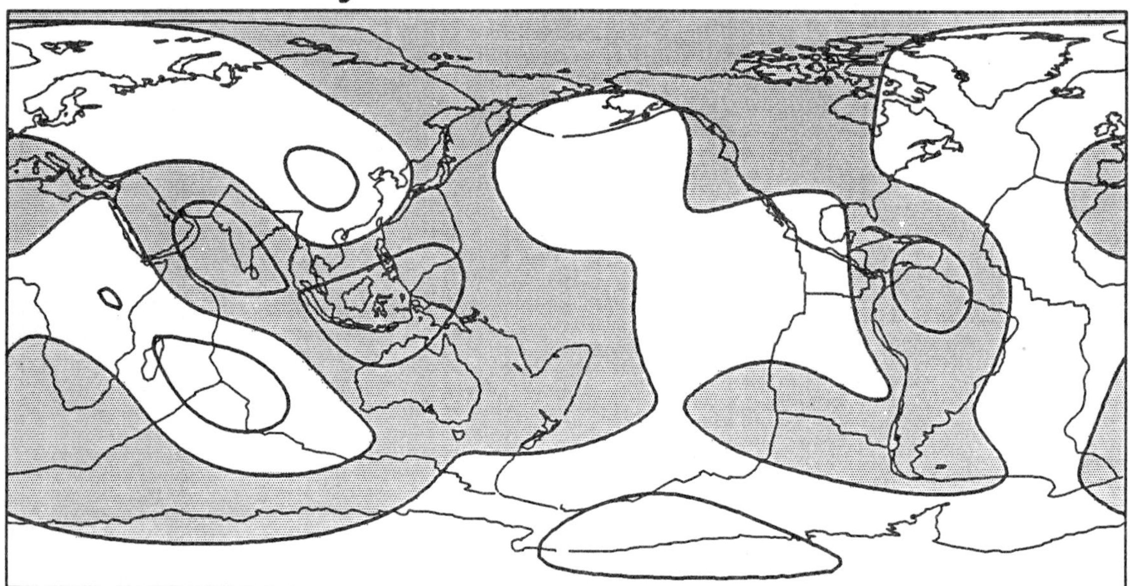

Fig. 1.7. The situation at the top of the lower mantle. One has the impression that some fragments of Pangaea have drifted over or caused lower mantle downwellings (Americas, India, Siberia) or drifted toward downwellings (Australia). Subduction zones are mainly over fast mantle while oceanic ridges are mainly over slow lower mantle (after Tanimoto submitted).

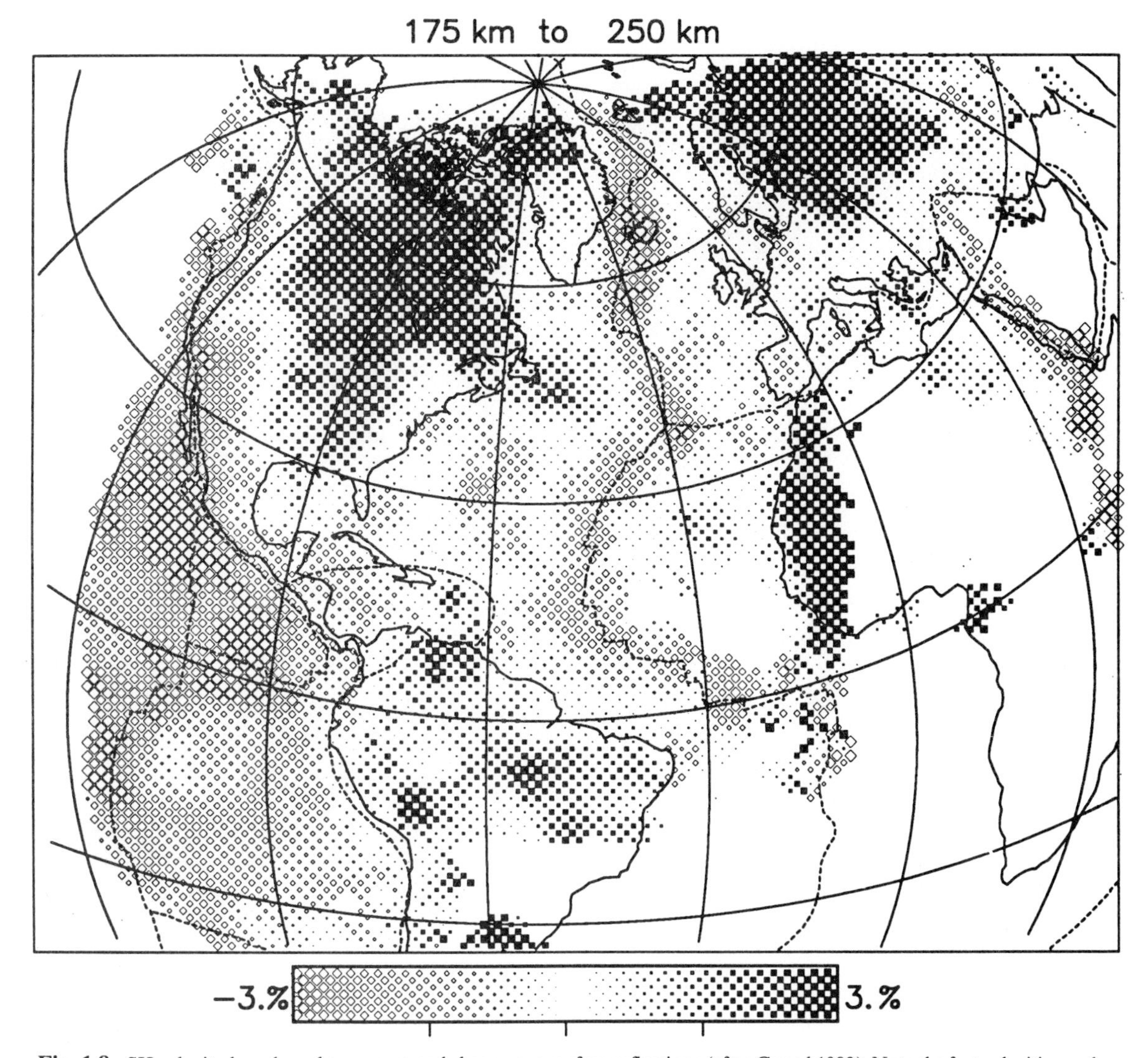

Fig. 1.8. SH velocity based on shear waves and shear wave surface reflections (after Grand 1989). Note the fast velocities under shields and slow velocities associated with continental-tectonic and young ocean (175–250 km depth).

power is in the long wavelengths. Body waves through the lower mantle emerge at distances between about 25° and 90° arc-distance. Because of the large-scale coherence of the lower mantle velocity structure, this means that seismic waves can travel in certain directions through entirely slow or entirely fast lower mantle. In other words seismic anomalies can accumulate, at least in certain directions. For seismic events in the central Pacific the lower mantle is generally slow for rays travelling to circumpacific stations. On the other hand rays starting from circumpacific sources will exhibit a pronounced asymmetry in their azimuthal anomaly pattern. Rays going to North America and the southwest from events in the Kuriles will be fast compared to rays going in other directions. Generally fast lower mantle occurs beneath most subduction zones and continents. Neglect of these variations has contributed to hypotheses regarding the deep structure of continents and deep slab penetration. Teleseismic waves and ScS waves spend most of their travel time in the lower mantle

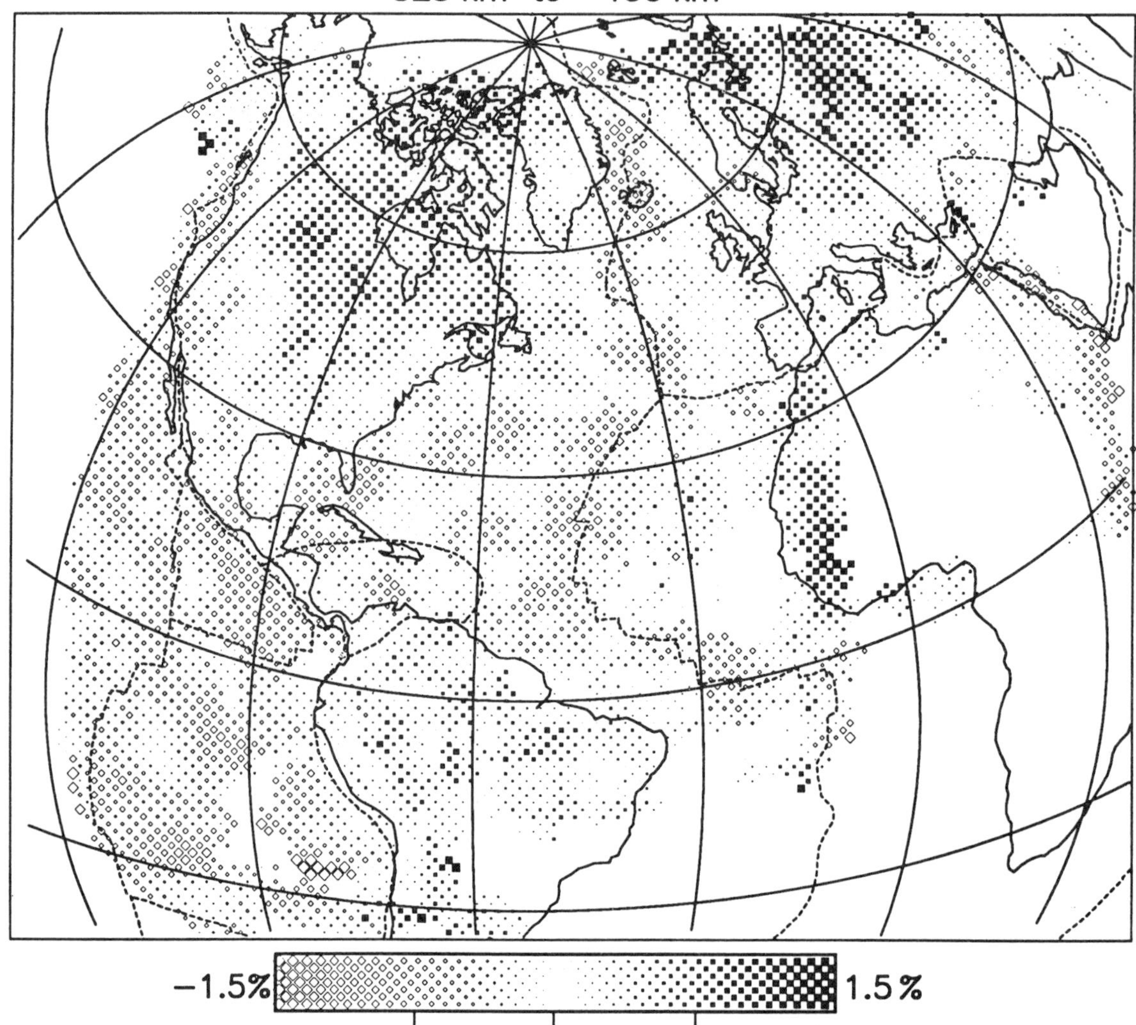

Fig. 1.9. Same as Fig. 1.8 but for 325–400 km depth. Note the much lower level of lateral variation. One could argue from this map that shields are deeply rooted, or that mid-ocean ridges are broad and deeply rooted but all we really know is that significant lateral variations persist to order 400 km (after Grand, manuscript in preparation).

and, even though the lateral variations there are small, the accumulated effect can dominate the observed variation.

The second thing that is apparent from the lower mantle tomographic images is the general equatorial symmetry. Equatorial regions are generally slow, polar regions are generally fast and there are two north–south fast bands encompassing eastern Asia and the Americas. Thus, a fast band surrounds the Pacific or, alternatively, Africa and parts of the Atlantic and Indian Oceans. The lower mantle helps control the moments of inertia of the Earth and we expect the regions with mass excesses to be arranged symmetrically about the equator. Subduction zones are the predominant mass excesses in the upper mantle and their symmetry about the equator is fairly obvious. The largest geoid anomaly, near New Guinea, is almost precisely on the equator. This plus the South American subduction-related geoid high control, to a large extent, the orientation of the spin axis of the Earth. The symmetry of the lower mantle suggests that convec-

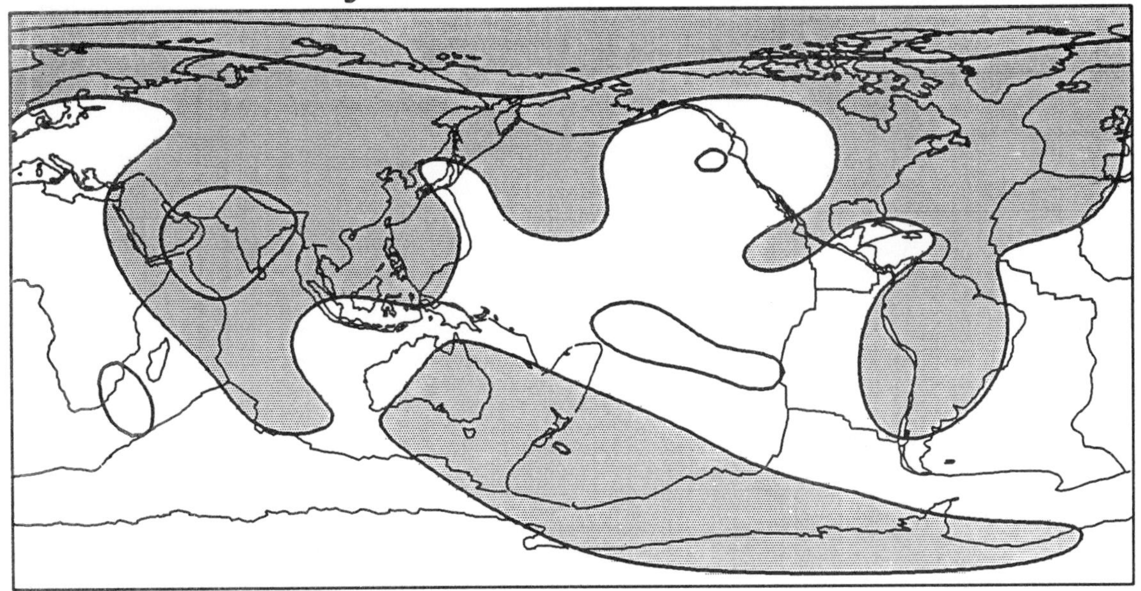

Fig. 1.10. SH velocity in the lower mantle (1555–1816 km). The contour interval is 0.5 per cent (after Tanimoto submitted).

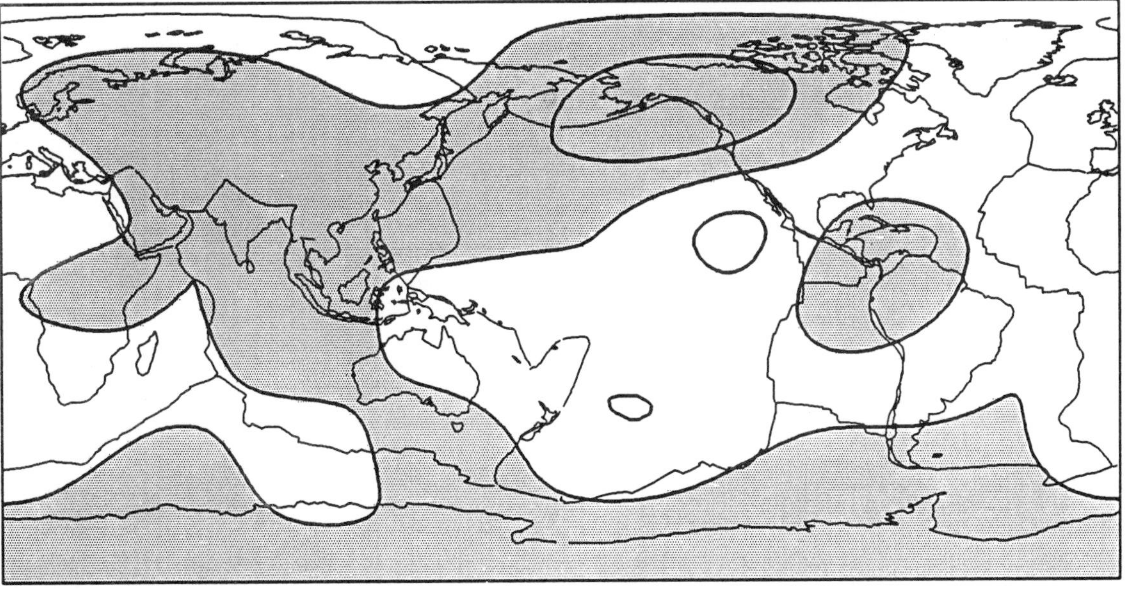

Fig. 1.11. SH velocity (2088–2359 km). The contour interval is 0.5 per cent (after Tanimoto submitted).

tion in the lower and upper mantles is coupled. One possibility is that subduction preferentially occurs over lower mantle downwellings. It is certainly true that there is little subduction in the central Pacific or in the hemisphere centred on Africa (which is antipodal to the mid-pacific). These are regions where both the geoid and the tomography suggest that hot mantle upwellings are occurring. The preference of hot spots for these zones suggests that the upper mantle is being heated by these lower mantle upwellings, even if no material transfer is taking place. The locations of hot spots and subduction zones are consistent with thermal coupling between the upper and lower mantle. They do not require transfer of material across the upper mantle–lower mantle boundary.

Hot upwellings in the lower mantle, in principle, deform the core–mantle boundary and any other chemical boundaries in the mantle. These upwarpings of dense interfaces are apparently responsible for the mass excesses which contribute to the broad geoid anomalies over the Pacific and African regions (Hager *et al.* 1985). These geoid highs are more-or-less centred on the equator and, along with the subduction zones, control the rotation axis of the Earth.

The third thing we notice about the lower mantle tomography is its relation to surface tectonics. The relationships are fairly subtle and we have already mentioned some of them. The locations of mid-ocean ridges bear little relation to the locations of fast and slow regions in the lower mantle. On the other hand we know that ridges are very mobile and we might not expect, at any given time, that ridges should bear any relation to the lower mantle, even for whole mantle convection. Lower mantle upwellings may only influence the locations of upper mantle upwellings in the most general terms, i.e. the hemispheres in which hot upper mantle is most likely to occur. Mid-ocean ridges tend to occur over both slow and fast regions of the lower mantle.

If we look at the locations of continents, relative to the fast, presumably cold, parts of the lower mantle, and if we keep in mind the drift history of the continents since the breakup of Pangaea, we can develop some insights about the recent history of the mantle.

- Throughout much of the depth range of the lower mantle there are two predominant bands of faster than average seismic velocity. One of these includes the Americas and one extends from central Africa to the subduction zones in the south-west Pacific. These bands are presumably colder than average mantle and are therefore dense downwellings.

- In general, the American plates, the Australian plate, and the African plate have been moving toward these downwellings over the past 100–200 Ma. The north Polar regions and the southern-most Pacific are less pronounced regions of inferred downwellings. The northern continents and Antarctica, respectively, have been moving generally toward these regions.

- Most of the African plate is underlain by seismically slow lower mantle, presumably the site of a broad upwelling. Africa, of course, was the centre of Pangaea. The continents surrounding it have apparently broken up and moved from the upwelling toward the downwellings. Most of the continents today are embedded in regions overlying inferred lower mantle downwellings and one can speculate that the present cycle of continental dispersal is almost complete. Subduction zones are mostly tied to the leading edges of continents and these also preferentially occur over lower mantle downwellings. One can also speculate that subduction can only initiate, or continue, in regions of lower mantle downwellings. Alternatively, it is possible that regions of long continuous subduction are where the top of the lower mantle is most effectively cooled and where lower mantle downwellings initiate.

- In this scenario the supercontinent of Pangaea insulated the lower mantle both by thermal blanketing and by the absence of subduction in the centre. The periphery of the supercontinent is where subduction could cool the upper mantle, causing large-scale lower mantle downwellings.

- The lower mantle is probably high-viscosity compared to the upper mantle and convection there may be sluggish. It is likely that convective patterns in the lower mantle are long-lived and that they control the thermal structure of the upper mantle. The long-wavelength geoid, a measure of the distribution of mass anomalies in the Earth, is primarily controlled by density variations in the lower mantle (Hager *et al.* 1985). This part of the geoid may have been relatively stable for more than 200 Ma and may have caused the breaking and dispersal of a long-lived supercontinent (Anderson 1982; see also Menzies and Kyle, Chapter 8, this volume). The hot upwelling inferred for the centre of Pangaea apparently still exists. The general lack of correlation of the long-wavelength geoid with present-day surface tectonics can be understood if the mobility of the surface plates, and their boundaries, relative to the persistence of lower mantle features, is taken into account.

- The stability of the Earth's rotation axis is another measure of the persistence of mass anomalies in the upper and lower mantle. Periods of rapid true polar wander may reflect rearrangements of the lower mantle convective patterns or of upper mantle subduction zones, or both (Anderson 1982).

- The tomographic pattern for the lower mantle is not invariant with depth but slower than average regions persist over most of the lower mantle under the south Atlantic–Africa–Southern Indian Ocean region and the central Pacific region. These are presumably hotter than average, buoyant, and upwelling. If the lower mantle is high-viscosity, then these hot upwellings may be very sluggish and, to the upper mantle, may appear to be relatively permanent hot patches. In this situation the upper mantle can be thermally coupled to the lower mantle, rather than shear coupled.

- The top of the lower mantle has a somewhat different tomographic pattern than the deeper part. The top of the lower mantle is slow under the north Atlantic, the north Pacific, the Antarctic plate, and northern Europe to south-east Asia. The lack of continuity between the anomalously hot parts of the upper mantle and lower mantle argues against simple models of whole mantle convection or deeply rooted mantle plumes. Some of the faster, and presumably coldest, parts of the transition region and the top of the lower mantle are where hot spots are most prominent (Atlantic Ocean, south-west Indian Ocean). The middle part of the lower mantle has fast seismic velocities under the two biggest hot spots, Iceland and Hawaii. On the other hand most hot spots occur in regions where the integrated velocity over the upper 650 km of the mantle is slower than average. This happens also to be true for the lowermost 1000 km of the mantle. The lack of radial continuity in the seismic velocity structure of the mantle could also mean the absence of a steady-state regime. There is no consistent pattern in the tomography that would encourage belief in whole mantle convection or any transfer of material between the upper and lower mantle.

One has to be careful about correlations between surface features and deep mantle features. For example, if one looks at the maps near the base of the mantle one might be tempted to conclude that continents have roots deeper than 2600 km. Most of the continents (except Africa) indeed overlie a faster than average lowermost mantle. The correlation, however, starts to break down at depths shallower than 2000 km. The weighted average of lower mantle densities, as expressed in the geoid, suggests that, on average, broad regions centred on the Pacific and Africa are upwellings and the surrounding regions, in which are embedded the most travelled continents, are downwellings. The top of the lower mantle under most of Eurasia is slow. The present may be the start of another period of continental insulation and protection of the underlying mantle from subduction cooling. Note that subducted plates spreading out along the base of the upper mantle can cool the top of the lower mantle and can also induce instabilities by deforming the boundary of the lower mantle. The reverse is also true. Subduction may only occur in the upper mantle where lower mantle downwellings permit it. In any event, correlations of seismic velocities across the upper-lower mantle boundary have at least three explanations, only one of which requires transfer of material across the (deformable) boundary. A technical difficulty is that there tends to be smearing of anomalies because of the low vertical resolution of seismic data. A sharp reversal of velocity anomaly would be difficult to resolve. High-amplitude upper mantle anomalies tend to be smeared into the top of the lower mantle.

Figure 1.12 shows the velocities in the lowermost mantle, just above the core. Note that most of the continents are in high-velocity areas and the slow regions are mostly oceanic. If this map were at a shallower depth there are some that might claim this as evidence for deep continental roots. In fact, at this depth (2630–2891 km) there is a better correlation of fast velocity with continents than there is for upper mantle maps deeper than about 300 km. (In the upper mantle, western North America, central Asia, Siberia, north-east Africa, and Antarctica are over slow areas and most of the Atlantic, western Pacific, and south-west Indian Ocean are fast.) The velocities of near vertical ScS waves also record the differences between continents and oceans (Okal and Anderson 1975; Sipkin and Jordan 1975). Some of this effect appears to come from the lower mantle. The correlation of ScS times with surface tectonics has been used to argue for deep continental roots (Jordan 1975), assuming that the lower mantle did not contribute to any correlated variations.

It might not be completely accidental that the tomographic images near the core–mantle boundary (and, in fact, throughout most of the lower mantle) mimic the distributions of oceans and continents. The long-wavelength geoid also looks much like Fig. 1.12, with geoid highs centred over Africa and the central Pacific. The African one has much the same shape as Pangaea (Anderson 1982). One interpretation of these correlations is that there is interaction between super-continents and mantle convection. A large supercontinent can insulate the mantle both by capping it with a thick insulating lid and isolating it from subductive cooling.

1.13. PLATE RECONSTRUCTIONS

Seismology, of course, maps the present-day physical properties of the interior. However, there is the strong suggestion that past events can also be dimly perceived.

MDLSH Layer 11 2630 - 2891 km

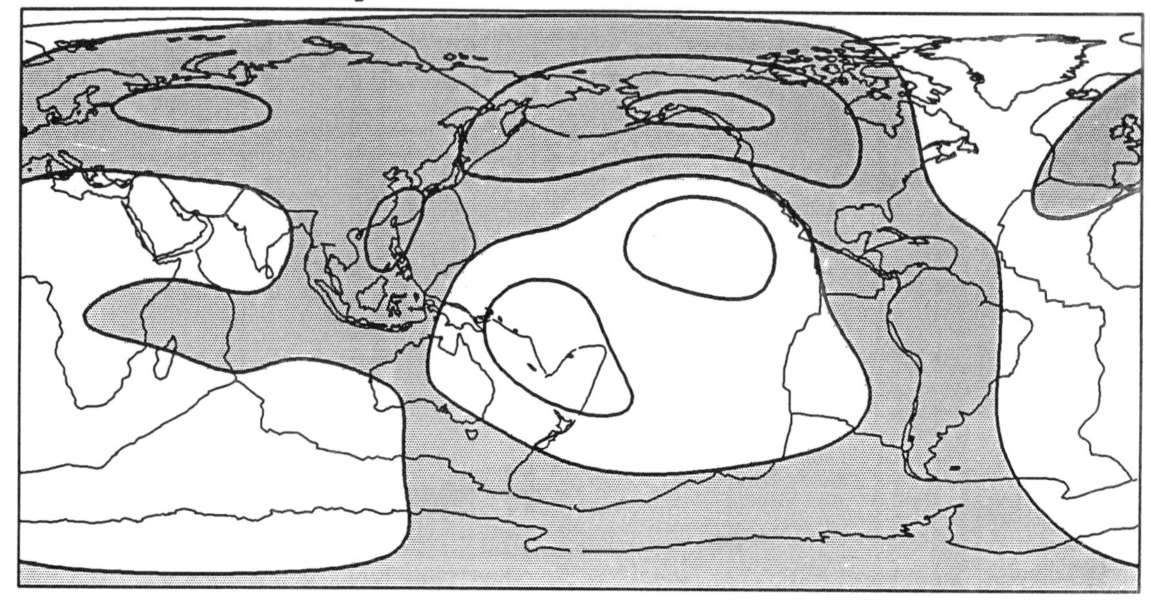

Fig. 1.12. SH velocity in lowermost mantle. Note that most of the continents lie above fast and, possibly, cold and dense regions of the lowermost mantle (after Tanimoto submitted).

Below 300 km there is only weak correlation with present surface tectonics. However, the fastest parts of the transition region correlate with regions in which oceanic lithosphere has been overridden by continents in the past 200 Ma. The slowest regions have been covered by continents for about 150 Ma, using almost any continental reconstruction map. This suggests that thick cold oceanic lithosphere has settled into the transition region under the drifting continents and under newly opened oceans (Fig. 1.13).

The present continents, by and large, have drifted into long-wavelength geoid lows formed by cold regions of the lower mantle. Most subduction zones also occur above cold regions of the lower mantle, a situation probably not unrelated to the present distribution of continents. Oceanic ridges tend to be on the edges of lower mantle hot regions. In fact, they tend to parallel the geoid contours, lying close to the zero contour of long-wavelength geoid maps.

The present relationship of surface fatures to the deep mantle cannot be a permanent one if lower mantle convection is more sluggish than plate motions. We suggest that this situation only prevails at the end of a period of continental breakup and dispersal. The continents and their peripheral subduction zones drift towards geoid lows which in turn are caused by lower mantle downwellings. Subduction may preferentially initiate or end up above cold regions of the lower mantle. The supercontinent of Pangaea could have been centred over a hot upwelling region of the lower mantle with the circum-Pangaeatic subduction, and the edges of Pangaea, over the present South Atlantic and the south-west Indian Ocean, with Tethyan subduction occurring under what is now Saudi Arabia through India and the northern Indian Ocean. Saying that subduction occurs preferentially above cold regions of the lower mantle is different from saying that cold or fast regions of the lower mantle are due to subduction. In fact, there are other faster regions of the lower mantle (under North America, Siberia, south Pacific, northern Pacific) not always related to present or recent subduction.

In summary, the tomography of the lower mantle indicates that there are two broad low-velocity regions, one centred under the Pacific hemisphere and one centred under Africa, extending from the north Atlantic to the Indian Ocean. The high-velocity regions surround the major oceans and the surface above these regions includes most of the continental fragments that made up the periphery of Pangaea. The faster than

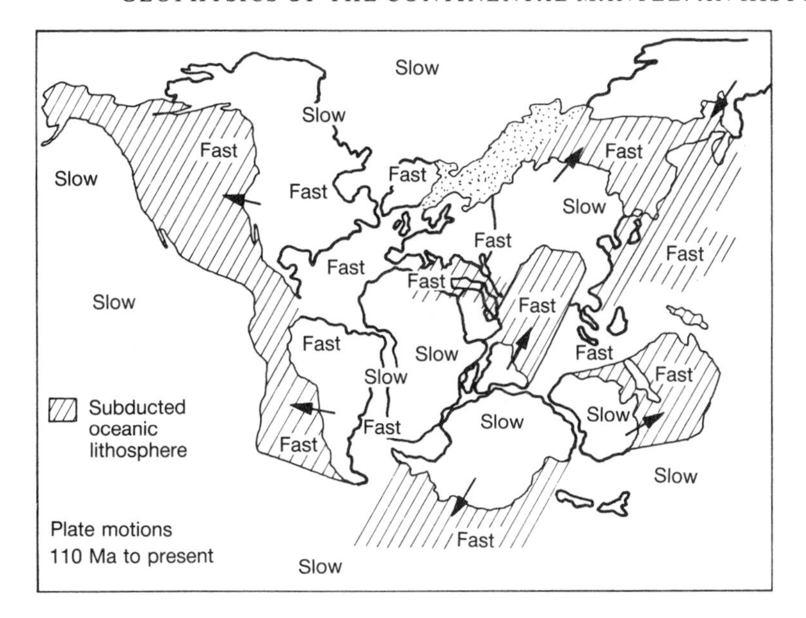

Fig. 1.13. Motion of the continents over the past 110 Ma (hatched regions). 'Fast' and 'slow' refer to seismic velocities in the transition region. This correlation is better than with present day tectonics.

average parts of the lower mantle are presumably colder than average and, in a convecting mantle, would represent the cold downwellings. It is natural to assume that continents will drift away from hot upwellings and toward cold downwellings. In fact, most of the continents may be coming to rest in the near future since any further motion will have them intruding into parts of the mantle which are above the hot upwellings in the lower mantle. No one has suggested that continental roots extend into the lower mantle but the correlation between continents and lower mantle velocities is as strong as with deep upper mantle velocities.

The main reason for suggesting ultradeep (>400 km) continental roots (Jordan 1975; Sipkin and Jordan 1975) was the ScS travel time distribution, interpreted with the assumptions of mantle isotropy and lower mantle homogeneity. Shear waves which reflected off the core (ScS) were observed to be much faster if they reflected under continents (actually, the data was confined to South America and Asia). These waves have very little depth resolution and the source of their travel time anomaly can be anywhere between the source and the core. It now appears that the integral effect through the lower mantle beneath continents is to speed the ScS waves. The remaining effect is primarily due to propagation above 220 km (Okal and Anderson 1975). I favour a 150–200 km thick cratonic plate even though, at present, there are features in the deeper mantle that correlate with surface tectonic features.

1.14. FORMATION OF THE CONTINENTAL LITHOSPHERE

One can easily tabulate various themes and variations.

1. In the beginning was oceanic crust and lithosphere. The proto-continental lithosphere was formed by doubling and redoubling of oceanic lithosphere under circumstances which favoured shallow-angle subduction. A variant of this is the formation of thin proto-continental lithosphere which then thickens by underplating of oceanic lithosphere. In either case eclogite pods or lenses would occur which represent former oceanic crust. These may be a future source of continental flood basalts and eclogite xenoliths in kimberlites.
2. Subduction of basalt/eclogite- and harzburgite-bearing oceanic plate through the upper mantle eventually leads to gravitational stratification of the upper mantle, with the denser eclogite concentrating in the transition region and buoyant depleted peridotite concentrating in the shallow mantle. When the deeper, more fertile, source melts it rises to form new oceanic plate, pushing aside the depleted peridotite which forms or thickens the continental lithosphere. Basaltic melts rising at midplate environments interact more with the shallow mantle, lithosphere, and crust and form ocean island and continental flood basalts.

3. Diapirs rising from the middle or deep mantle impinge on the base of previously formed continental lithosphere, cool, and fractionate. The more buoyant crystals (olivine and orthopyroxene) rise and denser crystals (garnet) or dense melts sink. The continental lithosphere is thereby underplated by buoyant refractory crystals.
4. Harzburgites and depleted peridotites can sink into fertile mantle if they are cold enough or if they are dragged down by mantle downwellings or eclogite sinkers. However, when they warm up or become delaminated from their denser portions they will rise to the top of the mantle, or the base of any overlying continental lithosphere. This will add relatively cold olivine-rich material to the continental keel.
5. The original differentiation of the Earth may have led to refractory-rich layers, more buoyant basalt-rich layers, and denser eclogite-rich layers. Convection could have led to lateral concentrations of the more refractory material.
6. Island arcs, mantle wedges and oceanic plateaus, including crust and upper mantle, may accrete at the edges of continents.

There is not yet enough high-resolution deep seismic sounding reflection work to discriminate between these various possibilities and, in fact, a combination of mechanisms may have occurred. In general, the shield lithosphere has high seismic velocity and seems to be relatively transparent to seismic waves. There is some evidence for high-velocity layers in the continental lithosphere. These may be due to changes in orientation of olivine-rich material or to eclogite pods. It is not yet clear if there is enough eclogite in the subcontinental mantle to form massive outpourings of flood basalts. An alternative is that depleted MORB-like material becomes fractionated and contaminated as it ponds beneath, intrudes through, and extrudes on continental mantle and crust.

1.15. EVOLUTION OF THE MANTLE

Most theories of the evolution of the mantle start with an Earth-sized planet and a homogeneous mantle. Whole mantle convection produces partial melting in the upper mantle where the adiabat crosses the solidus and melt/crystal separation forms the crust/depleted mantle part of the lithosphere in the basalt/refractory peridotite portions of the original mantle (usually about the 1:4 ratio adopted for pyrolite). Upon subduction the basalt and harzburgite are stirred back into the mantle and the cycle repeats. In some variants the continental crust was removed from the upper mantle at an earlier stage but the upper and lower mantles remain similar enough to allow the return of subducted slabs to the lower mantle and the extraction of at least some material from the lower mantle.

At the other extreme are models in which melting occurs during planetary accretion and there is continuous melt/crystal separation, primarily near the surface of the growing planet, during accretion. The separation is efficient and irreversible because of the large difference in melting temperature and density of the products of differentiation and because remixing would have to happen at different pressures than the original separation. In this scenario we can have a stratified mantle with the denser products of differentiation in the lower mantle and the more buoyant products (low-pressure melts) near the surface and intermediate products (olivine and high-pressure melts) at intermediate depths. Melts would either rise to the surface as basalt or, if deeper, would sink because of their high compressibility or because they crystallized to dense eclogite. In this scenario the incompatible and easily fusable (crustal) elements would be continuously fractionated into the upper mantle during the accretional differentiation of the Earth. The continental crust and enriched shallow mantle would represent processing of the whole mantle, not just the upper mantle.

If the lower mantle is intrinsically denser than the various parts of the upper mantle, enough so that temperature variations cannot overcome the density contrast, then subducted slabs will remain in the upper mantle. The lower mantle, of course, can still convect and, in fact, may control convection in the upper mantle. Thermal coupling (hot or over hot; cold under cold) will make it appear as if slabs penetrate the boundary, thereby confusing the more gullible seismologists. Such coupling appears likely in convecting layered systems (Nataf *et al.* 1988).

Most discussions of mantle convection and mantle reservoirs adopt the first scenario or, at most, a homogeneous upper mantle variant. There may, in fact, be several compositional layers in the upper mantle, in addition to the continental lithosphere. There is no reason to believe that the whole upper mantle is trace-element and isotopically depleted fertile peridotite, or that it is thoroughly stirred. In fact, the best seismic solution for the mineralogy of the transition region (400–650 km depth) is an olivine-poor (< 50 per cent) piclogite (Anderson 1987; Duffy and Anderson 1989). The shallower mantle is likely to be peridotite but is also likely to collect enriched material by subduction of sediments and altered oceanic crust from above (see also

Kyser, Chapter 7, this volume) and trapped melts from below. With time the shallow mantle is likely to become isotopically variable (see also Menzies, Chapter 4, and Richardson, Chapter 3, this volume) and a source of contamination for deeper magmas attempting to rise through it to midplate volcanoes (see also Menzies and Kyle, Chapter 8, this volume).

Probably all basalts are blends of melts from various depths and various degrees of partial melting (polybaric melting). For example, the adiabatic ascent of a hot region of the mantle will exhibit increasing degrees of partial melting as the depth decreases and melts emerging at the surface may represent blends of the small melt fractions at depth and the larger melt fractions near the surface. Lateral transfer of small melt fractions (see also Watson *et al.*, Chapter 6, this volume) at the wings or periphery of the upwelling toward the more extensively molten interior may also occur. In addition, as melts invade the cooler shallow mantle or crust they will undergo variable degrees of cooling, crystal fractionation, and contamination prior to eruption. Melts erupting through thick crust or lithosphere are likely to have experienced both deep and shallow crystal fractionation. There is probably no melt that represents a given amount of partial melting at a given depth from a parent rock of given composition. They represent the integrated effects of polybaric and polythermal melting and fractionation. There is therefore no melt that represents a 'primitive' or 'primary' magma that can be simply related to its source by a single-stage process. It is also unlikely that any source region is 'primitive' in the sense that it is simply related to chondritic or solar composition. There is certainly no evidence that the lower mantle is a primitive reservoir or, in fact, that any material sampled so far has recently come from the lower mantle. The assumption that the lower mantle is an important reservoir is partially based on the assumptions that the upper mantle is homogeneous, that MORB comes from the upper mantle, and that the entire upper mantle is LIL depleted. It is particularly misleading to talk about 'the depleted LVZ or asthenosphere'.

1.16. CONCLUDING STATEMENT

Geophysics provides snapshots of the present geometry and physical state of the mantle. In order to interpret these snapshots it must be recognized that the surface features of the Earth are mobile and transient and it must be appreciated that thermal inertia gives the mantle a very long memory. It takes of the order of a hundred million years for a 100 km thick subducted slab to warm appreciably and, in the process, to cool the adjacent mantle. Hot upwellings also have a long time constant and their general locations are insensitive to the reconfigurations of surface plates and plate boundaries. Plate motions and the locations of continents, however, do feed back into mantle convection. A large stationary continent can cause the underlying mantle to heat up and restricts the possible locations of subduction. Moving continents tend to have subduction on their leading edges and, therefore, generate cold mantle under their keels. The mantle under disappearing oceans such as the Tethys should also be cold because of the subducting of cold oceanic lithosphere. On the other hand, if an edge of a continent should override a segment of an oceanic ridge (clearly, not now a mid-ocean ridge), the newly continental mantle (sublithospheric) will probably remain hot for some time. The piece of the continent in question is likely to experience tectonism, uplift, and magmatism and either mountain building or rifting depending on the general stress state. Mantle magmas making their way to the surface are likely to be cooled, fractionated, and contaminated relative to MORB, which previously had a much freer ride prior to eruption. This possibility has the virtue of economy of reservoirs, the geochemical characteristics of hot-spot magmas being acquired in the shallow mantle (asthenosphere) or lithosphere by depleted magmas from a single deep reservoir. Of course, any fertile fraction of continental lithosphere (eclogite pods?) may also be mobilized by the sublithospheric heating being discussed but only small amounts of contamination, or wall rock reactions, are required to turn MORB tholeiites into CFB tholeiites.

ACKNOWLEDGEMENTS

I greatly appreciate the conversations I have had with Steve Grand and Toshiro Tanimoto and they have my gratitude for allowing me the use of figures from their unpublished papers. This research was supported by NSF-EAR-8509350, California Institute of Technology Contribution Number 4775.

REFERENCES

Anderson, Don L. (1964). Recent evidence concerning the structure of the upper mantle from the dispersion of long-period surface waves. In *Proceedings of the VESIAC Conference on Variations of the Earth's Crust and Upper Mantle*. Institute of Science and Technology, The University of Michigan.

Anderson, Don L. (1966). Recent evidence concerning the structure and composition of the Earth's mantle. *Physics and Chemistry of the Earth*, **6**, 1–131.

Anderson, Don L. (1967). Phase changes in the upper mantle. *Science*, **157**, 1165–73.

Anderson, Don L. (1979). The deep structure of continents. *Journal of Geophysical Research*, **84**, 7555–60.

Anderson, Don L. (1982). Hotspots, polar wander, mesozoic convection and the geoid. *Nature*, **297**, 391–3.

Anderson, Don L. (1987). Thermally induced phase changes, lateral heterogeneity of the mantle, continental roots and deep slab anomalies. *Journal of Geophysical Research*, **92**, 13968–80.

Anderson, Don L. (1989*a*). Composition of the Earth. *Science*, **243**, 367–70.

Anderson, Don L. (1989*b*). *Theory of the Earth*. Blackwell Scientific Publications, Boston, Massachusetts.

Anderson, Don L. and Bass, J. D. (1984). Mineralogy and composition of the upper mantle. *Geophysical Research Letters*, **11**, 637–40.

Anderson, Don L. and Bass, J. (1986). The transition region of the Earth's upper mantle. *Nature*, **320**, 321–8.

Anderson, Don L. and Dziewonski, A. (1982). Upper mantle anisotropy; evidence from free oscillations. *Geophysical Journal of the Royal Astronomical Society*, **69**, 383–404.

Anderson, Don L. and Julian B. (1969). Shear velocities and elastic parameters of the mantle. *Journal of Geophysical Research*, **74**, 3281–6.

Anderson, Don, L. and Kovach, R. L. (1964). Attenuation in the mantle and rigidity of the core from multiply reflected core phases. *Proceedings of the National Academy of Sciences, USA*, **51**, 168–72.

Anderson, Don L. and Sammis, C. (1970). Partial melting in the upper mantle. *Physics of the Earth and Planetary Interiors*, **3**, 41–50.

Anderson, Don L. and Toksöz, M. N. ((1963). Surface waves on a spherical Earth. *Journal of Geophysical Research*, **68**, 3483–500.

Archambeau, C. and Anderson, Don L. (1964). Inversion of surface wave dispersion data. *I.U.G.G. Reports, Publ. du Bur. Central Seismologigue Internat., Ser. A., Travaux Scientiligues, Fascicule 23*.

Bass, J. and Anderson, Don L. (1984). Composition of the upper mantle. *Geophysical Research Letters*, **11**, 237–40.

Boyd, F. R. and Gurney, J. J. (1986). Diamonds and the African lithosphere. *Science*, **232**, 472–7.

Brune, J. and Dorman, J. (1963). Seismic waves and Earth structure in the Canadian Shield. *Bulletin of the Seismological Society of America*, **53**, 167–209.

Bullen, K. (1953). *Introduction to the theory of seismology*. Cambridge University Press, London.

Chapman, D. and Furlong, K. (1977). Continental heat flow–age relationships. *EOS, Transactions of the American Geophysical Union*, **58**, 1240.

Crough, T. and Thompson, J. (1976). Numerical and approximate solutions for lithosphere thickening and thinning. *Earth and Planetary Science Letters*, **31**, 397–402.

Der, Z. A., McElfresh, T. W., and O'Donnell, A. (1982). An investigation of the regional variations and frequency dependence of anelastic attenuations in the mantle under the United States in the 0.5–4 Hz band. *Geophysical Journal of the Royal Astronomical Society*, **69**, 67–99.

Der, Z. A., Lees, A. C., and Cormier, V. F. (1986). Frequency dependence of Q in the mantle underlying the shield areas of Eurasia. Part III: The Q model. *Geophysical Journal of the Royal Astronomical Society*, **87**, 1103–12.

Dorman, J., Ewing, M., and Oliver, J. (1960). Study of shear-velocity distribution in the upper mantle by mantle Rayleigh waves. *Bulletin of the Seismological Society of America*, **50**, 87–115.

Duffy, T. and Anderson, Don L. (1989). Seismic velocities in mantle minerals and the mineralogy of the upper mantle. *Journal of Geophysical Research*, **94** (B2), 1895–912.

Dziewonski, A. and Anderson, Don L. (1981). Preliminary reference Earth model. *Physics of the Earth and Planetary Interiors*, **25**, 297–356.

Ebinger, C. J., Bechtel, T. D., Forsyth, D. W., and Bouin, C. O. (1989). Effective elastic plate thickness beneath the East African and Afar plateaus and dynamic compensation of the uplifts. *Journal of Geophysical Research*, **94**, 2883–901.

Ewing, M. and Press, F. (1954). An investigation of mantle Rayleigh waves. *Bulletin of the Seismological Society of America*, **44**, 121–47.

Fuchs, K. (1986). Reflections from the subcrustal lithosphere. In *Reflection seismology: the continental crust*, Geodynamic Series, Vol. 14, pp. 67–76. American Geophysical Union, Washington, DC.

Gough, I. (1989). Magnetometer array studies. Earth structure and tectonic processes. *Reviews of Geophysics*, **27**, 141–58.

Gough, D. I., Fordjor, C. K., and Bell, J. S. (1983). A stress province boundary and tractions on the North American plate. *Nature*, **305**, 619–21.

Grand, S. and Helmberger, D. V. (1984). Upper mantle shear structure of North America. *Geophysical Journal of the Royal Astronomical Society*, **76**, 399–438.

Gutenberg, B. (1959). The asthenosphere low velocity layer. *Annali di Geofisica*, **12**, 439–60.

Hager, B. H., Clayton, R., Richards, M., Comer, R., and Dziewonski, A. (1985). Lower mantle heterogeneity, dynamic topography and the geoid. *Nature*, **313**, 541–5.

Helmberger, D. and Engen, G. (1974). Upper mantle shear structure. *Journal of Geophysical Research*, **79**, 4017–28.

Herrin, E. and Taggart, J. (1962). Regional variations in P_h velocity. *Bulletin of the Seismological Society of America*, **52**, 1037–46.

Ibraham, A. and Nuttli, O. (1967). Travel-time curves and upper mantle structure from long-period S-waves. *Bulletin of the Seismological Society of America*, **57**, 1063–92.

Jeffreys, H. (1958). On the interpretation of Pd. *Geophysics Journal*, **1**, 191–7.

Johnson, L. (1967). Array measurements of P velocities in the upper mantle. *Journal of Geophysical Research*, **72**, 6309–24.

Jones, A. G. (1982). Observations of the electrical asthenosphere beneath Scandinavia. *Tectonophysics*, **90**, 37–55.

Jordan, T. (1975). The continental tectosphere. *Reviews of Geophysics and Space Physics*, **13**, 1–12.

Karner, G. D. and Watts, A. B. (1983). Gravity anomalies and flexure of the lithosphere at mountain ranges. *Journal of Geophysical Research*, **88**, 10449–77.

Knopoff, L. (1972). Observation and inversion of surface-wave dispersion. In The upper mantle (ed. A. R. Ritsema). *Tectonophysics*, **13** (1–4), 497–519.

Kono, Y. and Amano, M. (1978). Thickening model of the continental lithosphere. *Geophysical Journal of the Royal Astronomical Society*, **54**, 405–16.

Kovach, R. and Anderson, Don L. (1962). Long-period Love waves in a heterogeneous spherical Earth. *Journal of Geophysical Research*, **617**, 5243–55.

Kovach, R. and Anderson, Don L. (1964). Higher mode surface waves and their bearing on the structure of the Earth's mantle. *Bulletin of the Seismological Society of America*, **54**, 161–82.

Kovach, R. and Robinson, E. (1969). Upper mantle structure in the Basin and Range province, western North America, from the apparent velocities of S waves. *Bulletin of the Seismological Society of America*, **59**, 1653–65.

Lehmann, I. (1962). The travel times of the longitudinal waves of the Logan and Blanca atomic explosions. *Bulletin of the Seismological Society of America*, **52**, 519–26.

Lehmann, I. (1964). On the travel times of P as determined from nuclear explosions. *Bulletin of the Seismological Society of America*, **54**, 123–39.

MacDonald, G. J. F. (1963). The deep structure of continents. *Reviews of Geophysics*, **1**, 587–665.

McKenzie, D. (1969). Speculations on the consequences and causes of plate motions. *Geophysical Journal of the Royal Astronomical Society*, **18**, 1–32.

McNutt, M. K. and Menard, H. W. (1982). Constraints on yield strength in the oceanic lithosphere derived from observations of flexure. *Geophysical Journal of the Royal Astronomical Society*, **59**, 363–94.

McNutt, M. K., Diament, M., and Kogan, M. G. (1988). Variations of elastic plate thickness at continental thrust belts. *Journal of Geophysical research*, **93**, 8825–38.

Nakada, M. and Hashizume, M. (1983). Upper mantle structure beneath the Canadian Shield derived from higher modes of surface waves. *Journal of the Physics of the Earth*, **31**, 387–405.

Nakanishi, I. and Anderson, Don L. (1983). Measurements of mantle wave velocities and inversion for lateral heterogeneity and anisotropy. *Journal of Geophysical Research*, **88**, 10267–83.

Nakanishi, I. and Anderson, Don L. (1984). Aspherical heterogeneity of the mantle from phase velocities of mantle waves. *Nature*, **307**, 117–21.

Nataf, H. C., Nakanishi, I., and Anderson, Don L. (1984). Anisotropy and shear velocity heterogeneities in the upper mantle. *Geophysical Research Letters*, **11,** 109–12.

Nataf, H. C., Nakanishi, I., and Anderson, Don L. (1986). Measurements of mantle wave velocities. *Journal of Geophysical Research*, **91,** 7261–307.

Nataf, H. C., Moreno, S., and Cardin, Ph. (1988). What is responsible for thermal coupling in layered convection? *Journal de la Physique, France*, **49,** 1707–14.

Niazi, M. and Anderson, Don L. (1965). Upper mantle structure of western North America from apparent velocities of P waves. *Journal of Geophysical Research*, **70,** 4633–40.

Nuttli, O. (1963). Seismological evidence pertaining to the structure of the Earth's upper mantle. *Reviews of Geophysics*, **1,** 351–400.

Okal, E. and Anderson, Don L. (1975). A study of lateral inhomogeneities in the upper mantle by multiple ScS travel-time residuals. *Geophysical Research Letters*, **2,** 313–16.

Oxburgh, R. and Parmentier, M. (1978). Thermal processes in the formation of continental lithosphere. *Philosophical Transactions of the Royal Society, London*, **A288,** 415–29.

Pollack, H. and Chapman, D. (1977). On the regional variations of heat flow, geotherms and lithosphere thickness. *Tectonophysics*, **38,** 279–96.

Polyak, B. and Smirnov, Y. (1968). Relationship between terrestrial heat flow and the tectonics of continents. *Geotectonics*, **4,** 205–13.

Poupinet, G. (1979). On the relations between P-wave travel time residuals and the age of the continental plate. *Earth and Planetary Science Letters*, **43,** 149–61.

Regan, J. and Anderson, Don L. (1984). Anisotropic models of the upper mantle. *Physics of the Earth and Planetary Interiors*, **35,** 227–63.

Ringwood, A. (1969). Composition and evolution of the upper mantle. In *The Earth's crust and upper mantle*, Geophysics Monograph, No. 13 (ed. P. V. Hart). pp. 1–17. American Geophysical Union, Washington, DC.

Romanowicz, B. (1979). Seismic structure of the upper mantle beneath the United States by three dimensional inversion of body wave and arrival times. *Geophysical Journal of the Royal Astronomical Society*, **57,** 479–506.

Sato, Y. (1955). Analysis of dispersed surfaces waves by means of Fourier transform. *Bulletin of the Earthquake Research Institute, Tokyo*, **33,** 34–50.

Sclater, J. G., Jaupart, J., and Galson, D. (1980). The heat flow through oceanic and continental crust and the heat loss of the Earth. *Reviews of Geophysics and Space Physics*, **18,** 269–311.

Silver, P. and Chan, W. (1988). Implications for continental structure and evolution from seismic anistropy. *Nature*, **335,** 34–9.

Sipkin, S. and Jordan, T. (1975). Lateral heterogeneity of the upper mantle determined from the travel times of ScS. *Journal of Geophysical Research*, **80,** 1474–84.

Toksöz, M. and Anderson, Don L. (1966). Phase velocities of long-period surface waves and structure of the upper mantle. *Journal of Geophysical Research*, **71,** 1649–58.

Toksöz, M., Chinnery, M., and Anderson, Don L. (1967). Inhomogeneities in the Earth's mantle. *Geophysical Journal of the Royal Astronomical Society*, **13,** 31–59.

Watts, A. B. and Cox, K. G. (1989). The Deccan Traps: An interpretation in terms of progressive lithospheric flexure in responses to a migrating load. *Earth and Planetary Science Letters*, **93,** 85–97.

Yuen, D. and Fleitout, L. (1985). Thinning of the lithosphere and small-scale convective instabilities. *Nature*, **313,** 125–8.

2

Petrology and geochemistry of the continental mantle: an historical perspective

Martin A. Menzies

2.1. INTRODUCTION

Research into the petrology and geochemistry of the continental mantle has been active for more than a hundred years since the discovery of diamonds in the gravels of the Orange River, South Africa in 1866. This was to herald the beginning of a new era in petrology with the expectation that kimberlite-borne peridotites might represent the dense highly magnesian material believed to exist beneath the crust. The exact petrology of the outer 200 km of the mantle, however, has remained a contentious issue. Attempts to correlate earthquake-wave propagation data with known rock types soon led to the realization that, whilst the properties of the outer shell (i.e. sial) compared favourably with the properties of granite, the underlying material (i.e. sima) had the properties of dunite. Although eclogite and basalt were proposed as major constituents of the mantle, comparisons of rock compositions and seismic velocities pointed to a predominance of peridotite above 200 km and possibly eclogite at greater depths in the upper mantle. While most of the peridotites occurring as xenoliths had experienced partial melting events and were consequently different from undifferentiated mantle compositions, some did compare favourably with chondritic models for the upper mantle. Moreover continual interaction of the cold, stable lithosphere with the underlying hot, unstable asthenosphere has led to chemical enrichment of much of the lithosphere due to the upward migration and stagnation of potassic melts from the asthenosphere. In general, however, mantle rocks comprise peridotites and pyroxenites that are believed to represent: (1) potential source rocks for basaltic magmas (e.g. lherzolite); (2) residua remaining after partial melting processes (e.g. harzburgite); and (3) polybaric derivatives of silicate melts or hydrous fluids (e.g. pyroxenite). An integration of geophysical, petrological, and geochemical data now reveals that sub-Proterozoic and sub-Phanerozoic lithosphere is similar in physical and chemical properties to the lithosphere beneath old ocean basins while the sub-Archaean lithosphere is much thicker and has different physical and chemical properties.

2.2. LITHOSPHERE AND ASTHENOSPHERE

Fisher (1889) used the term 'Earth's crust' to denote a crystalline shell overlying a 'vitreous medium' but Barrell (1914) was possibly the first person to use the term lithosphere (*lithos* = stone; *sphairă* = ball) to mean a rigid layer that could support loads and that overlaid the weaker, less rigid asthenosphere (*asthenes* = weak). Although the term crust soon became synonymous with lithosphere, this tended to conceal the fact that lithosphere comprises crust and mantle portions and that a gradual transition exists between the lithosphere and asthenosphere.

The lithosphere comprises a mechanical, and in some cases, a chemical boundary layer (Rayleigh number $\ll$critical) and the upper portion of the thermal boundary layer (local Rayleigh number = critical) (Fig. 2.1). Moreover, since the thermal boundary layer is neither rigid nor vigorously convecting it serves as a transition from the rigid mechanical boundary layer (lithosphere) to the adiabatic interior (asthenosphere). A multitude of terms have been applied to the lithosphere and the asthenosphere some of which are shown in Fig. 2.1.

2.3. PETROLOGY OF THE CONTINENTAL MANTLE

Cohen (1879) first described eclogite or griquaite nodules from the kimberlite pipes of South Africa and Stelzner (1894) suggested they were early segregations of a kimberlitic magma, an idea reiterated by Beck (1899). During this period Dunn (1881) proposed that diamonds had a very shallow origin formed by the crystallization of carbon derived from the Karoo shales whilst Lewis (1888) speculated that the host rock to

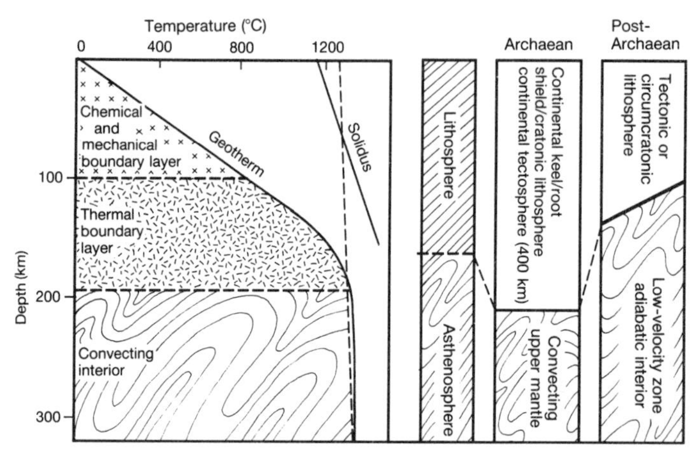

Fig. 2.1. The thermal structure of the lithosphere assuming a potential temperature of 1280°C in the convecting interior of the earth (after McKenzie and Bickle 1988). The adiabatic upwelling curve (dashed) and the dry peridotite solidus (solid) are shown. The lithosphere comprises a mechanical boundary layer (MBL) and, particularly in the Archaean, a chemical boundary layer (CBL) that overlie a thermal boundary layer (TBL) which acts as a transition into the adiabatic interior or asthenosphere. Isotopic heterogeneities are more likely to survive within the MBL and CBL than in the TBL as the latter would continuously equilibrate with small volume melts or fluids from the asthenosphere. The plethora of terminology used to refer to Archaean and post-Archaean lithosphere is shown to the right. Note that, whilst terms like keel, root, and tectosphere specifically apply to the Archaean lithospheric mantle and not the post-Archaean lithospheric mantle, terms like adiabatic interior and convecting upper mantle equally apply to the sublithospheric mantle in Archaean and post-Archaean times.

diamonds was probably a deep-seated porphyritic form of peridotite. More importantly, Lewis (1887) and Bonney (1897, 1899, 1900) reported the presence of diamonds in eclogite and peridotite xenoliths and diamond–garnet megacrysts from Griqualand West, South Africa, a discovery that would fuel speculation about the nature of subcrustal regions. Was it predominantly eclogite or peridotite? Fermor (1913) proposed that an eclogitic layer existed beneath the crust, his infra-plutonic zone, whilst Wagner (1914) suggested that a deep-seated peridotitic layer was responsible for the production of kimberlite. Furthermore, he speculated that eclogite xenoliths were shattered fragments of eclogitic schlieren scattered throughout the deep-seated peridotite matrix. The presence of the same minerals in xenoliths and the matrix of the kimberlite host, encouraged Wagner (1914) to conclude that most peridotitic xenoliths were cognate and that diamond was an original constituent of kimberlite. It is worth noting that in 1914, Wagner was able to list at least 20 papers that discussed different hypotheses regarding the 'cognate' or 'accidental' origin for xenoliths, a contentious issue that would remain unresolved for over half a century.

What emerged after almost 40 years of debate was that eclogite and peridotite were the strongest contenders for the dominant rock type in the continental mantle. Washington (1925) was one of the first to propose a crust–mantle cross-section (Fig. 2.2) where the outer SIAL (i.e. rich in silica and aluminium) portion of the Earth was interpreted as granite (<20 km), granodiorite–granite (20–40 km), and basalt–gabbro (40–60 km) and the SIMA (i.e. rich in silica and magnesium) portion of the Earth as peridotite. In contrast, Daly (1926) believed that much of the crust had to be basaltic to account for the volumes of basaltic magma erupted at the surface. Studies of kimberlite-borne xenoliths, however, produced a detailed crust–mantle section that was at variance with the ideas of Daly (1926, 1933). Wagner (1928) noted a vertical variation in the petrology of the crust and a subcrustal peridotitic mantle that contained 'schlieren' of eclogite and pyroxenite. He concluded that 'down to the depth at least from which kimberlite [was] derived the deeper substratum [consisted] of peridotite and not of eclogite'. Bowen (1928) also stated a preference for a peridotitic mantle on the basis of similarities in measured velocities in dunite and those reported from earthquake data (Anderson, Chapter 1, this volume). Bowen (1928), however, made a fundamental breakthrough in suggesting that a more barren peridotite would result from the extraction of a basaltic magma from the Earth's mantle thus linking the origin of basalts with mantle peridotites (Fig. 2.2). The next 60 years of petrological and geochemical research tended to reiterate this assertion and essentially concluded that most xenolith rock types were related by melt formation and extraction processes, i.e. source (lherzolite)–residue (harzburgite)–melt(pyroxenite). With regard to eclogites Bowen (1928) declined to

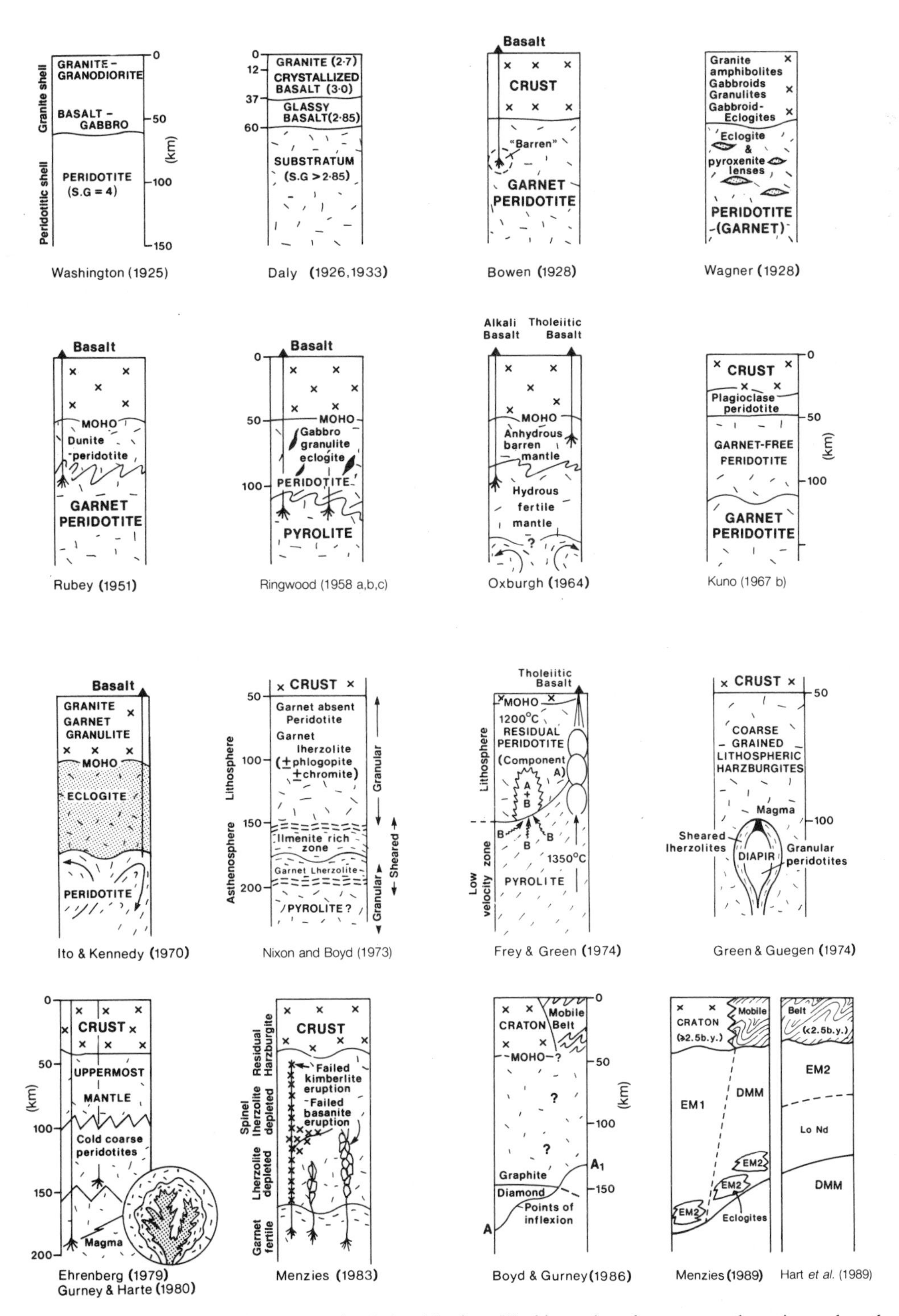

Fig. 2.2. The changing perception of crust–mantle relationships from Washington's early crust–mantle sections, where the SIAL and SIMA acronyms are believed to best match a granite–basalt crust and a peridotite mantle, to the most recent geochemical acronyms used to identify geochemical domains in the lithospheric mantle beneath Archaean cratons and post-Archaean mobile belts. See the text for the historical relevance of these crust–mantle sections.

comment on whether or not an eclogitic layer existed above or below the Earth's peridotitic layer but he maintained that a basaltic composition existed within the constituent minerals of eclogite (viz. garnet and pyroxene). This provided an important basis for later studies of the gabbro–eclogite transition and the modern idea that eclogites may be recycled basalt (i.e. oceanic lithosphere). Over the next 50 years the idea that peridotite was the dominant rock type in the upper mantle gained wide acceptance (e.g. Sobolev 1977), and Rubey (1951) was possibly the first person to suggest petrological and geochemical zonation in the subcontinental mantle with the residue from extraction of basalt *underlain* by more fertile garnet peridotite (Fig. 2.2).

Most investigators accepted that peridotitic xenoliths were deep-seated in origin when the primary mineral assemblages observed in xenoliths and orogenic peridotites compared favourably with experimentally constrained pressure–temperature stability fields. Those stability fields for peridotite facies rocks were plagioclase peridotite (< 30 km), spinel peridotite (< 70 km), and garnet peridotite (> 70 km) (Boyd and England 1960; O'Hara 1967*a*,*b*, 1968). Several contrasting interpretations of the crust–mantle boundary, however, were forthcoming from experimental work completed in the 1960s. Kuno (1967*b*) produced a lithospheric cross-section for the island of Japan based on xenolith and seismic data where major changes in the peridotite facies of the mantle (plagioclase peridotite to garnet peridotite) were consistent with known experimental work (Fig. 2.2). In contrast, Ito and Kennedy (1970) suggested that the crust–mantle boundary or the Moho was essentially the gabbro–eclogite transition with the implication that the subcrustal mantle was eclogite and not peridotite (Fig. 2.2).

Whether or not xenoliths were precipitates of basaltic or kimberlitic magmas or actual fragments of the subcrustal mantle received a lot of attention (Wilshire and Binns 1961; O'Hara and Mercy 1963; Carter 1965; Challis 1965; Frantsesson 1968). Ernst (1965) noted that the metamorphic textures in xenoliths were incompatible with the suggestion that they were magmatic accumulates, and others noted some fundamental differences in the mineral chemistry of basalt-borne and kimberlite-borne xenoliths and believed that kimberlite xenoliths were deeper in origin (Forbes and Kuno 1967; Kuno 1967*a*,*b*). Their preference was for a model where peridotites had a mantle origin and were residua after partial melting processes. Furthermore, O'Hara (1967*b*) and Forbes and Kuno (1967) stressed the importance of xenoliths and concluded that they were mantle-derived and must be related to the genesis of basaltic magmas. In contrast, Davidson (1967) believed that eclogites and peridotites were crustal in origin and Frantsesson (1968) compared the petrology of kimberlite-borne xenoliths with cumulate peridotites and pyroxenites formed by high-level fractionation processes.

The significance of the textural variability observed in high-pressure xenoliths was unravelled with the help of experimental data. Shearing of mantle peridotite or syntectonic recrystalization of olivine was shown to be an important, possibly dominant flow mechanism in the convecting upper mantle or asthenosphere (Ave'Lallement and Carter 1970; Carter and Ave'Lallement 1970). Sheared textures were observed to grade into granular textures on the scale of a single hand specimen and it was concluded that the granular texture found in xenoliths was the 'oldest' texture (Cox *et al.* 1973; Harte *et al.* 1975; Mercier and Nicolas 1975; Harte 1977; Coisy and Nicolas 1978). Furthermore, thermobarometric data for sheared and granular xenoliths (Boyd 1973; Boyd and Nixon 1973, 1975) revealed an inflexion in the geotherm marking a change from low-temperature/low-pressure (granular) xenoliths to higher-temperature/higher-pressure (sheared) xenoliths. On the basis of the coincidence of these xenoliths with the continental geotherm, coarse granular peridotites were interpreted as fragments of the lithosphere (cold and equilibrated). In contrast, the deviation of the sheared xenoliths toward the oceanic geotherm was taken as evidence of an asthenospheric influence during their formation, possibly in the form of shear-heating at the base of the lithosphere (Nixon and Boyd 1973). Consequently, the crust–mantle section of Nixon *et al.* (1973), based on kimberlite-borne xenoliths and the thermobarometric data of Boyd (1973), was one of the first to subdivide the mantle into lithosphere (granular) and asthenosphere (sheared) (Fig. 2.2). This vertical variation should be compared with the earlier ideas of Rubey (1951) and Oxburgh (1964). Green and Guegen (1974) reinterpreted the thermobarometric data of Boyd (1973) and proposed a diapiric origin for the association of granular and sheared xenoliths (Fig. 2.2), noting that the high strain rate of sheared xenoliths cannot be maintained for long periods of time (Goetze 1975). In their model the outer portion of the diapir deformed plastically and had a porphyroclastic or mosaic–porphyroclastic texture and the core remained intact and had a coarse granular texture. Wilshire and Pike (1975) adopted this model to explain the presence of gabbroic and pyroxenitic bands in lherzolitic xenoliths and massifs in terms of polybaric partial melting processes and the segregation of melt products. The cross-cutting relationships in composite xenoliths and orogenic peridotite bodies were consistent with crystallization of basaltic magmas in the spinel–facies mantle. Further refinements of the diapiric model

(Ehrenberg 1979; Dawson 1980; Gurney and Harte 1980; Harte 1983) were undertaken in an attempt to explain chemical modifications in the lithosphere that would result from upwelling of silicate melts from the asthenosphere (Fig. 2.2). Sheared xenoliths were known to have formed at a higher temperature than granular xenoliths and the model of cold lithosphere–hot asthenosphere interaction (Nicolas *et al.* 1987) was compatible with the enrichment in magmatophile elements (e.g. Fe, Ti, and K) in the sheared hot xenoliths and the refractory nature of the granular cold xenoliths. The presence of a metasomatic melt or fluid may be of fundamental importance to the recrystallization process if, as recently suggested, the recrystallization of peridotite minerals is fluid-assisted (Drury and Van Roermund 1989).

Studies of the specific gravity of kimberlite-borne and basalt-borne xenoliths revealed that the residue left after partial melting was less dense than the parental material (O'Hara 1975; Boyd and McCallister 1976; Jordan 1979), an observation that would help constrain the spatial relationships of mantle peridotites. Undepleted and hotter lherzolites were found to be denser than depleted and cooler harzburgites and dunites thus explaining the predominance of harzburgites in kimberlite-borne xenolith suites (Archaean lithosphere). The ubiquity of less refractory spinel lherzolites beneath Proterozoic and Phanerozoic continental crust betrayed a fundamental difference in the formation of Archaean and post-Archaean lithospheric mantle (Richter 1988). In general, underplating of the continental and oceanic crust with refractory peridotite appeared to be an inevitable consequence of partial melting processes in the Earth's mantle (Anderson, Chapter 1, this volume). Such refractory or depleted peridotite would grade at depth into fertile or undepleted asthenosphere still capable of producing basalt. This zonal distribution had many similarities to that previously suggested by Rubey (1951) (Fig. 2.1).

Petrological and seismological models of the uppermost mantle (Ringwood 1975; Anderson 1977; Jordan 1975*a*,*b* 1988; Anderson, Chapter 1, this volume) are in general agreement but it should be noted that the various petrological solutions to seismic data can only be distinguished above 220 km (lithosphere). At depths of 200–400 km (lithosphere–asthenosphere) under the oceans and the continents the seismic properties of the mantle are consistent with the presence of either undifferentiated peridotite or eclogite (Anderson 1979*a* 1987). There is a possible discontinuity at 200 km which has been variably interpreted as: (1) a chemical discontinuity between eclogite and peridotite; (2) a boundary between fertile and depleted material; or (3) the base of the lithospheric mantle (Anderson, Chapter 1, this volume).

It is now generally accepted that xenolith suites contain a mixed population that comprises mantle and crustal lithologies and polybaric derivatives of tholeiitic and alkaline magmas (Wass and Irving 1976; Dawson 1980; Wilshire *et al.* 1985; Nixon 1987). Moreover, the presently accepted geometry of the continental mantle is that the cold, granular, equilibrated, less dense lithosphere (mechanical/chemical boundary layer) overlies a substratum of hot, deformed, denser asthenosphere (adiabatic interior) and that a transition zone exists beneath these two extremes (Fig. 2.1).

2.4. GEOCHEMISTRY OF THE CONTINENTAL MANTLE—MANTLE XENOLITHS

2.4.1. Undifferentiated mantle

Ultramafic rocks of all types provided a basis for models of the undifferentiated Earth. Ringwood (1958*a–c*, 1966) revolutionized our understanding of the evolution of the Earth's mantle with the introduction of a theoretical undifferentiated mantle composition (i.e. pyrolite) based on mixing appropriate amounts of peridotite (70 per cent) and basalt (30 per cent) (Table 2.1). The basis of this work was the assumption that an initial undifferentiated lherzolitic composition would leave a residual/refractory harzburgite after extraction of a basaltic melt (cf. Bowen 1928). Ringwood (1958*a–c*) proposed that beneath the Moho a dunitic and peridotitic mantle extended to 100–150 km (Fig. 2.2) and contained segregations of differentiated material (i.e. eclogite, pyroxene granulite, or gabbro) which had failed to reach the surface as eruptive products. This in turn was underlain by undifferentiated mantle (e.g. pyrolite). The concept of 'pyrolite' gained such immediate acceptance that petrologists and geochemists undertook a search for natural material that best matched 'pyrolite' in terms of major and minor elements, either in the form of orogenic peridotites, basalt-borne xenoliths or kimberlite-borne xenoliths (e.g. Green 1964; Melson *et al.* 1967; Hutchison 1974; Maaloe and Aoki 1977). The Lizard orogenic peridotite (Green 1964) and basalt-borne xenoliths from the Massif Central, France (Hutchison 1974) were believed to be representative of undepleted continental mantle while the St. Paul's rocks (Melson *et al.* 1967) and the Tinaquillo peridotite constituted heterogeneous oceanic mantle (Table 2.1). Whether such peridotites were related to basaltic magmatism very much depended on

Table 2.1. Primitive upper mantle compositions.

	Weight per cent according to reference								
	a	b	c	d	e	f	g	h	i
SiO_2	43.60	45.16	45.0	44.71	45.6	45.1	46.2	46.0	45.8
TiO_2	0.04	0.71	0.09	0.16	0.2	0.22	0.23	0.18	0.1
Al_2O_3	2.40	3.54	3.5	2.46	3.3	4.14	4.75	4.06	3.3
FeO	8.80	8.04	8.0	8.15	8.1	7.82	7.7	7.55	8.0
MnO	0.10	0.14	0.11	0.18	0.15	0.13	0.13	—	0.16
MgO	41.50	37.47	39.0	41.0	38.5	38.0	35.5	37.8	38.3
CaO	2.50	3.08	3.25	2.42	3.1	3.54	4.36	3.21	2.8
Na_2O	0.32	0.57	0.28	0.29	0.4	0.36	0.40	0.33	0.2
Cr_2O_3	0.40	0.43	0.41	0.42	0.4	0.45	0.43	0.47	0.4
NiO	0.34	0.20	0.25	0.26	—	0.27	0.23	—	0.2

a, Hess (1964). Orogenic peridotite/serpentinite.
b, Ringwood (1966). Hypothetical mixture of basalt (high Ti) and peridotite.
c, Hutchison (1974). Basalt-borne spinel lherzolites.
d, Maaloe and Aoki (1977). Basalt-borne and kimberlite-borne spinel and garnet lherzolites.
e, Ringwood (1979). Hypothetical mixture and basalt (low Ti) and peridotite.
f, Wanke (1981). Basalt-borne spinel lherzolites/meteorites.
g, Palme and Nickel (1985). Basalt-borne spinel lherzolites/meteorites.
h, Zindler and Hart (1986). Bulk silicate earth composition.
i, Boyd (1989). Hypothetical mixture of kimberlite-borne low-temperature peridotite and Yilgarn komatiite.

their isotopic ratios. Ten years after the introduction of the concept of 'pyrolite', strontium isotopic studies confirmed that some of the kimberlite-borne and basalt-borne peridotite xenoliths were related to basaltic magmas, either as residua, accumulates, or pristine material (Gast 1968; Leggo and Hutchison 1968). On the basis of low concentrations of titanium and alkaline earths, Ringwood (1975) questioned whether any of these garnet or spinel peridotites were appropriate candidates for undifferentiated mantle. It became apparent, however, that the Ti and K contents of pyrolite were too high due to the use of an enriched tholeiite in the calculation of the pyrolite composition. This led to a revision of the pyrolite composition (Ringwood 1979; Table 2.1). Ten years later Boyd (1989) calculated the composition of the undifferentiated Archaean mantle with the use of a kimberlite-borne low-temperature granular peridotite and a Yilgarn komatiite. The basic assumption was that the highly magnesian character of low-temperature peridotites had resulted from the removal of high-temperature magnesian melts. This latest undifferentiated mantle composition compares favourably with pyrolite (Table 2.1).

The composition of the undifferentiated Earth had been estimated by mixing of melt and residue in specific proportions (Ringwood 1975) and by compiling averages of xenolith chemical analyses (Hutchison 1974; Maaloe and Aoki 1977). A different approach was adopted by several investigators who compared the elemental abundances and elemental ratios of meteorites with xenolith data (Palme *et al.* 1978; Jagoutz *et al.* 1979; Wanke 1981; Palme and Nickel 1985). Palme *et al.* (1978) demonstrated that refractory element ratios in basalt-borne spinel lherzolites were chondritic in cases where magnesian compositions were close to that of the undifferentiated Earth (i.e. MgO = 38 per cent) and significantly different from chondritic element ratios if the xenolith in question was refractory (i.e. MgO > 38 per cent). While Jagoutz *et al.* (1979) reported that compatible refractory elements occurred in approximately chondritic abundances, Palme and Nickel (1985) found that the Ca/Al ratio in basalt-borne spinel lherzolites was non-chondritic and constant between many suites of lherzolites (Table 2.1). They took this to mean that the upper mantle had undergone a global differentiation event involving removal of garnet at times as early as the Archaean (cf. Anderson 1979*a*; Richardson, Chapter 3, this volume). High-pressure fractionation of majorite—a pyroxene with the garnet structure—has more recently been invoked as a mechanism to account for the geochemistry of granular kimberlite-borne xenoliths from South Africa (Herzberg *et al.* 1988). This hypothesis is at odds, however, with the major and trace element geochemistry of granular peridotites, a feature that is more consistent with an origin as a refractory residue of partial melting (Boyd and Mertzman 1987; Menzies, Chapter 4, this volume).

From these comparisons of xenolith data with meteorite data it is now accepted that several basalt-borne xenoliths closely match the composition of primordial mantle within 10–20 per cent for most elements (Palme *et al.* 1978; Jagoutz *et al.* 1979; Sachtleben and Seck 1981; Palme and Nickel 1985).

2.4.2. Orogenic and alpine peridotites

Hess (1955, 1962) suggested a form of mantle convection in the undifferentiated mantle (Table 2.1) under ocean ridges and this helped focus attention on alpine and orogenic massifs which represent the largest exposures of plagioclase-, spinel-, and garnet-bearing peridotite of mantle origin. It soon became apparent that the petrological similarities between orogenic peridotites and basalt-borne or kimberlite-borne xenoliths were of significance with regard to the internal nature of the Earth (Ross *et al.* 1954; Wilshire and Pike 1975). Orogenic peridotites were believed to be accumulates from basaltic magmas extracted from depleted mantle or fragments of refractory mantle (e.g. Green 1967) because of their depletion in light rare earth elements relative to chondrite (Haskin *et al.* 1966; Frey 1969, 1970). While petrological and geochemical studies of orogenic peridotites related them to basaltic magmas, isotopic data made such an hypothesis untenable unless one invoked disequilibrium partial melting. The high $^{87}Sr/^{86}Sr$ ratios in alpine peridotites compared to the low $^{87}Sr/^{86}Sr$ ratios in mid-ocean ridge basalts spawned various complex genetic models (Roe 1964; Stueber 1965; Stueber and Murthy 1966) that were extended to oceanic peridotites (Bonatti 1968, 1971). A layer of residual 'alpine type peridotite' was postulated to exist beneath the crust of the Atlantic Ocean, in addition presumably, to the residue left after production of mid-ocean ridge basalt. These ideas changed dramatically with the advent of plate tectonic theory and the belief that alpine massifs were obducted fragments of oceanic crust and mantle (Dewey and Bird 1971; Nicolas and Jackson 1972). Moreover, the realization that high-strontium isotopic ratios in rocks with low-strontium contents might result from secondary processes (Hart 1970) aroused some concern about the significance of isotopic analyses of serpentinized peridotites. Strontium (Brueckner 1974; Menzies and Murthy 1978) and neodymium isotopic analyses (Richard and Allegre 1980; Reisberg and Zindler 1986) of the mineral constituents of orogenic peridotites revealed that, whilst the whole-rock peridotite could indeed have an enhanced isotopic composition (due to interaction with the fluids responsible for serpentinization), the constituent clinopyroxene had retained a primary isotopic composition in part due to the high concentration of Sr and its resistance to alteration. It is now generally accepted that orogenic peridotites have complex origins and that some massifs (e.g. Lanzo, Italy) represent the closest approximation to the source of mid-ocean ridge basalts (cf. Hurley 1967) and, as such, contain evidence of mantle processes that are closely linked to the genesis of tholeiitic magmas (Bodinier *et al.* in press).

2.4.3. Metasomatized xenoliths

Oxburgh (1964) first suggested that amphibole (or phlogopite) might represent a suitable upper mantle repository for heat-producing elements like potassium (Fig. 2.2). He noted the barren nature of many xenoliths and the vast enrichment in potassium in amphibole-bearing xenoliths from the south-western USA (Lausen 1927). More significantly Oxburgh (1964) suggested that the subcrustal mantle was zoned from anhydrous and refractory mantle, immediately beneath the crust, to hydrous and fertile mantle at depth. The proposed mantle zonation had implications for mantle convection, and he commented that 'convection cells [must have operated] at depths too great to disturb the layering postulated'. Several years later the importance of incompatible elements and hydrous phases in peridotitic rocks was elegantly demonstrated by some of the first high-precision trace-element studies of hydrous peridotites (Griffin and Murthy 1969) and experimental work that showed amphibole and mica to be stable at mantle pressures and temperatures (Lambert and Wyllie 1968; Yoder and Kushiro 1969; Hill and Boettcher 1970).

The concept of mantle metasomatism or the addition of hydrous or anhydrous phases to the mantle had its foundation in these early studies (Erlank 1973; Frey and Green 1974; Boyd and Nixon 1975; Harte *et al.* 1975; Lloyd and Bailey 1975). In a very significant paper Frey and Green (1974) recognized two distinct mantle components within basalt-borne xenoliths (Fig. 2.2), a lithospheric residue (component A) formed by tholeiite extraction, and an asthenospheric alkalic melt (component B). The latter behaved as a migratory fluid phase and brought about a contiguous enrichment in the light rare earth elements in otherwise refractory spinel lherzolites. Irving (1974, 1980) later demonstrated the presence of silicate melts, or the derivatives thereof, in the mantle in the form of igneous-textured pyroxenite veins or stockworks and thus reiterated an earlier suggestion of Ringwood (1958*a–c*) that the pyroxenitic and eclogitic lenses in the mantle were melts that had stagnated *en route* to the surface (Fig. 2.2). Mantle metasomatism was proposed as a precursor to

(Boettcher and O'Neil 1980; Menzies and Murthy 1980*a,b*; Wass and Rogers 1980), and, as a consequence of (Roden *et al.* 1984; Menzies *et al.* 1985), alkaline magmatism in the upper mantle. This confirmed the model of Frey and Green (1974) and was a necessary outcome of the models of Rubey (1951) and Ringwood (1962*a,b*) where mantle-derived magmas stagnated and crystallized within the upper mantle (Fig. 2.2). The chemical heterogeneity observed in lithospheric xenoliths and basaltic rocks was thus interpreted as the result of the migration of fluids (K-rich) and silicate melts (Fe–Ti rich) (Hawkesworth *et al.* 1984; Menzies and Hawkesworth 1987*a*). Metasomatism provided a mechanism by which to transform the petrology, chemistry (Kyser, Chapter 7, this volume; Menzies, Chapter 4, this volume), and oxidation state of the lithosphere (Haggerty, Chapter 5, this volume). While in a general sense metasomatism was perceived as the result of asthenosphere–lithosphere interaction (Menzies *et al.* 1987; McKenzie 1989), detailed investigations of basalt-borne xenoliths (Griffin *et al.* 1988; O'Reilly and Griffin 1988) indicated that metasomatism was a complex form of enrichment. In contrast to xenoliths, orogenic peridotites provided sufficient exposure such that the effects of chromatographic fractionation around metasomatic veins could be studied in detail (Fabries *et al.* 1989; Bodinier *et al.* 1990). In all these enrichment models it can be shown that alkaline or tholeiitic melts, or derivatives thereof, were the dominant metasomatic component. Similarly, the importance of potassic or carbonate rich fluids/melts in metasomatism has recently been demonstrated by experimental (Watson *et al.*, Chapter 6, this volume) and theoretical considerations (McKenzie 1989).

Despite the fact that enrichment processes and mantle metasomatism were embraced in the 1980s as a panacea for all problems relating to basalt genesis, considerable debate centred around whether or not hydrous phases in mantle xenoliths were primary or secondary, the latter being due to infiltration of the xenolith by the host magma. Dawson and Powell (1969), Carswell (1975), and Francis (1976) were among the first to report primary mica in garnet and spinel lherzolites. The compositional characteristics and textural relationships of these amphiboles and micas proved that they had been introduced 'prior to entrainment' and were therefore of mantle origin. This was heralded as a very important contribution in that it verified that metasomatism could occur at mantle pressures and temperatures. The metasomatism could, however, have involved magmas similar in composition to the host magma. Indeed, whether or not the metasomatism observed in xenoliths involved the infiltration of kimberlitic or basaltic melts or their derivatives at mantle pressures and temperatures was debated (Hawkesworth *et al.* 1983; Kramers *et al.* 1983; Erlank *et al.* 1987; Menzies *et al.* 1987). The polybaric (100–150 km) nature of metasomatic processes in the cratonic lithosphere was known from thermobarometric data for metasomatized garnet peridotites from Bultfontein, South Africa (Waters and Erlank 1988). Moreover, micro-inclusions in diamonds (Navon *et al.* 1988) revealed the presence of potassic metasomatic melts or fluids at 180–200 km which may have had an origin in the asthenosphere or lower mantle as proto-kimberlite (Menzies 1988). The presence of small volume melts in the hot asthenosphere (McKenzie 1989) and their effectiveness as metasomatic agents (Watson and Brenan 1987; Watson *et al.*, Chapter 6, this volume) makes it highly probable that most of the metasomatism observed in lithospheric xenoliths was ultimately of sub-lithospheric origin.

Other isotopic studies of basalt-borne and kimberlite-borne xenoliths concentrated on whether or not the lithospheric mantle was isotopically identical to oceanic basalts. Erlank and Shimizu (1977) analysed richterite-bearing peridotites and demonstrated the presence of high $^{87}Sr/^{86}Sr$ ratios in the mantle. They maintained that such mantle had evolved in response to a higher Rb/Sr ratio than that commonly associated with the source of oceanic basalts. Earlier several investigators had concluded that crustal contamination during entrainment had been responsible for the high $^{87}Sr/^{86}Sr$ ratios in whole-rock peridotites from South Africa (e.g. Barrett 1975). Sr and Nd isotopic studies of clinopyroxene and garnet separates from ultramafic xenoliths, however, showed this not to be the case and established the presence of isotopically enriched lithospheric mantle ($^{87}Sr/^{86}Sr = 0.702$–0.836) beneath the cratonic regions of Africa (Menzies and Murthy 1980*a*; Cohen *et al.* 1984; Richardson *et al.* 1984, 1985). In addition, oxygen, carbon, and hydrogen isotopic studies (e.g. Boettcher and O'Neil 1980; Kyser *et al.* 1981, 1982; Mattey *et al.* 1985; Kyser, Chapter 7, this volume) indicated that many of the trace element enrichments in the subcratonic lithosphere may have involved recycling processes. Interestingly, the isotopic heterogeneities of the subcontinental mantle were adequate to account for the characteristics of oceanic basalts provided one could find a mechanism of incorporating the continental mantle into the source region of ocean island basalts (e.g. McKenzie and O'Nions 1983; Hawkesworth *et al.* 1986).

2.4.4. On-craton, off-craton systematics

With the use of regression analysis Maaloe and Aoki

(1977) provided an estimate of the composition of the undifferentiated Earth (Table 2.1) and demonstrated that basalt-borne spinel peridotites were less refractory than kimberlite-borne garnet lherzolites. Jagoutz *et al.* (1979) further demonstrated that several basalt-borne spinel lherzolites had a full complement of basaltic elements and that only highly incompatible elements were depleted. Taken together these data meant that the mantle beneath the stable, Archaean craton of South Africa was chemically different from that beneath the adjacent more unstable, Proterozoic mobile belts. This pointed to a marked lateral change in the lithosphere in that the composition of the lithosphere beneath the Proterozoic crust was thought to be very similar to pyrolite or undepleted mantle and the lithosphere beneath Archaean crust was thought to be devoid of any magmatophile elements. The idea of petrologically and chemically distinct lithospheres beneath crustal provinces of different age was further investigated by Boyd *et al.* (1985), Boyd (1987) and Boyd and Mertzmann (1987). They suggested that cold granular highly refractory peridotites were restricted to the Archaean cratons of the world and that they were probably produced by high degrees of partial melting perhaps during production of komatiitic melts (Boyd 1989). This was not at variance with the apparent abundance of subcratonic garnet 'lherzolites' because many of these peridotites are actually garnet harzburgites and their lherzolitic mineralogy was produced by exsolution of orthopyroxene, clinopyroxene, and garnet from Ca–Al rich orthopyroxene (Dawson *et al.* 1980; Cox *et al.* 1987). Dawson *et al.* (1980) demonstrated exsolution of garnet and clinopyroxene from high Al–Ca pyroxene in fertile harzburgite and Cox *et al.* (1987) inferred the same by consideration of the spatial relationships between pyroxenes and garnet. Furthermore, Cox *et al.* (1987) proposed that the cratonic residua had formed by high-level (100–250 km) melting of almost dry peridotite during the production of tholeiitic or komatiitic magmas. In contrast, the hot sheared peridotites were thought to represent accreted oceanic lithosphere added by subduction processes to the edge of Archaean cratonic nucleii (Boyd 1987; Boyd and Mertzmann 1987; Kyser, Chapter 7, this volume).

Studies of the suboceanic mantle provided a fundamentally important comparative data base for unravelling the complexities of the mantle beneath continents. The chemistry of oceanic peridotites varies systematically from preoceanic rifts, passive margins, and mature ocean basins to supra-subduction environments (Dick and Fisher 1984; Dick *et al.* 1984; Michael and Bonatti 1985; Bonatti and Michael 1989). Bonatti and Michael (1989) investigated the chemical heterogeneity of the Phanerozoic lithosphere and noted that the lithospheric mantle is more depleted in a supra-subduction environment than in a mature ocean or a preoceanic setting. Dick and Bullen (1984) had previously suggested that active margin magmatism could have caused a marked depletion of the peridotitic protolith due to the involvement of water in the melting process. Higher temperatures were not necessarily needed to produce a highly refractory residue. This should be borne in mind when interpreting the highly refractory nature of the Archaean lithospheric mantle; perhaps that is a result of the Archaean mantle being 'highly processed' during lithospheric accretion and active margin processes.

Variations in the geological (Gurney 1972) and geophysical (Richter 1988) properties of the continental lithospheric mantle beneath Archaean and post-Archaean crust reveal fundamentally different refractory lithospheres that may have evolved in response to different processes. Isotopically discrete lithospheres are also apparent beneath Archaean and post-Archaean crust (Basu and Tatsumoto 1980; Menzies and Murthy 1980*a,b*; Menzies 1983; Richardson *et al.* 1985; Erlank *et al.* 1987; Menzies and Hawkesworth 1987*b*; Menzies and Halliday 1988; Roden *et al.* 1988) but, in the case of sub-Archaean lithospheric mantle, the processes are more complex than can be inferred from major and minor element chemistry (Menzies, Chapter 4, this volume; Richardson, Chapter 3, this volume).

2.4.5. Eclogites and subducted lithosphere

Eclogites have not been the focus of attention mainly because seismological and petrological constraints indicated that the shallow lithospheric mantle was olivine-rich and consequently research has tended to concentrate on peridotites. It should be noted, however, that one cannot use seismic data alone to distinguish between eclogitic or peridotitic assemblages in the upper mantle (Anderson 1987) and that seismic anisotropy is only marginally in favour of peridotite. Perhaps at depth garnatite is a fundamentally more important rock type (Anderson 1981). The paucity of eclogites in basalt-borne and kimberlite-borne xenolith suites and the prolific amount of peridotites has also contributed to the lack of data on eclogites. The idea that basalt may be produced by near total melting of a 'basaltic' parent (cf. Bowen 1928) has more recently helped focus attention on eclogites (Hofmann and White 1982; Ringwood 1982). These recent models argue for the involvement of an olivine eclogite in the genesis of mid-ocean ridge basalt or subducted oceanic crust in the production of continental and oceanic basalts.

Terms like 'magmatic' or 'metamorphic' eclogites,

based on the classification of Norwegian eclogites (Eskola 1921), have confused eclogites of shallow crustal origin (Bobrievich and Sobolev 1957) with eclogites of deeper mantle origin. Although Milashev (1960) and Milashev *et al.* (1963) reported a petrological continuum between peridotites and eclogites, a continuous range in compositin is less convincing as intermediate compositions are rare (Bobrievich and Sobolev 1962; Carswell and Dawson 1970). Many of the world's eclogites are broadly basaltic in composition (Rogers 1977; Robey 1981), a feature noted by Bowen (1928), and consequently most petrological models have related the formation of eclogites to basalts. A number of eclogites, however, are different from basaltic protoliths particularly in terms of their potassium and titanium contents (Dawson 1980). Despite this, most of the models proposed for the origin of eclogites involve basaltic derivatives: differentiates of peridotitic magmas (Wagner 1914); derivatives of partial melts of garnet peridotite (O'Hara and Yoder 1967; MacGregor and Carter 1970); subducted oceanic lithosphere (Ringwood and Green 1966; Helmstaedt and Doig 1975; MacGregor and Manton 1986); ancient remnants of lithosphere (Kramers 1979); subducted Archaean anorthositic crust (Jagoutz *et al.* 1984); basaltic melts modified by accumulation, contamination, and metamorphism (Arculus *et al.* 1988); or relicts of primary differentiation of the Earth (Anderson 1981; McCulloch 1982). Detailed studies of eclogites from the environs of the Kaapvaal craton, South Africa have uncovered intrinsic chemical and mineralogical distinctions between eclogites from on and off the craton (Gurney *et al.* 1988; Richardson, Chapter 3, this volume), features that may reflect different thermal regimes. Off-craton and craton-margin eclogites are unlike anything found on the craton and may have a lower crustal origin. Moreover, the garnet–granulite eclogite type xenoliths entrained by non-diamondiferous off-craton kimberlites are not sampled by on-craton kimberlites. Bonney (1899) and Wagner (1914) reported that on-craton eclogites were garnet–clinopyroxene rocks with accessory phases and diamonds, a feature that provides independent evidence of their high-pressure origin as does the presence of coesite (Gurney *et al.* 1988 and references therein). Shervais *et al.* (1988) noted that eclogites from South Africa, Namibia, and Angola could be classified using the system initially outlined by Coleman *et al.* (1965) for Franciscan eclogites produced by Phanerozoic subduction. Eclogites may thus represent metamorphosed lithosphere and polybaric derivatives of mantle-derived melts.

Eclogites and eclogitic inclusions in diamonds from Australia, North America, and Africa have a considerable range in Sr, Nd, and Pb isotopic composition that overlaps with oceanic basalts and averages for the upper and lower crust. These data also indicate late Archaean–early Proterozoic ages (2.6–1.1 billion years) for eclogite inclusions (Kramers 1979; McCulloch 1982; Jagoutz *et al.* 1984; Richardson 1987; Jagoutz 1988; Arculus *et al.* 1989). Some eclogites of early Proterozoic (2.1 billion years) age are believed to be derivatives of recycled oceanic crust, a factor that may point to plate tectonic processes active several billion years ago (Shervais *et al.* 1988). Similarly, eclogites from Premier, South Africa have affinities with asthenospheric mid-ocean ridge basalt sources (Richardson 1987) supporting the earlier suggestion of Ringwood (1962*b*) that eclogitic or pyroxenitic segregations in the upper mantle may represent 'failed eruptions' of asthenosphere-derived material. In contrast, eclogites from Argyle, Australia may be derivatives or hybrids of processes in the asthenosphere and lithosphere similar to those involved in the source regions of lamproites and micaceous kimberlites. It seems appropriate that, with the plethora of data on peridotitic xenoliths, our attention shoud have more recently turned to eclogites —amongst the first xenoliths to be described in detail by Bonney (1899) nearly 100 years ago.

2.4.6. Age of the lithospheric mantle

Linear relationships on Rb/Sr and $^{87}Sr/^{86}Sr$ ratio diagrams for mantle xenoliths were interpreted as mantle ages ranging from Phanerozoic to Archaean, 0.2–3.4 billion years (Manton and Tatsumoto 1969; Kleeman and Cooper 1970; Stueber and Ikramuddin 1974; Basu and Murthy 1977; Erlank *et al.* 1987). Kleeman and Cooper (1970) provided Pb–Pb data for lherzolite xenoliths from Australia that indicated an age of 2.0–2.5 billion years and the Roberts Victor pipe in South Africa contained xenoliths with ages of 1.7–2.4 billion years (Manton and Tatsumoto 1969). A spinel lherzolite xenolith from Baja California gave a relatively precise mineral isochron that indicated the presence of Proterozoic mid-ocean ridge basalt mantle beneath the south-western USA (Basu and Murthy 1977). The existence of such ages in the lithosphere was later substantiated by Sm–Nd isotopic analysis (Stosch *et al.* 1980). Erlank *et al.* (1987) interpreted the isotopic systematics for a suite of metasomatized kimberlite-borne xenoliths from South Africa as indicative of a late Phanerozoic metasomatic event synchronous with Karoo volcanism. Ironically the majority of isotopic data for metasomatic or hydrous minerals in basalt-borne and kimberlite-borne xenoliths gave the age of entrainment indicating that the minerals were either introduced during the episode of volcanism that ultimately transported them to the surface, or, that entrainment reset the isotopic systems during transport (Allsopp and Barrett 1975; Erlank *et al.* 1987). Hydrous minerals with Proterozoic to Archaean ages, however, have been reported in peridotite xenoliths from Africa (Cohen *et al.* 1984; Waters 1987).

Other age information was forthcoming from mantle megacrysts and much of it provided important constraints on the age and origin of the lithosphere. For example, a considerable amount of 'age' data exists for diamonds because of the need to solve whether or not diamonds were phenocrystal or xenocrystal (Harris 1987). Diamond inclusion Sm–Nd data indicates ages of 3.2–3.3 billion years for minerals of the peridotite paragenesis and 1.2–1.5 billion years for minerals of the eclogite paragenesis (Richardson *et al.* 1984; Richardson 1987; Chapter 3, this volume). These data contrast with the Archaean to Proterozoic ages reported for whole-rock eclogites and peridotites and their mineral constituents (Kramers 1979; McCulloch 1982; Jagoutz *et al.* 1984; Richardson 1987; Jagoutz 1988; Arculus *et al.* 1989; Walker *et al.* 1989).

2.5. GEOCHEMISTRY OF THE CONTINENTAL MANTLE—BASALTIC ROCKS

2.5.1. Introduction

Bowen (1928) noted that 'at 37–60 km [one] encounter[s] material of a peridotitic character . . . and if this is true the only source of basaltic magma . . . lies in the peridotite zone, from which it must be produced by selective fusion.' Moreover, he was one of the first to suggest that extraction of basalt from peridotite led to the peridotite being more 'barren' (Fig. 2.2). Although a very acceptable idea towards the end of the twentieth century, these ideas were advanced at a time when some of his colleagues believed in basaltic layers deep in the Earth as a source for basaltic magmas (Daly 1926, 1933; Fig. 2.1).

Experimental petrology greatly advanced our understanding of the origin of basalts. Kuno (1960) linked the genesis of different magmas to depth and some of the earliest basalt–peridotite experiments (Yoder and Tilley 1962) studied the formation of melts to depths of 90 km. O'Hara (1965, 1968) noted that melts in equilibrium with peridotite were different from so-called 'primary magmas' formed at depth and the need for fractionation of olivine became apparent. Green and Ringwood (1967) presented a comprehensive scheme for the production of tholeiitic and alkaline basaltic magmas in terms of polybaric partial melting of undepleted mantle peridotite (=pyrolite) and fractional crystallization. Soon the importance of CO_2 in mantle processes was realized and the stability of mantle carbonate was linked to the genesis of kimberlites and carbonatities (Eggler 1975, 1976, 1978; Wyllie 1987; Wallace and Green 1988; Green and Wallace 1988). In contrast, the compositions of basaltic magmas produced by melting of dry or volatile-free spinel lherzolite were perhaps best estimated from the experimental work of Takahashi and Kushiro (1983) and Fujii and Scarfe (1985). Throughout the 1970s and early 1980s most models of basalt genesis were based on the assumption that large degrees of melting were necessary to produce basaltic magmas from a peridotitic source and that small degree melts were unlikely to be easily mobilized. However, recent theoretical and experimental considerations indicate that small volume melts are mobile at mantle pressures and temperatures and, therefore, earlier models may need to be revised (McKenzie 1984, 1985; Watson and Brenan 1987; Brenan and Watson 1988; McKenzie and Bickle 1988; Hunter and McKenzie 1989; McKenzie 1989; Watson *et al.*, Chapter 6, this volume).

Gast (1960, 1968) and Schilling (1971) pointed out the complementary geochemical relationship between oceanic tholeiites (depleted in large ion lithophile (LIL) and light rare earth (LRE) elements) and alkaline basalts (enriched in LIL and LRE elements). Gast (1968) proposed that the source of mid-ocean ridge basalt (MORB) had undergone a partial melting episode prior to extraction of MORB and he related the origin of alkaline (i.e. ocean island) and tholeiitic (i.e. mid-ocean ridge) basalts to variable degrees of melting of a heterogeneous mantle source (cf. McKenzie and Bickle 1988). Basalts within ocean basins that were not produced at ocean ridges were believed to be produced by hot-spot magmatism (Wilson 1965; Morgan 1971) and it was soon realized that ridge basalts were derived from shallow depleted asthenosphere and ocean island basalts were supplied by deeper mantle plumes (Schilling 1973; Hofmann and Hart 1975; Sun and Hanson 1975).

2.5.2. Continental alkaline volcanic rocks and mantle domains

Lewis (1888, 1897) studied the matrix of diamond (blue ground) and concluded that it was a porphyritic peridotite of igneous eruptive character. Bonney (1899) prefaced an article on the parent rock of the diamond by stating that so much had been written on the occurrence of diamond by 1899 that brevity was the best introduciton (Bonney 1899 and references therein). Despite that much of the next 50 years saw a rapid increase in the data base concerning the geology and mineralogy of kimberlites (Wagner 1914; Williams 1932). Wagner

(1914) believed diamonds to have formed from kimberlitic magmas and noted that some pipes were barren of diamonds. He suggested that this might be due to a lack of carbon or carbon compounds in the deep-seated peridotite zone which he believed to be the ultimate source of kimberlite. Wagner was also the first to note that diamondiferous kimberlites were erupted in stable geological terrains, a point that was re-investigated over 50 years later (Frantsesson 1968; Frantsesson and Prokopchuk 1968; Dawson 1970). They noted that most diamondiferous kimberlites in southern Africa and Siberia were confined to the ancient cratons and that their abundance decreased into the surrounding mobile belts.

Probably because of the association of kimberlites with old crust enriched in radiogenic strontium the high $^{87}Sr/^{86}Sr$ ratios in kimberlites were believed to be due to: (1) incorporation of micro-xenoliths of crust (Powell 1966); (2) contamination with Sr derived from overlying limestones and shales (Brookins 1967); or (3) post-emplacement alteration (Berg and Allsopp 1972). The underlying assumption at this time was that all mantle-derived rocks had a source similar to oceanic basalts and as such should be isotopically identical. If their isotopic composition was different, then this was taken as evidence of crustal contamination and must constitute 'high level' interaction with the crust. This was accepted despite the fact that many 'altered' kimberlites had apparently lower $^{87}Sr/^{86}Sr$ ratios than fresh kimberlites (Paul 1979; Demaiffe and Fieremans 1981). Whilst a source similar to oceanic basalts for basaltic kimberlites is consistent with much of the modern isotopic data (Kramers 1977; Basu and Tatsumoto 1980; Kramers *et al.* 1981), not all kimberlites are genetically the same. In a landmark contribution, Smith (1983) demonstrated that basaltic or Group I kimberlites had a mantle source with the Sr, Nd, and Pb isotopic characteristics of ocean island basalts (OIB) and were therefore interpreted as asthenospheric melts. In contrast, micaceous or Group II kimberlites had Sr, Nd, and Pb isotopic characteristics similar to basalt-borne and kimberlite-borne xenoliths and were therefore different from oceanic basalts. These data were interpreted to mean that micaceous kimberlites had a significant source contribution from the subcontinental lithospheric mantle. The Group I and II terminology introduced by Smith (1983) is now more commonplace than the 'basaltic–micaceous kimberlite' terminology of Wagner (1914). The model of depleted asthenosphere (OIB)-enriched lithosphere interaction proposed by Smith (1983) was re-interpreted by McCulloch *et al.* (1983) who believed that the data was better explained by interaction of asthenospheric (MORB) melts and enriched lithospheric mantle. The restriction of micaceous diamondiferous kimberlites to the craton and the dominant contribution from lithospheric mantle in their magmagenesis supports the contention that an enriched Archaean keel exists beneath the cratons.

Geochemical data for alkaline and tholeiitic basalts provide us with important constraints on the global distributions of mantle reservoirs. Many continental alkaline basalts (e.g. basanites, nephelinites, and alkali olivine basalts), like basaltic kimberlites, are chemically indistinguishable from ocean island basalts (Allegre *et al.* 1981, 1982), a fact that requires parts of the upper mantle to have been a common reservoir beneath both ocean basins and continents. Moreover, this reservoir must have been isolated for billions of years (Sun and Hanson 1975; Tatsumoto 1978; Sun 1980; Chase 1981). Whilst most investigators agreed that an enriched reservoir was required for alkaline ocean island and continental basalts and a depleted reservoir for mid-ocean ridge basalts the exact geometry of these reservoirs remained a hottly debated topic. Gast (1968) and Schilling (1971) proposed that the source for alkali basalts was deep and the source of MORB was shallow, Tatsumoto (1978) proposed the opposite with a shallow enriched reservoir and a deep depleted reservoir. Anderson (1979*a,b*) queried a more fundamental assumption—that the upper mantle was necessarily peridotite. He believed that, as a result of early differentiation within the Earth, a layer of eclogite lay beneath a layer of peridotite. MORB was derived from this eclogite and the source was replenished by subduction.

To this day the exact geometry of MORB and OIB sources, the latter feeding continental and alkaline intraplate volcanoes, remains the source of some debate (Allegre *et al.* 1981, 1982; Thompson *et al.* 1984; Zindler and Hart 1986). It is generally accepted that much of the asthenosphere is MORB-like and that OIB melts may be generated as small volume melts in the lithosphere, the asthenosphere, or in the deeper mantle. Detailed studies of the geochemistry of oceanic basalts (Zindler and Hart 1986) have produced evidence for several chemically distinct reservoirs: DMM or depleted MORB mantle, PREMA or prevalent mantle for ocean islands (e.g. Hawaii), HIMU or mantle with high μ (i.e. U/Pb), EM2 or enriched mantle 2, and EM1 or enriched mantle 1. An estimate of the composition of the bulk silicate earth from the work of Zindler and Hart (1986) compares favourably with that of Wanke (1981) (Table 2.1).

Within the continental regions the correlation between crustal age and kimberlite type helped define chemical provinces in the mantle beneath the Archaean

Kaapvaal craton and the surrounding Proterozoic mobile belt in South Africa. Similarly, compilations of isotopic data for continental alkaline volcanic rocks from the western USA reveal isotopic provinciality (Leeman 1982). Menzies (1989) used xenolith-bearing volcanic rocks as a probe of the geometry and chemistry of lithospheric mantle domains and was able to assign a domain identity to subcrustal regions (Fig. 2.2). In summary, the isotopic characteristics of the lithosphere beneath the mobile belts is similar to that beneath the ocean basins (Hart *et al.* 1989) and the lithosphere beneath stable cratons is to some degree isotopically unique. A close correlation exists between areas of high heat flow (>2 HFU), thin crust (<30 km), and the eruption of continental volcanic rocks whose source is in the asthenosphere, whereas continental volcanic rocks with a significant source contribution from the lithospheric mantle tend to be restricted to regions of low regional heat flow (<2 HFU) and thick crust (>40 km). Chemical provinciality has also been recorded in the lithosphere beneath Scotland (Thirlwall 1982) and has similarly been interpreted as a result of lateral and vertical variability in the lithospheric mantle (Menzies and Halliday 1988). It should be noted that a shallow lithospheric source for continental alkaline basalts (Leeman 1982; Perry *et al.* 1987) from the western USA and Canada contrasts with the classic demonstration of an asthenospheric source for continental alkaline volcanic rocks from West Africa and the Gulf of Guinea (Fitton and Dunlop 1985). Indeed these data seem to indicate that under different circumstances alkaline melts can be generated in the convecting asthenosphere (or deeper) and in the lithosphere. If the chemical provinciality observed in continental volcanic rocks indicates the involvement of lithospheric mantle domains in their genesis, then continental volcanic rocks can be used to literally 'map' mantle domains (Fitton *et al.* 1988; Leat *et al.* 1988; Ormerod *et al.* 1988; Menzies 1989; Menzies and Kyle, Chapter 8, this volume). Similarly, magma distribution and the variability in composition can be used to monitor the vertical movement of the asthenosphere–lithosphere boundary (McKenzie and Bickle 1988). It should be stressed that not all continental alkaline and tholeiitic volcanic rocks are suitable candidates for 'mapping' of the subcrustal lithosphere. Ewart *et al.* (1988) demonstrated that high-level processes involving continental crust may produce similar chemical provinciality in erupted basalts.

2.5.3. Continental flood basalts

The petrology of continental flood basalts was dealt with in many of the early texts (Washington 1922; Kuno 1969) and, more recently, Thompson (1977) deduced that certain flood basalt provinces were extruded at rates equivalent to mid-ocean ridge basalts. To account for these large volumes he proposed that continental basalts had an origin similar to oceanic basalts and that any modification of their trace element or isotopic ratio was due to digestion of variable quantities of crust during upwelling or accumulation of the magmas in the crust (e.g. Cox 1980). Whilst this is undoubtedly the case for several continental flood basalt provinces, including the British Tertiary, many others retain evidence of mantle heterogeneity. Leeman (1975, 1977) and Brooks and Hart (1978) proposed that the continental regions were underlain by lithospheric mantle that was very different from that beneath the ocean basins. Leeman (1977) believed in the existence of old enriched subcontinental mantle characterized by high Rb/Sr rtios and high $^{87}Sr/^{86}Sr$ ratios and Brooks and Hart (1978) suggested that flood basalts inherited age information from their lithospheric mantle source, thus explaining the linear arrays of isotopic and trace element data. Hawkesworth *et al.* (1983, 1986, 1988) have more recently invoked the participation of Proterozoic lithospheric mantle in the production of the Karoo and the Parana flood basalt provinces. This was substantiated by recent Sm–Nd isotopic studies (Ellam and Cox 1989) which point to a Proterozoic age in the source region of some of the Karoo volcanic rocks. Pb isotope data for continental tholeiites from the western USA (Leeman 1975, 1977) also point to a Proterozoic lithospheric mantle source. The fact that Proterozoic rather than Archaean ages are forthcoming from these continental flood basalt provinces is of importance. The major, minor, and trace element chemistry of Archaean lithospheric mantle is more refractory than post-Archaean lithospheric mantle which is closer to pyrolite in composition. It is therefore more likely that such lherzolitic material could melt to produce vast quantities of tholeiitic magma than the refractory harzburgites found beneath the Archaean cratons. Whether continental flood basalts are derived from lithospheric mantle or asthenosphere with high-level contamination with crust remains unresolved in many cases (e.g. Hawkesworth and Norry 1983 and references therein; Moorbath *et al.* 1984 and references therein; MacDougall 1989 and references therein; Leeman and Fitton 1989 and references therein; Menzies and Kyle, Chapter 8, this volume).

2.6. CONCLUDING STATEMENT

Volcanic rocks and xenoliths provide a unique insight

into the petrology and geochemistry of the asthenosphere and lithosphere but it must be remembered that lherzolites and metasomatized xenoliths constitute a small proportion of the lithosphere. The continental lithospheric mantle may be largely comprised of a depleted protolith that remains in a 'pristine' state at some distance from magma conduits within the mechanical boundary layer and well outwith the thermal boundary layer where the ingress of small volume melts from the asthenosphere produces a 'secondary' assemblage. The geochemistry of this depleted protolith holds the key to the evolution of the Archaean lithosphere, perhaps the only part of the upper mantle whose major element chemistry can be explained by extraction of crustal material (e.g. komatiites). Similarly, the precise significance of the lithosphere beneath Proterozoic and Phanerozoic terrains has not been thoroughly investigated. It can be demonstrated that Proterozoic and Phanerozoic lherzolites are a suitable source for tholeiitic melts with the isotopic characteristics of mid-ocean ridge basalts. The existence of such material beneath post-Archaean crust requires some explanation in that MORB mantle was depleted in the first billion years of Earth's history. Moreover, the recent recognition of important correlations between seismic tomography and the geoid and tectonic features like subducted slabs may help focus our attention on active margin processes and the petrology and geochemistry of eclogites. Can active margin processes produce highly magnesian peridotites like those found beneath the Archaean crust? Finally, the importance of small volume melts and metasomatic processes are only now being realized with the study of wetting angles. The hitherto unspecified nature of metasomatic melts can now be more accurately defined (i.e. carbonate melts) but the exact relationship of such chemical transfer mechanisms to the origin of alkaline magmas is unknown as small volume melts can be generated within the asthenosphere and the lithosphere.

ACKNOWLEDGEMENTS

Joe Boyd is thanked for his comments and his 'direction' after reading an earlier version of this chapter. Barry Dawson provided a very useful review and pointed me toward literature that I had inadvertently omitted. Above all the staff at the RHBNC library are thanked for their eagerness to seek out the most obscure references with the minimum of information or time. Christine, Craig, and Kevin are thanked for their graphic and photographic assistance.

REFERENCES

Allegre, C. J., Dupre, B., Lambert, B., and Richard, P. (1981). The subcontinental versus oceanic debate; I. Lead–neodymium–strontium isotopes in primary alkali basalts from a shield area; the Ahaggar volcanic suite. *Earth and Planetary Science Letters*, **52,** 85–92.

Allegre, C. J., Dupre, B., Lambert, B., Richard, P., Rousseau, D., and Brooks, C. (1982). The subcontinental versus suboceanic debate; II. Nd–Sr–Pb isotopic comparison of continental tholeiites with mid-ocean ridge tholeiites, and the structure of continental lithosphere. *Earth and Planetary Science Letters*, **57,** 25–34.

Allsopp, H. L. and Barrett, D. R. (1975). Rb–Sr determinations of South African Kimberlite pipes. *Physics and Chemistry of the Earth*, **9,** 605–17.

Anderson, Don L. (1977). The 650 km discontinuity. *Geophysical Research Letters*, **3,** 347–9.

Anderson, Don L. (1979*a*). Chemical stratification of the mantle. *Journal of Geophysical Research*, **84,** 6297–8.

Anderson, Don L. (1979*b*). The deep structure of continents. *Journal of Geophysical Research*, **72,** 4181–8.

Anderson, Don L. (1981). Hotspots, basalts and the evolution of the mantle. *Science*, **213,** 82–9.

Anderson, Don L. (1987). The depths of mantle reservoirs. In *Magmatic processes: physicochemical principles* (ed. B. O. Mysen), pp. 3–12. Geochemical Society, Pennsylvania.

Arculus, R. J., Ferguson, J., Campbell, B. W., Smith, D., McCulloch, M. T., Jackson, I., Hensel, H. D., Taylor, S. R., Knutson, J., and Gust, D. A. (1988). Trace element and isotopic characteristics of eclogites and other xenoliths derived from the lower continental crust of southeastern Australia and southwestern Colorado Plateau U.S.A. In *Eclogites and Eclogite-facies rocks* (ed. D. Smith), pp. 335–386, Elsevier, Amsterdam.

Ave'Lallement, H. G. and Carter, N. L. (1970). Syntectonic recrystallisation of olivine and modes of flow in the upper mantle. *Geological Society of America, Bulletin*, **81,** 2203–20.

Barrell, J. (1914). The strength of the Earth's crust, *Journal of Geology*, **22,** 425–683.

Barrett, D. R. (1975). The genesis of kimberlite and associated rocks: strontium isotope evidence. *Physics and Chemistry of the Earth*, **9,** 637–54.

Basu, A. R. and Murthy, V. R. (1977). Ancient lithospheric xenolith in alkali basalt from Baja, California. *Earth and Planetary Science Letters*, **34**, 246–53.

Basu, A. R. and Tatsumoto, M. (1980). Nd isotopes in selected mantle-derived rocks and minerals and their implications for mantle evolution. *Contributions to Mineralogy and Petrology*, **75**, 43–54.

Beck, R. (1899). Neues von den afrikanischen Diamantlagerstatten. *Ebenda, S.* 417–19.

Berg, G. W. and Allsopp, H. L. (1972). Low $^{87}Sr/^{86}Sr$ ratios in fresh South African kimberlites. *Earth and Planetary Science Letters*, **16**, 27–30.

Bobrievich, A. P. and Sobolev, V. S. (1957). Eclogitization of pyroxene crystalline schists from the Archaean complex. *Zapiski Vsesoyuznogo Mineralogischeskogo Obshchestvo*, **86**, 3–17.

Bobrievich, A. P. and Sobolev, V. S. (1962). The kimberlitic association of the northern part of the Siberian Platform. *Petrografiya Vostochnoi Sibiri*, **1**, 341–416.

Bodinier, J. L., Menzies, M. A., and Thirlwall, M. F. (1991). Elemental and isotopic geochemistry of the Lanzo lherzolite massif—implications for the temporal evolution of the MORB source. *Journal of Petrology* (In press).

Bodinier, J. L., Vasseur, G., Vernieres, J., Dupuy, C., and Fabries, J. (1990). Mechanisms of mantle metasomatism: geochemical evidence from the Lherz orogenic peridotite. *Journal of Petrology* (in press).

Boettcher, A. L. and O'Neil, J. R. (1980). Stable isotope, chemical and petrographic studies of high pressure amphiboles and micas: evidence for metasomatism in the mantle source regions of alkali basalts and kimberlites. *American Journal of Science*, **280A**, 594–621.

Bonatti, E. (1968). Ultramafic rocks from the Mid-Atlantic ridge—petrologic and strontium isotopic evidence for an alpine type rock assemblage. *Earth and Planetary Science Letters*, **9**, 247–56.

Bonatti, E. (1971). Ancient continental mantle beneath oceanic ridges. *Journal of Geophysical Research*, **76**, 3825–31.

Bonatti, E. and Michael, P. J. (1989). Mantle peridotites from continental rifts to ocean basins to subduction zones. *Earth and Planetary Science Letters*, **91**, 297–311.

Bonney, T. G. (1897). On some rock specimens from Kimberley, South Africa. *Geological Magazine*, **4**, 448–502.

Bonney, T. G. (1899). The parent-rock of the diamond in South Africa. *Geological Magazine*, **6**, 309–21.

Bonney, T. G. (1900). Additional notes on boulders and other rock-specimens from Newlands diamond mines, Griqualand-West. *Proceedings of the Royal Society, London*, **67**, 475–84.

Bowen, N. L. (1928). *The evolution of the igneous rocks.* Princeton University Press, Princeton, New Jersey.

Boyd, F. R. (1973). A pyroxene geotherm. *Geochimica et Ćosmochimica Acta*, **37**, 2533–46.

Boyd, F. R. (1987). High- and low-temperature garnet peridotite xenoliths and their possible relation to the lithosphere-asthenosphere boundary beneath southern Africa. In *Mantle xenoliths* (ed. P. H. Nixon), pp. 403–12. John Wiley and Sons, Chichester.

Boyd, F. R. (1989). Compositional distinction between oceanic and cratonic lithosphere. *Earth and Planetary Science Letters*, **96**, 15–26.

Boyd, F. R. and England, J. L. (1960). Apparatus for phase-equilibrium measurements at pressures up to 50 kilobars and temperatures up to 1750°C. *Journal of Geophysical Research*, **65**, 741–8.

Boyd, F. R. and Gurney, J. J. (1986). Diamonds and the African lithosphere. *Science*, **239**, 472–7.

Boyd, F. R. and McCallister, R. H. (1976). Densities of fertile and sterile garnet peridotites. *Geophysical Research Letters*, **3**, 509–12.

Boyd, F. R. and Mertzman, S. A. (1987). Composition and structure of the Kaapvaal lithosphere, southern Africa. In *Magmatic processes: physiochemical principles* (ed. B. O. Mysen), pp. 13–24. Geochemical Society, Pennsylvania.

Boyd, F. R. and Nixon, P. H. (1973). Structure of the upper mantle beneath Lesotho. *Carnegie Institution of Washington Yearbook*, **72**, 431–45.

Boyd, F. R. and Nixon, P. H. (1975). Origins of the ultramafic nodules from some kimberlites of northern Lesotho and the Monastry Mine, South Africa. *Physics and Chemistry of the Earth*, **9**, 431–54.

Boyd, F. R., Gurney, J. J., and Richardson, S. H. (1985). Evidence for a 150–200 km thick Archaean lithosphere from diamond inclusion thermobarometry. *Nature*, **315**, 387–9.

Brenan, J. N. and Watson, E. B. (1988). Fluids in the lithosphere, 2: Experimental constraints on CO_2 transport in dunite and quartzite at elevated P–T conditions with implications for mantle and crustal decarbonation processes. *Earth and Planetary Science Letters*, **91**, 141–58.

Brookins, D. G. (1967). The strontium geochemistry of carbonates in kimberlites and limestones from Riley County Kansas. *Earth and Planetary Science Letters*, **2**, 235–40.

Brooks, C. and Hart, S. R. (1978). Rb–Sr mantle isochrons and variations in the chemistry of Gondwanaland's lithosphere. *Nature*, **271**, 220–3.

Brueckner, H. K. (1974). Mantle Rb/Sr and $^{87}Sr/^{86}Sr$ ratios from clinopyroxenes from Norwegian garnet peridotites and pyroxenites. *Earth and Planetary Science Letters*, **24**, 26–32.

Carswell, D. A. (1975). Primary and secondary phlogopites and clinopyroxenes in garnet lherzolite xenoliths. *Physics and Chemistry of the Earth*, **9**, 417–29.

Carswell, D. A. and Dawson, J. B. (1970). Garnet peridotite xenoliths in South African kimberlite pipes and their petrogenesis. *Contributions to Mineralogy and Petrology*, **25**, 163–84.

Carter, J. L. (1965). The origin of olivine bombs and related inclusions in basalts. Unpublished D.Phil. thesis. Rice University, Houston, Texas.

Carter, N. L. and Ave'Lallement, H. G. (1970). High temperature flow of dunite and peridotite. *Geological Society of America Bulletin*, **81**, 2181–202.

Challis, G. A. (1965). The origin on New Zealand ultramafic intrusions. *Journal of Petrology*, **6**, 322–64.

Chase, C. (1981). Oceanic island Pb: two stage histories and mantle evolution. *Earth and Planetary Science Letters*, **52**, 277–84.

Cohen, E. (1879). Uber einen Eklogit, welcher als Einschluss in den Diamantgruben von Jagersfontein, Orange Freistaat, Sudafrika vorkommt. *Neues Jahrbuch für Seismologie*, 864–72.

Cohen, R. S., O'Nions, R. K., and Dawson, J. B. (1984). Isotope geochemistry of xenoliths from East Africa: implications for development of mantle reservoirs and their interaction. *Earth and Planetary Science Letters*, **68**, 209–20.

Coisy, P. and Nicholas, A. (1978). Structure et géodynamique du manteau superieur sous le Massif Central (France d'après l'étude des enclaves des basaltes). *Bulletin Mineralogique*, **101**, 424–36.

Coleman, R. G., Lee, D. E., Beatty, L. B., and Brannock, W. W. (1965). Eclogites and eclogites: their differences and similarities. *Geological Society of America Bulletin*, **76**, 483–508.

Cox, K. G. (1980). A model for flood basalt vulcanism. *Journal of Petrology*, **21**, 629–50.

Cox, K. G., Gurney, J. J., and Harte, B. (1973). Xenoliths from the Matsoku pipe. In *Lesotho kimberlites* (ed. P. H. Nixon), pp. 76–100. Lesotho National Development Corporation, Maseru.

Cox, K. G., Smith, M. R., and Beswetherick, S. (1987). Textural studies of garnet lherzolites: evidence of exsolution origin from high-temperature harzburgites. In *Mantle xenoliths* (ed. P. H. Nixon), pp. 537–50. John Wiley and Sons, Chichester.

Daly, R. A. (1926). *Our mobile Earth*. C. Scribner's Sons, New York.

Daly, R. A. (1933). *Igneous rocks and the depths of the Earth*. McGraw-Hill, New York.

Davidson, C. F. (1967). The so-called "cognate xenoliths" of kimberlite. In *Ultramafic and related rocks* (ed. P. J. Wyllie), pp. 342–5. John Wiley and Sons, New York.

Dawson, J. B. (1970). The structural setting of African kimberlite magmatism. In *African magmatism and tectonics* (ed. T. N. Clifford and I. G. Gass), pp. 321–55. Oliver and Boyd, Edinburgh.

Dawson, J. B. (1980). *Kimberlites and their xenoliths*. Springer-Verlag, Berlin.

Dawson, J. B. and Powell, D. G. (1969). Mica in the upper mantle. *Contributions to Mineralogy and Petrology*, **22**, 233–7.

Dawson, J. B., Smith, J. V., and Hervig, R. L. (1980). Heterogeneity in upper mantle lherzolites and harzburgites. *Philosophical Transactions of the Royal Society, London*, **A297**, 323–31.

Demaiffe, D. and Fieremans, M. (1981). Strontium isotoe geochemistry of the Mbuji–Mayi and Kundelungu kimberlites, Zaire, Central Africa. *Chemical Geology*, **31**, 311–23.

Dewey, J. and Bird, J. M. (1971). Mountain belts and the new global tectonics. *Journal of Geophysical Research*, **75**, 2025–47.

Dick, H. J. B. and Bullen, T. (1984). Chromian spinel as a petrogenetic indicator in abyssal and alpine-type peridotites and spatially associated lavas. *Contributions to Mineralogy and Petrology*, **86**, 54–76.

Dick, H. J. B. and Fisher, R. L. (1984). Mineralogic studies of the residues of mantle melting: abyssal and alpine-type peridotites. In *Kimberlites, the mantle and crust–mantle relationships* (ed. J. Kornprobst), pp. 295–308. Elsevier, Amsterdam.

Dick, H. J. B., Fisher, R. L., and Bryan, W. B. (1984). Mineralogic variability of the uppermost mantle along mid-ocean ridges. *Earth and Planetary Science Letters*, **69**, 88–106.

Drury, M. R. and Van Roermund, H. L. M. (1989). Fluid assisted recrystallisation in upper mantle peridotite xenoliths from kimberlites. *Journal of Petrology*, **30**, 133–52.

Dunn, E. J. (1881). Notes on the diamond fields South Africa. *Cape Press XIII*, Cape Town.

Eggler, D. H. (1975). CO_2 as a volatile component of the mantle: the system Mg_2SiO_4–SiO_2–H_2O–CO_2. *Physics and Chemistry of the Earth*, **9,** 869–81.

Eggler, D. H. (1976). Does Co_2 cause partial melting in the low velocity layer of the mantle? *Geology*, **4,** 69–72.

Eggler, D. H. (1978). Stability of dolomite in a hydrous mantle with implications for the mantle solidus. *Geology*, **6,** 397–400.

Ehrenberg, S. N. (1979). Garnetiferous ultramafic inclusions in minette from the Navajo volcanic field. In *The mantle sample: inclusions in kimberlites and other volcanics* (ed. F. R. Boyd and H. O. A. Meyer), pp. 330–44. American Geophysical Union, Washington, DC.

Ellam, R. and Cox, K. G. (1989). A Proterozoic lithospheric source for Karoo magmatism: evidence from the Nuanetsi picrites. *Earth and Planetary Science Letters*, **92,** 207–18.

Erlank, A. J. (1973). Kimberlite potassic richterite and the distribution of potassium in the upper mantle. In *First International Kimberlite Conference, Cape Town*, pp. 103–6 (abstr.).

Erlank, A. J. and Shimizu, N. (1977). Strontium and strontium isotope distributions in some kimberlite nodules and minerals. In *Second International Kimberlite Conference, Santa Fe*, (abstr.).

Erlank, A. J., Waters, F. G., Hawkesworth, C. J., Haggerty, S. E., Allsopp, H. L., Rickard, R. S., and Menzies, M. A. (1987). Evidence for mantle metasomatism from the Kimberley Pipes, South Africa. In *Mantle metasomatism* (ed. M. A. Menzies and C. J. Hawkesworth), pp. 221–311. Academic Press, London.

Ernst, T. (1965). Do peridotitic inclusions in basalts represent mantle material? In *The Upper Mantle Symposium New Delhi* (eds. C. H. Smith and T. Sorgenfrei), pp. 180–5. Det Berlingske Bogtrykkeri, Copenhagen.

Eskola, P. (1921). On the eclogites of Norway. *Škrifter utg. af Videnskabsselskabet i Christiania*, **8,** 1–118.

Ewart, A., Chappell, B. W., and Menzies, M. A. (1988). An overview of the geochemical and isotopic characteristics of the eastern Australian Cainozoic volcanic provinces. In *Oceanic and continental lithosphere: similarities and differences* (ed. M. A. Menzies and K. G. Cox). pp. 225–74. Oxford University Press, Oxford.

Fabries, J., Bodinier, J. L., Dupuy, C., Lorand, J. P., and Benkerrou, C. (1989). Evidence for modal metasomatism in the orogenic spinel lherzolite body from Caussou, northeastern Pyrenees, France. *Journal of Petrology*, **30,** 199–228.

Fermor, L. L. (1913). Preliminary note on garnet as a geological barometer and on an infra-plutonic zone in the Earth's crust. *Records of the Geological Survey, India*, **63,** 41–7.

Fisher, O. (1889). *Physics of the Earth's crust.* MacMillan and Co, London.

Fitton, J. G. and Dunlop, H. (1985). The Cameroon Line, West Africa and its bearing on the origin of oceanic and continental alkali basalt. *Earth and Planetary Science Letters*, **72,** 23–38.

Fitton, J. G., James, D., Kempton, P. D., Ormerod, D. S., and Leeman, W. P. (1988). The role of lithospheric mantle in the generation of late Cenozoic basic magmas in the western United States. In *Oceanic and continental lithosphere: similarities and differences* (ed. M. A. Menzies and K. G. Cox), pp. 331–50. Oxford University Press, Oxford.

Forbes, R. B. and Kuno, H. (1967). Peridotite inclusions and basaltic host rocks. In *Ultramafic and related rocks* (ed. P. J. Wyllie), pp. 328–36. J. Wiley and Sons, New York.

Francis, D. M. (1976). Amphibole pyroxenite xenoliths: cumulate or replacement phenomena from the upper mantle, Nunivak Island, Alaska. *Contributions to Mineralogy and Petrology*, **58,** 51–61.

Frantsesson, E. V. (1968). *The petrology of the kimberlites.* Department of Geology, Australian National University.

Frantsesson, E. V. and Prokopchuk, B. I. (1968). Kimberlites—the tectonomagmatic facies of the alkaline ultramafic association of the platforms, *Vulkanizm i tecktogenez*, 159–164.

Frey, F. A. (1969). Rare earth abundances in a high-temperature peridotite intrusion. *Geochimica et Cosmochimica Acta*, **33,** 1429–47.

Frey, F. A. (1970). Rare earth abundances in Alpine ultramafic rocks. *Physics of the Earth and Planetary Interiors*, **3,** 323–30.

Frey, F. A. and Green, D. H. (1974). The mineralogy, geochemistry and origin of lherzolite inclusions in Victorian basanites. *Geochimica et Cosmochimica Acta*, **38,** 1023–59.

Fujii, T. and Scarfe, C. M. (1985). Composition of liquids coexisting with spinel lherzolite at 10 kb. and the genesis of MORBs. *Contributions to Mineralogy and Petrology*, **90,** 18–28.

Gast, P. W. (1960). Limitations on the composition of the upper mantle. *Journal of Geophysical Research*, **65,** 1287–97.

Gast, P. W. (1968). Trace element fractionation and the origin of tholeiitic and alkaline magma types. *Geochimica et Cosmochimica Acta*, **32,** 1057–86.

Goetze, C. (1975). Sheared lherzolites: from the point of view of rock mechanics. *Geology*, **3,** 172–3.

Green, D. H. (1964). The petrogenesis of the high temperature peridotite intrusion in the Lizard area, Cornwall. *Journal of Petrology*, **5,** 134–88.

Green, D. H. (1967). The stability fields of aluminous pyroxene peridotite and garnet peridotitie and their relevance in upper mantle structure. *Earth and Planetary Science Letters*, **3,** 151–60.

Green, D. H. and Ringwood, A. E. (1967). The genesis of basaltic magmas. *Contributions to Mineralogy and Petrology*, **15,** 103–90.

Green, D. H. and Wallace, M. E. (1988). Mantle metasomatism by ephemeral carbonatite melts. *Nature*, **336,** 459–62.

Green, H. W. II and Guegen, Y. (1974). Origin of kimberlite pipes by diapiric upwelling in the upper mantle. *Nature*, **249,** 617–20.

Griffin, W. L. and Murthy, V. R. (1969). Distribution of K, Rb, Sr and Ba in some minerals relevant to basalt genesis. *Geochimica et Cosmochimica Acta*, **33,** 1389–414.

Griffin, W. L., O'Reilly, S. Y., and Stabel, T. (1988). Mantle metasomatism beneath western Victoria Australia II: isotopic geochemistry of Cr diopside lherzolites and Al–augite pyroxenites. *Geochimica et Cosmochimica Acta*, **52,** 449–59.

Gurney, J. J. (1972). Plumbing the secrets of the Earth's mantle. *Int. Diamonds*, **2,** 42–8.

Gurney, J. J. and Harte, B. (1980). Chemical variations in upper mantle nodules from Southern African kimberlites. *Philosophical Transactions of the Royal Society, London*, **297,** 273–93.

Gurney, J. J., Moore, R. O., Otter, M. L., Kirkley, M. B., Hops, J. J., and McCandless, T. E. (1988). Southern African kimberlites. [Preprint available from the Department of Geochemistry, University of Cape Town, Rondesbosch, 7700, S.A.]

Harris, J. W. (1987). Recent physical, chemical and isotopic research of diamond. In *Mantle xenoliths* (ed. P. H. Nixon), pp. 477–500. J. Wiley and Sons, Chichester.

Hart, S. R. (1970). Chemical exchange between sea water and deep ocean basalts. *Earth and Planetary Science Letters*, **9,** 269–79.

Hart, W K., WoldeGabriel, G., Walter, R. C., and Mertzman, S. A. (1989). Basaltic volcanism in Ethiopia: constraints on continental rifting and mantle interactions. *Journal of Geophysical Research*, **94,** 7731–48.

Harte, B. (1977). Rock nomenclature with particular relation to deformation and recrystallisation textures in olivine-bearing xenoliths. *Journal of Geology*, **85,** 279–88.

Harte, B. (1983). Mantle peridotites and processes—the kimberlite sample. In *Continental basalts and mantle xenoliths* (ed. C. J. Hawkesworth and M. J. Norry), pp. 46–91. Shiva Publishing, England.

Harte, B., Cox, K. G., and Gurney, J. J. (1975). Petrography and geological history of upper mantle xenoliths from the Matsoku kimberlite pipe. *Physics and Chemistry of the Earth*, **9,** 447–506.

Haskin, L. A., Frey, F. A., Schmitt, R., and Smith, R. H. (1966). Meteoritic, solar and terrestrial rare-earth distributions. *Physics and Chemistry of the Earth*, **7,** 167–211.

Hawkesworth, C. J. and Norry, M. J. (1983). *Continental basalts and mantle xenoliths*. Shiva Publishing, England.

Hawkesworth, C. J., Erlank, A. J., Marsh, J. S., Menzies, M. A., and van Calsteren, P. (1983). Evolution of the continental lithosphere: evidence from volcanics and xenoliths in southern Africa. In *Continental basalts and mantle xenoliths* (ed. C. J. Hawkesworth and M. J. Norry), pp. 111–38. Shiva Publishing, England.

Hawkesworth, C. J., Rogers, N. W., van Calsteren, P. W. C., and Menzies, M. A. (1984). Mantle enrichment processes. *Nature*, **311,** 331–5.

Hawkesworth, C. J., Mantovani, M. S. M., Taylor, P. N., and van Calsteren, P. (1986). Evidence from the Parana of South Brazil for a continental contribution to DUPAL basalts. *Nature*, **322,** 356–9.

Hawkesworth, C. J., Mantovani, M. S. M., and Peate, D. (1988). Lithosphere remobilisation during Parana CFB magmatism. In *Oceanic and continental lithosphere: similarities and differences* (ed. M. A. Menzies and K. G. Cox), pp. 205–24. Oxford University Press, Oxford.

Helmstaedt, H. and Doig, R. (1975). Eclogite nodules from kimberlite pipes of the Colorado Plateau—samples of subducted Franciscan-type oceanic lithosphere. *Physics and Chemistry of the Earth*, **9,** 95–111.

Herzberg, C., Feigenson, M., Skuba, C., and Ohtani, E. (1988). Majorite fractionation recorded in the geochemistry of peridotites from South Africa. *Nature*, **332,** 823–6.

Hess, H. H. (1955). Serpentines, orogeny and epeirogeny. *Geological Society of America Special Paper* **62,** 391–408.

Hess, H. H. (1962). History of ocean basins. In *Petrologic studies: a volume to honor A. F. Buddington* (ed. A. E. J. Engel, H. L. James, and B. F. Leonard), pp. 64–93. Geological Society of America, Denver.

Hess, H. H. (1964). The oceanic crust, the upper mantle and the Mayaguez serpentinized peridotite. In *A study of the serpentinite* (ed. C. A. Burk), pp. 95–111. National Academy of Science–National Research Council, Washington, DC.

Hill, R. E. T. and Boettcher, A. L. (1970). Water in the earth's mantle: melting curves of basalt–water and basalt–water–carbon dioxide. *Science*, **167,** 980–1.

Hofmann, A. W. and Hart, S. R. (1975). An assessment of local and regional isotopic equilibrium in the mantle. *Earth and Planetary Science Letters*, **38,** 44–62.

Hofmann, A. W. and White, W. M. (1982). Mantle plumes from ancient oceanic crust. *Earth and Planetary Science Letters*, **57,** 421–36.

Hunter, R. H. and McKenzie, D. P. (1989). The geometry of carbonate melts in rocks of mantle composition. *Earth and Planetary Science Letters*, **92,** 347–56.

Hurley, P. M. (1967). Rb^{87}/Sr^{87} relationships in the differentiation of the mantle. In *Ultramafic and related rocks* (ed. P. J. Wyllie), pp. 372–5. J. Wiley and Sons, New York.

Hutchison, R. (1974). The formation of the earth. *Nature*, **250,** 556–8.

Irving, A. J. (1974). Pyroxene-rich ultramafic xenoliths in the Newer Basalts of Victoria, Australia. *Neues Jahrbuch für Mineralogie*, **120,** 147–67.

Irving, A. J. (1980). Petrology and geochemistry of composite ultramafic xenoliths in alkalic basalts and implications for magmatic processes within the mantle. *American Journal of Science*, **280A,** 389–426.

Ito, K. and Kennedy, G. C. (1970). The fine structure of the basalt-eclogite transition. In *Fiftieth anniversary symposia*, Mineralogical Society of America (ed. B. A. Morgan), special paper 3.

Jagoutz, E. (1988). Nd and Sr systematics in an eclogite xenolith from Tanzania: evidence for frozen mineral equilibria in the continental lithosphere. *Geochimica et Cosmochimica Acta*, **52,** 1285–93.

Jagoutz, E., Palme, H., Baddenhausen, H., Blum, K., Cendales, M., Dreibus, G., Spettrl, B., Lorenz, V., and Wanke, H. (1979). The abundances of major, minor and trace elements in the Earth's mantle as derived from primitive ultramafic nodules. *Proceedings of the Tenth Lunar Science Conference*, Geochimica et Cosmochimica Acta, Supplement 11, 2031–50.

Jagoutz, E., Dawson, J. B., Hoernes, S., Spettel, B., and Wanke, H. (1984). Anorthositic oceanic crust in Archean Earth. *Lunar and Planetary Science*, **XV,** 395–6 (abst.).

Jordan, T. H. (1975*a*). The continental tectosphere. *Reviews of Geophysics and Space Physics*, **13,** 1–12.

Jordan, T. H. (1975*b*). Lateral heterogeneity and mantle dynamics. *Nature*, **257,** 745–50.

Jordan, T H. (1979). Mineralogies, densities and seismic velocities of garnet lherzolites and the geophysical implications. In *The mantle sample: inclusions in kimberlite and other volcanics* (ed. F. R. Boyd and H. O. A. Meyers), pp. 1–14. American Geophysical Union, Washington, DC.

Jordan, T. H. (1988). Structure and formation of the continental tectosphere. In *Oceanic and continental lithosphere: similarities and differences* (ed. M. A. Menzies and K. G. Cox), pp. 11–38. Oxford University Press, Oxford.

Kleeman, J. D. and Cooper, J. A. (1970). Geochemical evidence for the origin of some ultramafic inclusions from Victorian basanites. *Phyics of the Earth and Planetary Interiors*, **3,** 302–8.

Kramers, J. D. (1977). Lead and strontium isotopes in Cretaceous kimberlites and mantle-derived xenoliths from Southern Africa. *Earth and Planetary Science Letters*, **34,** 419–31.

Kramers, J. D. (1979). Lead, uranium, strontium, potassium and rubidium in inclusion bearing diamonds and mantle derived xenoliths from southern Africa. *Earth and Planetary Science Letters*, **42,** 58–70.

Kramers, J. D., Smith, C. B., Lock, N., Harmon, R., and Boyd, F. R. (1981). Can kimberlite be generated from ordinary mantle? *Nature*, **291,** 53–6.

Kramers, J. D., Ruddick, J., and Dawson, J. (1983). Trace element and isotope studies of veined, metasomatic and 'MARID' xenoliths from Bultfontein South Africa. *Earth and Planetary Science Letters*, **65,** 90–106.

Kuno, H. (1960). High-alumina basalt. *Journal of Petrology*, **1,** 121–45.

Kuno, H. (1967*a*). Mafic and ultramafic nodules from Itinome-gata Japan. In *Ultramafic and related rocks* (ed. P. J. Wyllie), pp. 337–41. J. Wiley and Sons, New York.

Kuno, H. (1967*b*). Volcanological and petrological evidences regarding the nature of the Upper Mantle. In *The earth's mantle* (ed. T. F. Gaskell), pp. 89–110. Academic Press, New York.

Kuno, H. (1969). Plateau basalts. In *The earth's crust and upper mantle* (ed. P. J. Hart), pp. 495–501. American Geophysical Union, Washington, DC.

Kyser, T. K., O'Neil, J. R., and Carmichael, I. S. E. (1981). Oxygen isotope thermometry of basic lavas and mantle nodules. *Contributions to Mineralogy and Petrology*, **77,** 11–23.

Kyser, T. K., O'Neil, J. R., and Carmichael, I. S. E. (1982). Genetic relations among basic lavas and ultramafic nodules and evidence from oxygen isotopic compositions. *Contributions to Mineralogy and Petrology*, **81,** 88–102.

Lambert, I. B. and Wyllie, P. J. (1968). Stability of hornblende and a model for the low velocity zone. *Nature*, **219,** 1240–1.

Lausen, C. (1927). The occurrence of olivine bombs near Globe, Arizona. *American Journal of Science*, **14,** 293–306.

Leat, P. T., Thompson, R. N., Morrison, M. A., Hendry, G. L., and Dickin, A. P. (1988). Compositionally-diverse Miocene–Recent rift-related magmatism in NW Colorado: partial melting and mixing of mafic magmas from three different asthenospheric and lithospheric mantle sources. In *Oceanic and continental lithosphere: similarities and differences* (ed. M. A. Menzies and K. G. Cox), pp. 351–78. Oxford University Press, Oxford.

Leeman, W. P. (1975). Radiogenic tracers applied to basalt genesis in the Snake River Plain–Yellowstone National Park region—evidence for a 2.7-b.y.-old upper mantle keel. *Geological Society of America Abstracts with Programs*, **7,** 1165 [abstr.].

Leeman, W. P. (1977). Comparison of Rb/Sr, U/Pb and rare earth characteristics of sub-continental and sub-oceanic mantle regions. In *Magma genesis* (ed. H. J. B. Dick), pp. 149–68. Oregon Department of Geology Mineralogy and Industry, Bulletin 96.

Leeman, W. P. (1982). Tectonic and magmatic significance of strontium isotopic variations in Cenozoic volcanic rocks from the western United States. *Geological Society of America, Bulletin*, **93,** 487–503.

Leeman, W. P. and Fitton, J. G. (ed.) (1989). Magmatism and lithospheric extension. *Journal of Geophysical Research*, **94,** 7682–986.

Leggo, P. J. and Hutchison, R. (1968). A Rb–Sr isotope study of ultrabasic xenoliths and their basaltic host rocks from the Massif Central France. *Earth and Planetary Science Letters*, **5,** 71–5.

Lewis, H. C. (1887). On a diamondiferous peridotite and the genesis of diamond. *Geological Magazine*, **5,** 22–4.

Lewis, H. C. (1888). The matrix of the diamond. *Geological Magazine*, **5,** 129–31.

Lloyd, F. E. and Bailey, D. K. (1975). Light element metasomatism of the continental mantle: the evidence and the consequences. *Physics and Chemistry of the Earth*, **9,** 389–416.

Maaloe, S. and Aoki, K. (1977). The major element composition of the upper mantle estimated from the composition of lherzolites. *Contributions to Mineralogy and Petrology*, **63,** 161–73.

MacDougall, I. (ed.) (1988). *Continental flood basalts*. Kluwer Academic Publishers, Boston.

MacGregor, I. D. and Carter, J. L. (1970). The chemistry of clinopyroxenes and garnets of eclogites and peridotite xenoliths from the Roberts Victor Mine, southern Africa. *Physics of the Earth and Planetary Interiors*, **3,** 391–7.

MacGregor, I. D. and Manton, W. I. (1986). Roberts Victor eclogites: ancient oceanic crust. *Journal of Geophysical Research*, **91,** 14063–79.

Manton, W. I. and Tatsumoto, N. (1969). Isotopic composition of lead and strontium in nodules from the Roberts Victor Mines, South Africa. *Transactions of the American Geophysical Union*, **50,** 343.

Mattey, D. J., Menzies, M., and Pillinger, C. T. (1985). Carbon isotopes in lithospheric peridotites and pyroxenites. *Terra Cognita*, **5,** 147.

McCulloch, M. T. (1982). Identification of earth's earliest differentiates. *Fifth International Conference of Geochemical and Cosmochemical Isotope Geology*, pp. 244–6 (abstr.).

McCulloch, M. T., Jaques, A. L., Nelson, D. R., and Lewis, J. D. (1983). Nd and Sr isotopes in kimberlites and lamproites from Western Australia: an enriched mantle origin. *Nature*, **302,** 400–3.

McKenzie, D. (1984). The generation and compaction of partially molten rock. *Journal of Petrology*, **25,** 713–65.

McKenzie, D. (1985). The extraction of magma from the crust and mantle. *Earth and Planetary Science Letters*, **74,** 81–91.

McKenzie, D. (1989). Some remarks on the movement of small melt fractions in the mantle. *Earth and Planetary Science Letters*, **95,** 53–72.

McKenzie, D. and Bickle, M. J. (1988). The volume and composition of melt generated by extension of the lithosphere. *Journal of Petrology*, **29,** 625–80.

McKenzie, D. and O'Nions, R. K. (1983). Mantle reservoirs and ocean island basalts. *Nature*, **301,** 229–31.

Melson, W. G., Jarosewich, E., Bowen, V. T., and Thompson, G. (1967). St. Peter and St. Paul Rocks: a high-temperature, mantle-derived intrusion. *Science*, **155,** 1532–55.

Menzies, M. A. (1983). Mantle ultramafic xenoliths in alkaline magmas: evidence for mantle heterogeneity modified by magmatic activity. In *Continental basalts and mantle xenoliths* (ed. C. J. Hawkesworth and M. Norry), pp. 92–110. Shiva Publishing, England.

Menzies, M. A. (1988). Mantle melts in diamonds. *Nature*, **335**, 769–70.

Menzies, M. A. (1989). Cratonic, circum-cratonic and oceanic mantle domains beneath the western U.S.A. *Journal of Geophysical Research*, **94**, 7899–915.

Menzies, M. A. and Halliday, A. N. (1988). Lithospheric mantle domains beneath the Archean and Proterozoic crust of Scotland. In *Oceanic and continental lithosphere: similarities and differences* (ed. M. A. Menzies and K. G. Cox), pp. 275–302. Oxford University Press, Oxford.

Menzies, M. A. and Hawkesworth, C. J. (ed.) (1987*a*). *Mantle metasomatism*. Academic Press, London.

Menzies, M. A. and Hawkesworth, C. J. (1987*b*). Upper mantle processes and composition. In *Mantle xenoliths* (ed. P. H. Nixon), pp. 725–38. J. Wiley and Sons, Chichester.

Menzies, M. A. and Murthy, V. R. (1978). Strontium isotope geochemistry of alpine tectonite lherzolites: data compatible with a mantle origin. *Earth and Planetary Science Letters*, **38**, 346–54.

Menzies, M. A. and Murthy, V. R. (1980*a*). Enriched mantle: Nd and Sr isotopes in diopsides from kimberlite nodules. *Nature*, **282**, 634–6.

Menzies, M. A. and Murthy, V. R. (1980*b*). Nd and Sr isotope geochemistry of hydrous mantle nodules and their host alkali basalts: implications for local heterogeneities in metasomatically veined mantle. *Earth and Planetary Science Letters*, **46**, 323–34.

Menzies, M. A., Kempton, P., and Dungan, M. (1985). Interaction of continental lithosphere and asthenospheric melts below the Geronimo volcanic field, Arizona. *Journal of Petrology*, **26**, 663–93.

Menzies, M. A., Rogers, N., Tindle, A., and Hawkesworth, C. J. (1987). Metasomatic and enrichment processes in lithospheric peridotites, an effect of asthenosphere–lithosphere interaction. In *Mantle metasomatism* (ed. M. A. Menzies and C. J. Hawkesworth), pp. 313–61. Academic Press, London.

Mercier, J. -C.C. and Nicolas, A. (1975). Textures and fabrics of upper mantle peridotites as illustrated by basalt xenoliths. *Journal of Petrology*, **16**, 454–87.

Michael, P. J. and Bonatti, E. (1985). Peridotite composition from the North Atlantic: regional and tectonic variations and implications for partial melting. *Earth and Planetary Science Letters*, **73**, 91–104.

Milashev, V. A. (1960). Cognate inclusions in the "Obnazhennaya" kimberlite pipe (basin of River Olenek). *Zapiski Vsesoyuznogo Mineralogicheskogo Obshchestvo*, **89**, 284–99.

Milashev, V. A., Krutoyarsky, M. A., Rabkin, M. I., and Erlikh, E. N. (1963). The kimberlitic rocks and picritic porphyries of the northeastern part of the Siberian Platform. *Trudy Nauchno-issledóvatel'skogo Instituta Geologii Arktiki*, 126.

Moorbath, S., Thompson, R. N., and Oxburgh, E. R. (1984). *The relative contributions of mantle, oceanic crust and continental crust to magma genesis*. Royal Society, London.

Morgan, W. J. (1971). Convection plumes in the lower mantle. *Nature*, 230, 42–3.

Navon, O., Hutcheon, I. D., Rossman, G. R., and Wasserburg, G. J. (1988). Mantle-derived fluids in diamond micro-inclusions. *Nature*, **335**, 784–9.

Nicolas, A. and Jackson, E. D. (1972). Repartition en deux provinces des peridotites des chaines alpines logeant la Mediterranée: implications géotectonique. *Schweizer Mineralogisch Petrographisch Mitteilungen*, **52**, 479–95.

Nicolas, A., Lucazeau, F., and Bayer, R. (1987). Peridotite xenoliths in Massif Central basalts, France: textural and geophysical evidence for asthenospheric diapirism. In *Mantle xenoliths* (ed. P. H. Nixon), pp. 563–74. John Wiley and Sons, Chichester.

Nixon, P. H. (ed.) (1987). *Mantle xenoliths*. John Wiley and Sons, Chichester.

Nixon, P. H. and Boyd, F. R. (1973). Petrogenesis of the granular and sheared ultrabasic nodule suite in kimberlites. In *Lesotho kimberlites* (ed. P. H. Nixon), pp. 48–56. Lesotho National Development Corporation, Maseru.

Nixon, P. H., Boyd, F. R., and Boullier, A-M. (1973). The evidence of kimberlite and its inclusions on the constitution of the outer part of the Earth. In *Lesotho kimberlites* (ed. P. H. Nixon), pp. 312–18. Lesotho National Development Corporation, Maseru.

O'Hara, M. J. (1965). Primary magmas and the origin of basalts. *Scottish Journal of Geology*, **1**, 19–40.

O'Hara, M. J. (1967*a*). Crystal-liquid equilibria and the origins of ultramafic nodules in basic igneous rocks. In *Ultramafic and related rocks* (ed. P. J. Wyllie), pp. 346–9. J. Wiley and Sons, New York.

O'Hara, M. J. (1967*b*). Mineral paragenesis in ultrabasic rocks. In *Ultramafic and related rocks* (ed. P. J. Wyllie), pp. 393–401. J. Wiley and Sons, New York.

O'Hara, M. J. (1968). The bearing of phase equilibria studies in synthetic and natural systems on the origin and evolution of basic and ultrabasic rocks. *Earth Science Review*, **4**, 69–133.

O'Hara, M. J. (1975). Is there an Icelandic mantle plume? *Nature*, **253**, 708–10.

O'Hara, M. J. and Mercy, E. L. P. (1963). Petrology and petrogenesis of some garnetiferous peridotites. *Transactions of the Royal Society of Edinburgh*, **65**, 251–314.

O'Hara, M. J. and Yoder, H. S. (1967). Formation and fractionation of basic magmas at high pressures. *Scottish Journal of Geology*, **3**, 67–117.

O'Reilly, S. Y., and Griffin, W. C. (1988). Mantle metasomatism beneath western Victoria Australia, I: Metasomatic processes in Cr diopside lherzolitres. *Geochimica et Cosmochimica Acta*, **52**, 433–48.

Ormerod, D. S., Hawkesworth, C. J., Rogers, N., Leeman, W. P., and Menzies, M. A. (1988). Tectonic and magmatic transitions in the western Great Basin U.S.A. *Nature*, **333**, 349–53.

Oxburgh, E. R. (1964). Petrological evidence for the presence of amphibole in the Upper Mantle and its petrogenetic and Geophysical implications. *Geological Magazine*, **101**, 1–19.

Palme, H. and Nickel, K. G. (1985). Ca/Al ratio and composition of the earth's upper mantle. *Geochimica et Cosmochimica Acta*, **49**, 2123–32.

Palme, H., Baddenhausen, H., Blum, K., Cendales, M., Dreibus, G., Hofmeisiter, H., Kruse, H., Palme, C. H., Spettel, B., Vilsek, E., Wanke, H., and Kurat, G. (1978). New data on lunar samples and achondrites and a comparison of the least fractionated samples from the earth, the moon and the eucrite parent body. *Proceedings of the Lunar and Planetary Science Conference, 9th, Geochimica et Cosmochimica Acta* Supplement 10, 25–57.

Paul, D. K. (1979). Isotopic composition of strontium in Indian kimberlites. *Geochimica et Cosmochimica Acta*, **43**, 389–94.

Perry, F. V., Baldridge, W. S., and DePaolo, D. J. (1987). Role of asthenosphere and lithosphere in the genesis of late Cenozoic basaltic rocks from the Rio Grande rift and adjacent regions of the southwestern United States, *Journal of Geophysical Research*, **92**, 9193–213.

Powell, J. L. (1966). Isotopic composition of strontium in carbonatites and kimberlites. In *Mineralogical Society of India—International Mineralogical Society Papers*, Fourth General Meeting, Mineralogical Association, New Delhi, pp. 58–66.

Reisburg, L. and Zindler, A. (1986). Extreme isotopic variations in the upper mantle: evidence from Ronda. *Earth and Planetary Science Letters*, **81**, 29–45.

Richard, P. and Allegre, C. (1980). Nd and Sr isotope study of ophiolite and orogenic lherzolite petrogenesis. *Earth and Planetary Science Letters*, **47**, 65–74.

Richardson, S. H. (1987). Latter-day origin of diamonds of eclogitic paragenesis. *Nature*, **322**, 623–6.

Richardson, S. H., Gurney, J. J., Erlank, A. J., and Harris, J. W. (1984). Origin of diamonds in old enriched mantle. *Nature*, **310**, 198–202.

Richardson, S. H., Erlank, A. J., and Hart, S. R. (1985). Kimberlite-borne garnet peridotite xenoliths from old enriched subcontinental lithosphere. *Earth and Planetary Science Letters*, **75**, 116–28.

Richter, F. M. (1988). A major change in the thermal state of the earth at the Archean–Proterozoic bondary: consequences for the nature and preservation of continental lithosphere. In *Oceanic and continental lithosphere: similarities and differences* (ed. M. A. Menzies and K. G. Cox), pp. 39–52. Oxford University Press, Oxford.

Ringwood, A. E. (1958*a*). The constitution of the mantle–I. Thermodynamics of the olivine–spinel transition. *Geochimica et Cosmochimica Acta*, **13**, 303–21.

Ringwood, A. E. (1958*b*). The constitution of the mantle—II. Further data on the olivine–spinel transition. *Geochimica et Cosmochimica Acta*, **15**, 18–29.

Ringwood, A. E. (1958*c*). The constitution of the mantle—III. Consequences of the olivine–spinel transition. *Geochimica et Cosmochimica Acta*, **15**, 195–212.

Ringwood, A. E. (1962*a*). A model for the upper mantle. *Journal of Geophysical Research*, **67**, 857–67.

Ringwood, A. E. (1962*b*). A model for the upper mantle, 2. *Journal of Geophysical Research*, **67**, 4473–7.

Ringwood, A. E. (1966). Chemical evolution of the terrestrial planets. *Geochimica et Cosmochimica Acta*, **30**, 41–104.

Ringwood, A. E. (1975). *Composition and petrology of the earth's mantle*. McGraw-Hill, Sydney, 618 pp.

Ringwood, A. E. (1979). *Origin of the earth and moon*. Springer-Verlag, New York.

Ringwood, A. E. (1982). Phase transformations and differentiation in subducted lithosphere: implications for mantle dynamics, basalt petrogenesis and crustal evolution. *Journal of Geology*, **90**, 611–43.

Ringwood, A. E. and Green, D. H. (1966). An experimental investigation of gabbro–eclogite transformation and some geophysical consequences. *Tectonophysics*, **3**, 383–427.

Robey, J. A. (1981). Kimberlites of the Central Cape Province, R.S.A. Unpublished D.Phil. thesis. University of Cape Town.

Roden, M. F., Frey, F. A., and Francis, D. M. (1984). An example of the consequent mantle metasomatism in peridotite inclusions from Nunivak Island Alaska. *Journal of Petrology*, **25**, 546–77.

Roden, M. F., Irving, A. J., and Murthy, V. R. (1988). Isotopic and trace element composition of the upper mantle beneath a young continental rift; results from Kilbourne Hole New Mexico. *Geochimica et Cosmochimica Acta*, **52**, 461–73.

Roe, G. D. (1964). Rubidium–strontium analyses of ultramafic rocks and the origin of peridotites. Unpublished D.Phil. thesis. Massachusetts Institute of Technology.

Rogers, N. (1977). Granulite xenoliths from Lesotho kimberlites and the lower continental crust. *Nature*, **270**, 681–4.

Ross, C. J., Foster, M. D., and Myers, A. T. (1954). Origin of dunites and olivine-rich inclusions in basaltic rocks. *American Mineralogist*, **39**, 693–737.

Rubey, W. W. (1951). Geologic history of sea water. *Geological Society of America Bulletin*, **62**, 111–47.

Sachtleben, Th. and Seck, H. A. (1981). Chemical control of Al-solubility in orthopyroxene and its implications on pyroxene geothermometry. *Contributions to Mineralogy and Petrology*, **78**, 157–65.

Schilling, J-G. (1971). Sea-floor evolution: rare earth evidence. *Philosophical Transactions of the Royal Society, London*, **A268**, 663–700.

Schilling, J-G. (1973). Iceland mantle plume. *Nature*, **246**, 141–3.

Shervais, J. W., Taylor, L. A., Lugmair, G. W., Clayton, R. N., Mayeda, T. K., and Korotev, R. L. (1988). Early Proterozoic oceanic crust and the evolution of the subcontinental mantle: eclogites and related rocks from southern Africa. *Geological Society of America, Bulletin*, **100**, 411–23.

Smith, C. B. (1983). Pb, Sr and Nd isotopic evidence for sources of African Cretaceous kimberlites. *Nature*, **304**, 51–4.

Sobolev, N. V. (1977). *Deep seated inclusions in kimberlites and the problem of the composition of the upper mantle*. American Geophysical Union, Washington, DC.

Stelzner, R. (1894). Die Diamantgruben von Kimberley. Isis. Fresden, 1893. *Zeitschrift Für Praktische Geologie*, 153–7.

Stosch, H. G., Carlson, R. W., and Lugmair, G. W. (1980). Episodic mantle differentiation: Nd and Sr isotopic evidence. *Earth and Planetary Science Letters*, **47**, 263–71.

Stueber, A. M. (1965). A geochemical study of ultramafic rocks. Unpublished D.Phil. thesis. University of California, San Diego.

Stueber, A. M. and Ikramuddin, M. (1974). Rubidium, strontium and the isotopic composition of strontium in ultramafic nodules, minerals and host basalts. *Geochimica et cosmochimica Acta*, **38**, 207–16.

Stueber, A. M. and Murthy, V. R. (1966). Sr isotope and alkali element abundances in ultramafic rocks. *Geochimica et Cosmochimica Acta*, **33**, 543–53.

Sun, S-S. (1980). Lead isotopic study of young volcanic rocks from mid-ocean ridges, ocean islands and island arcs. *Philosophical Transactions of the Royal Society, London*, **A297**, 409–45.

Sun, S-S. and Hanson, G. N. (1975). Origins of Ross Island basanitoids and limitations upon the heterogeneity of mantle sources for alkali basalts and nethelinites. *Contributions to Mineralogy and Petrology*, **52**, 77–106.

Takahashi, E. and Kushiro, I. (1983). Melting of a dry peridotite at high pressures and basalt magma genesis. *American Mineralogist*, **68**, 859–79.

Tatsumoto, N. (1978). Isotopic composition of lead in oceanic basalts and its implication to mantle evolution. *Earth and Planetary Science Letters*, **38**, 63–87.

Thirlwall, M. F. (1982). Systematic variation in chemistry and Nd–Sr isotopes across a Caledonian calc-alkaline volcanic arc: implications for source materials. *Earth and Planetary Science Letters*, **58**, 27–50.

Thompson, R. N. (1977). Columbia/Snake River—Yellowstone magmatism in the context of western U.S.A. Cenozoic geodynamics. *Tectonophysics*, **39**, 621–36.

Thompson, R. N., Morrison, M. A., Hendry, G. L., and Parry, S. (1984). An assessment of the relative roles of crust and mantle in magma genesis: an elemental approach. *Philosophical Transactions of the Royal Society, London*, **A310**, 111–46.

Wagner, P. A. (1914). *The diamond fields of Southern Africa*. Transvaal Leader, Harrison Street, Johannesburg.

Wagner, P. A. (1928). The evidence of the kimberlite pipes on the constitution of the outer part of the Earth. *South African Journal of Science*, **25**, 127–48.

Walker, R. J., Carlson, R. W., Shirley, S B., and Boyd, S. R. (1989). Os, Sr, Nd, and Pb isotope systematics of southern Africa peridotite xenoliths: implications for the chemical evolution of subcontinental mantle. *Geochimica et Cosmochemica Acta*, **53,** 1583–695.

Wallace, M. E. and Green, D. H. (1988). An experimental determination of primary carbonatite magma composition. *Nature*, **335**, 343–6.

Wanke, H. (1981). Constitution of terrestrial planets. *Philosophical Transactions of the Royal Society*, *London*, **A303,** 287–302.

Washington, H. S. (1922). Deccan traps and other plateau basalts. *Geological Society of America*, *Bulletin*, 33, 765–804.

Washington, H. S. (1925). The chemical composition of the Earth. *American Journal of Science* (*Fifth Series*), **9,** 351–78.

Wass, S. Y. and Irving, A. J. (1976). *XENMEG: a catalogue of occurrences of xenoliths and megacrysts in volcanic rocks of eastern Australia*. Australian Museum, Sydney.

Wass, S. Y. and Rogers, N. W. (1980). Mantle metasomatism—precursor to continental alkaline volcanism. *Geochimica et Cosmochimica Acta*, **44,** 1811–23.

Waters, F. G. (1987). A geochemical study of metasomatised peridotite and MARID nodules from the Kimberley Pipes, South Africa. Unpublished D.Phil. thesis. University of Cape Town.

Waters, F. G. and Erlank, A. J. (1988). Assessment of the vertical extent and distribution of mantle metasomatism below Kimberley, South Africa. In *Oceanic and continental lithosphere: similarities and differences* (ed. M. A. Menzies and K. G. Cox), pp. 185–204. Oxford University Press, Oxford.

Watson, E. B. and Brenan, J. M. (1987). Fluids in the lithosphere, 1: Experimentally-determined wetting characteristics of CO_2–H_2O fluids and their implications for fluid transport, host-rock physical properties, and fluid inclusion formation. *Earth and Planetary Science Letters*, **85,** 497–515.

Williams, A. F. (1932). *The genesis of the diamond*. Ernest Benn, London.

Wilshire, H. G. and Binns, R. A. (1961). Basic and ultrabasic xenoliths from volcanic rocks of New South Wales. *Journal of Petrology*, **2,** 185–208.

Wilshire, H. G. and Pike, J. E. N. (1975). Upper mantle diapirism: evidence from analogous features in alpine peridotites and ultramafic inclusions in basalt. *Geology*, **3,** 467–70.

Wilshire, H. G., Meyer, C. E., Nakata, J. K., Calk, L. C., Shervais, J. W., Nielson, J. E., and Schwarzmann, E. C. (1985). Mafic and ultramafic rocks of the western United States. *United States Geological Survey Open File Report*, **85**.

Wilson, J. T. (1965). Submarine fracture zones, aseismic ridges and the ICSU line: proposed western margin of the East Pacific ridge. *Nature*, **207,** 907–11.

Wyllie, P. J. (1987). Transfer of cratonic carbon into kimberlites and rare earth carbonatites. In *Magmatic processes: physicochemical principles*, The Geochemical Society Special Publication 1 (ed. B. O. Mysen), pp. 107–19. Geochemical Society, Pennsylvania.

Yoder, H. S. and Kushiro, I. (1969). Melting of a hydrous phase: phlogopite. *American Journal of Science*, **267A,** 558–82.

Yoder, H. S. and Tilley, C. E. (1962). Origin of basalt magmas: an experimental study of natural and synthetic systems. *Journal of Petrology*, **3,** 342–532.

Zindler, A. and Hart, S. R. (1986). Chemical geodynamics. *Annual Reviews of Earth and Planetary Science*, **14,** 493–571.

3

Age and early evolution of the continental mantle

Stephen H. Richardson

3.1. INTRODUCTION

Inclusion-bearing diamonds provide the most important means for determining the early evolutionary history of continental mantle. Diamond is unique in that it effectively isolates inclusions of other minerals from diffusive exchange at mantle temperatures and pressures. Syngenetic inclusions encapsulated by diamond have thus been closed systems since diamond crystallization and preserve a record of their age and initial conditions of formation (see also Haggerty, Chapter 5, this volume). In contrast, the constituent minerals of diamond host rocks were open systems at mantle temperatures and pressures and record ambient conditions immediately prior to sampling by kimberlite.

In southern Africa, diamondiferous kimberlites occur within a 3.5 billion year crustal rock record. Furthermore, kimberlites erupted within the boundaries of the Kaapvaal craton are diamondiferous, while those in adjacent Proterozoic mobile belts are barren. Southern Africa is thus a type area for the study of diamonds as time capsules from old continental mantle.

Diamond is only a trace constituent of kimberlite (or lamproite), with abundances varying from zero to an order-of-magnitude maximum of 1 p.p.m. In turn, only a small fraction of diamonds (typically <1 per cent) contains macroscopic monomineralic and occasionally multimineralic inclusions belonging to peridotitic or eclogitic parageneses. In addition, the diamond host assemblages are generally disaggregated during transport to the surface by kimberlite and not available for study as articulated xenoliths. There are thus various practical difficulties in extracting age and thermobarometric information from such mantle-derived materials and these will be mentioned in the course of this chapter.

3.2. DIAMOND INCLUSIONS AND HOST ASSEMBLAGES

Macroscopic silicate, oxide, and sulphide inclusions in diamonds are typically of the order of 10–100 μm in size and 1–10 μg in weight. The dominant peridotitic inclusion minerals are olivine, orthopyroxene, chromite, and purple chrome-pyrope garnet while the eclogitic inclusion minerals are primarily orange pyrope-almandine garnet and pale bluish-green omphacitic clinopyroxene. Sulphides are often numerically the most abundant inclusion phase but there are no optical mineralogical criteria for assigning them to either of the above inclusion suites unless accompanied by other identifiable inclusion minerals. None the less, diamond inclusion assemblages parallel the mineralogy of the two most important categories of mantle xenolith in kimberlites, garnet peridotite, and eclogite. Indeed, diamondiferous eclogite xenoliths are relatively common at certain kimberlite localities. However, there are significant differences in chemical composition between diamond inclusions and xenolith minerals, particularly for the peridotite suite. For example, the commonest peridotitic garnets in diamonds have significantly higher Mg and Cr and lower Ca contents than their peridotite xenolith counterparts, which are generally saturated with Ca. The former subcalcic garnets and corresponding olivines and orthopyroxenes are thus distinctly harzburgitic, while the latter are broadly lherzolitic. Lherzolitic diamond inclusions (including bright green diopside) do occur but they are generally rare.

Macrocrystic garnets (up to a few millimetres in size) of subcalcic composition have also been identified in kimberlite heavy mineral concentrates, where the bulk of kimberlite-borne mantle material is to be found as disaggregated mineral grains. The distribution of these discrete subcalcic garnets in southern African kimberlites generally mirrors that of diamonds in being restricted to the Archaean craton (Gurney and Switzer 1973; Boyd and Gurney 1986), implying an intimate genetic association (i.e. these are unencapsulated garnets from disaggregated diamond host assemblages).

The two intracratonic kimberlite occurrences (Fig. 3.1) where this association has been most extensively studied are Kimberley, which is 90 Ma old (Davis 1977), and Finsch, which is 118 Ma old (Smith 1983). These kimberlites are respectively representative of two

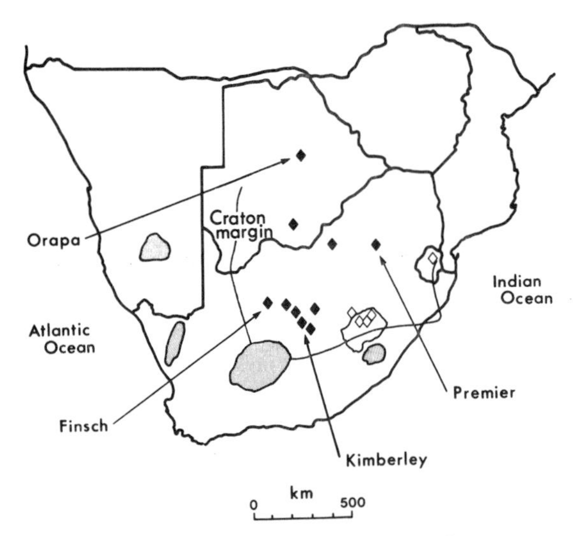

Fig. 3.1. Sketch map of the distribution of kimberlites in southern Africa (modified from Gurney and Harte 1980, fig. 1) highlighting the location of Kimberley (which includes the Bultfontein, De Beers, Dutoitspan, and Wesselton pipes) and Finsch in relation to an estimate of the southern margin of the Archaean Kaapvaal craton. Other kimberlites mentioned in the text (Premier, Orapa) are also indicated. Symbols: black diamonds, diamondiferous kimberlites; white diamonds, marginally diamondiferous kimberlites; shaded areas, clusters of non-diamondiferous kimberlites. Other lines are political boundaries.

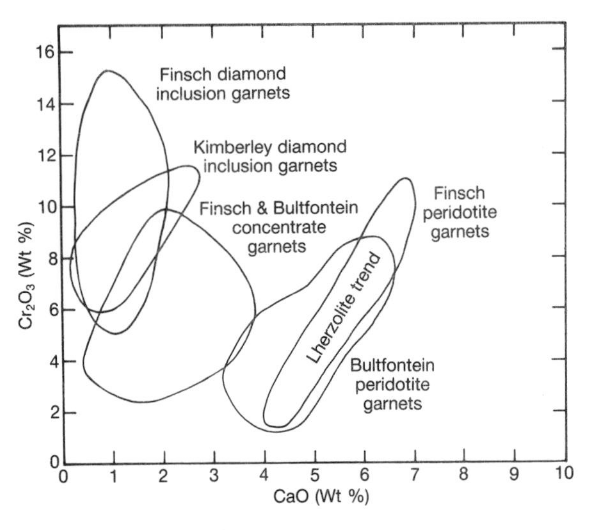

Fig. 3.2. Distribution of CaO and Cr_2O_3 in subcalcic garnet inclusions in Finsch (Gurney *et al.* 1979) and Kimberley (Richardson, unpublished) diamonds, selected subcalcic garnets from Finsch and Bultfontein (Kimberley) heavy mineral concentrates (Richardson *et al.* 1984), and garnets in Finsch (Shee *et al.* 1982) and Bultfontein (Lawless 1978) peridotite xenoliths. Positive correlations described by peridotite xenolith garnets have become known as the lherzolite trend (Sobolev *et al.* 1973) as they reflect saturation with a diopside component, regardless of the modal abundance of diopside. Subcalcic garnets lie to the left of this trend, indicating diopside-free host assemblages.

classic types of kimberlite (basaltic or Group I, and micaceous or Group II) recognized on the basis of petrographic, chemical, and isotopic characteristics (Barrett and Berg 1975; Smith 1983) though this distinction appears to have no relevance to diamond genesis.

Peridotitic diamond inclusions, of which 10–20 percent are garnets, comprise more than 90 per cent of inclusions in Kimberley diamonds (Harris *et al.* 1984) and more than 98 per cent of those in Finsch diamonds (Harris and Gurney 1979). A comprehensive electron microprobe study of diamond inclusion major element compositions by Gurney *et al.* (1979) revealed that more than 95 per cent of peridotitic garnets in Finsch diamonds have extremely high Mg and Cr and low Ca contents (Fig. 3.2) and are thus harzburgitic in composition. Associated olivine and orthopyroxene inclusion compositions are also extremely magnesian with average values of Mg/(Mg+Fe) of 0.941 and 0.949, respectively. Harzburgitic garnets in Kimberley diamonds have similar subcalcic compositions to their Finsch counterparts though the range of Cr contents is a little less extreme (5–15 wt per cent Cr_2O_3 for Finsch; 6–12 wt per cent Cr_2O_3 for Kimberley; Fig. 3.2). Lherzolitic garnet and diopside inclusions are rare at these localities, emphasizing the residual major element character of the peridotitic inclusion assemblage worldwide (Meyer and Tsai 1976; Sobolev 1977; Harris and Gurney 1979; Meyer 1987).

Subcalcic garnets from disaggregated harzburgitic diamond host assemblages are to be found in relative abundance in the kimberlite heavy mineral concentrate from both Finsch and Kimberley. At Finsch, such garnets comprise around 100 p.p.m. of the kimberlite as a whole (Gurney and Switzer 1973) as compared with a diamond abundance of 0.2 p.p.m. (Shee *et al.* 1982). Their major element compositions scatter between and partially overlap the fields for diamond inclusion garnets and non-diamondiferous peridotite xenolith garnets (Fig. 3.2), consistent with the open-system behaviour exhibited by unencapsulated garnets during residence in the mantle.

The trace element and radiogenic isotope geochemistry of diamond inclusions and their host assemblages is even more intriguing. Trace element abundance data for individual 50-μm inclusions can be obtained using an ion microprobe (Shimizu and Richardson 1987) or a proton microprobe (Griffin *et al.* 1988). For radiogenic isotope studies, garnet and clinopyroxene are the most important diamond inclusion minerals since they are the major carrier phases of Nd and Sr and are easily assigned to peridotitic or eclogitic parageneses on the basis of their distinctive colours. Sulphides are the major carrier phases of Pb but, as previously mentioned, paragenetic identification remains a problem (cf. Kramers 1979). Unfortunately, with Nd and Sr concentrations of only a few p.p.m., the average 10-μg garnet or clinopyroxene inclusion contains less than 50 pg of each of these elements, for which appropriately accurate and sufficiently precise isotopic ratios are not obtainable with conventional chemical extraction and mass spectrometric techniques. Therefore, to acquire these data it has been necessary to work with either composites of many small cogenetic inclusions or single large (millimetre-sized) inclusions, which are exceedingly rare, in order to obtain the required nanogram or more of the elements of interest. This is not a problem with diamond host assemblage mineral grains recovered from kimberlite heavy mineral concentrate since these may be up to a few millimetres in size and several milligrams in weight.

The trace element characteristics of individual inclusion and concentrate garnet grains from Kimberley and Finsch have been investigated by Shimizu and Richardson (1987) using an ion microprobe. A general enrichment of light rare earth elements (LREE) relative to heavy rare earth elements (HREE) in subcalcic garnets was documented (Fig. 3.3), which is the opposite of that normally observed in garnets from a variety of sources. Thus, inclusion garnet Sm/Nd ratios are significantly lower than the bulk earth (chondritic) value. The similarities in REE patterns for subcalcic diamond inclusion and concentrate garnets are also clearly demonstrated. Overall, subcalcic garnets are endowed with trace element abundances which belie their residual major element compositions. Concentrations of Nd (3–15 p.p.m.) and Sr (0.6–12 p.p.m.) determined by isotope dilution mass spectrometry (Richardson *et al.* 1984) range up to significantly higher values than those for peridotite xenolith garnets from Kimberley (1–4 p.p.m. Nd; 0.5–1 p.p.m. Sr; Richardson *et al.* 1985). The LREE enrichment, coupled with high Cr (5–15 wt. % Cr_2O_3) and low Ti (40–500 p.p.m.) contents is hard to reconcile with crystallization of these garnets in equilibrium with known silicate or carbonate magmas (cf. Shimizu and Richardson 1987; see also Watson *et al.*, Chapter 6, this volume). Rather, a subsolidus metasomatic origin for harzburgitic diamonds is indicated, as also suggested by thermobarometry (cf. subsequent section). Furthermore, during the

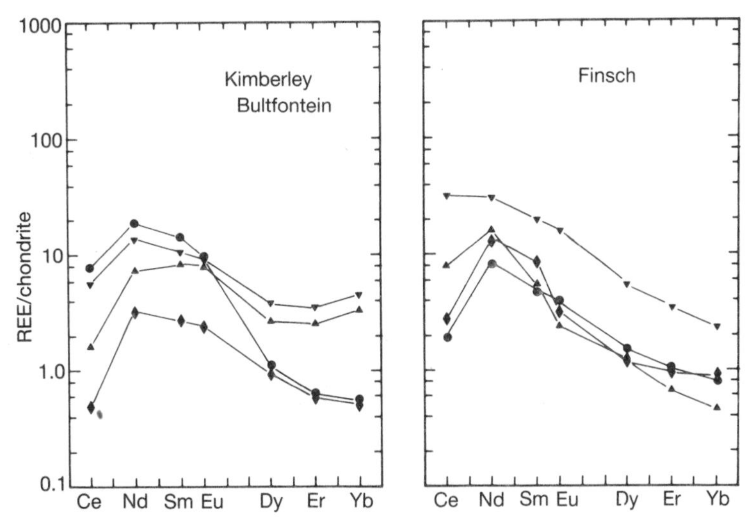

Fig. 3.3. Chondrite-normalized REE abundance patterns for individual subcalcic garnet grains (modified from Shimizu and Richardson 1987, fig. 1). Diamond symbols denote subcalcic garnet inclusions from Kimberley and Finsch diamonds while other symbols denote subcalcic garnets from Bultfontein (Kimberley) and Finsch heavy mineral concentrates. (Reprinted with permission from Geochimica et Cosmochimica Acta **51**, 755–8, Copyright 1987, Pergamon Press plc.)

growth of garnet together with diamond, olivine, orthopyroxene, and chromite, the bulk of the REE present in the local system was probably incorporated into garnet. This is of particular relevance to Sm–Nd isotope systematics since it implies that garnet (and diamond) crystallization did not significantly fractionate REE ratios (including Sm/Nd).

Combined $^{143}Nd/^{144}Nd$ and $^{87}Sr/^{86}Sr$ ratios for subcalcic garnets from Kimberley and Finsch are illustrated on a Nd–Sr isotope correlation diagram (Fig. 3.4) constructed for 90 Ma, the approximate time of kimberlite emplacement. Corresponding age corrections for residence at the surface are small for $^{143}Nd/^{144}Nd$, since Sm/Nd radios are low (0.1–0.5), and generally insignificant for $^{87}Sr/^{86}Sr$, since Rb/Sr ratios are negligible (0.0001–0.01).

The first characteristic of the data in Fig. 3.4 is that subcalcic diamond inclusion garnets, concentrate garnets, and host kimberlites have distinct isotopic signatures. The inclusion garnets have very low $^{143}Nd/^{144}Nd$ ratios indicative of an ancient Nd isotopic component and syngenetic host diamonds are thus xenocrysts in kimberlites. The concentrate garnets display a large spread in $^{143}Nd/^{144}Nd$ ratios, which are negatively correlated with extremely high $^{87}Sr/^{86}Sr$ ratios. Such concentrate garnets are also necessarily xenocrysts in kimberlites but are clearly isotopically differentiated from inclusion garnets. Yet these encapsulated and unencapsulated garnets show correlated differences in isotopic signature between localities (Fig. 3.4).

The second characteristic is that garnet $^{87}Sr/^{86}Sr$ ratios are unsupported by low intrinsic Rb contents (Richardson *et al.* 1984, 1985). Therefore, the garnets must have inherited such $^{87}Sr/^{86}Sr$ ratios by crystallization from and, in the case of unencapsulated garnets, subsequent diffusive equilibration with host assemblages which had higher Rb/Sr ratios for an appropriate period of time.

3.3. AGE CONSTRAINTS ON PERIDOTITIC DIAMOND FORMATION

Since garnet is the only major carrier phase of Nd and Sr in harzburgitic diamonds, useful isochron relationships

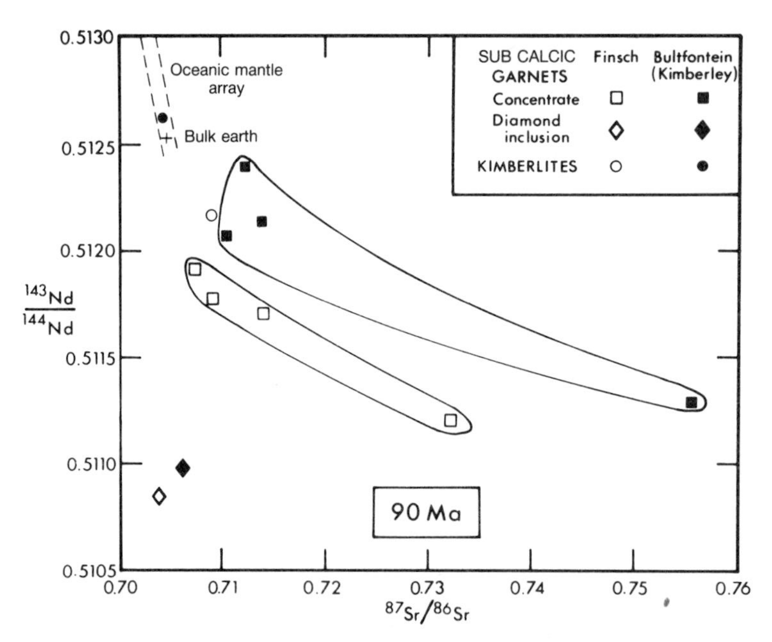

Fig. 3.4. Nd–Sr isotope correlation diagram constructed at 90 Ma (approximate time of Kimberley kimberlite emplacement) for subcalcic garnets (modified from Richardson *et al.* 1984, fig. 2 and references therein). A slightly older age of 118 Ma for the Finsch kimberlite (Smith 1983) does not significantly affect the presentation. Subcalcic concentrate garnets are single grains from Bultfontein (Kimberley) and Finsch heavy mineral concentrates, while inclusion garnets are composites of many specimens from Kimberley and Finsch diamonds. The oceanic mantle array, a bulk earth value, and host kimberlite values are shown for reference. (Reprinted with permission from *Nature* **310,** 198–202, Copyright 1984, Macmillan Magazines Ltd.)

are not obtainable. However, the data do lend themselves to model age interpretation. Sm–Nd model age relationships are best illustrated in a $^{143}Nd/^{144}Nd$ evolution diagram (Fig. 3.5). Intersections of garnet and reference curves yield Sm–Nd model ages for diamond precursor derivation from the hypothetical bulk earth reservoir. Ages thus obtained are 3410 ± 80 Ma for the Kimberley inclusion garnet composite and 3320 ± 30 Ma for the best of two Finsch inclusion garnet composites (Richardson *et al.* 1984). These model ages are within analytical error of each other and approximately synchronous harzburgitic diamond precursor formation over a wide area beneath the Kaapvaal craton is suggested.

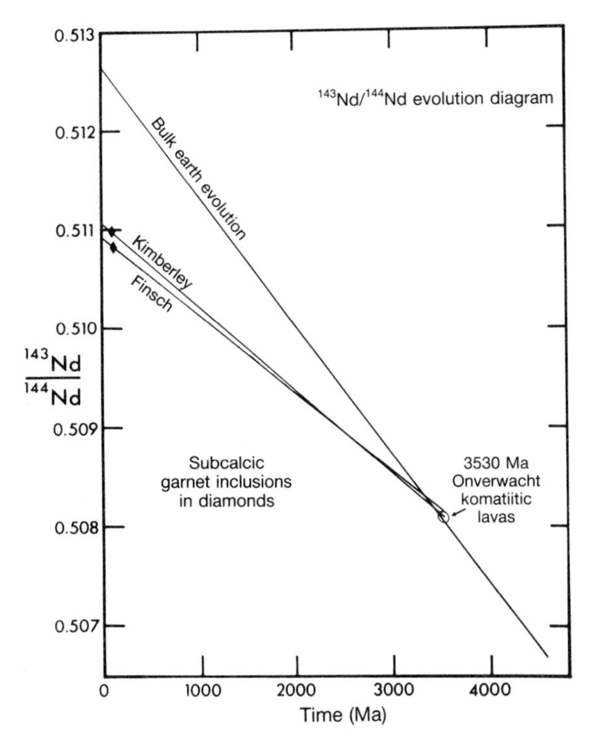

Fig. 3.5. $^{143}Nd/^{144}Nd$ evolution diagram showing evolution curves for subcalcic garnets in Kimberley and Finsch diamonds (modified from Richardson *et al.* 1984, fig. 3 and references therein). Curves are constructed from measured $^{143}Nd/^{144}Nd$ and $^{147}Sm/^{144}Nd$ ratios, using $\lambda_{Sm} = 6.54 \times 10^{-12}$ year^{-1}. Diamond symbols at 90 and 118 Ma mark the approximate times of diamond transport to surface by host kimberlites. A bulk earth evolution curve, based on average chondritic Nd isotopic evolution (Jacobsen and Wasserburg 1980) and the Onverwacht komatiite datum (Hamilton *et al.* 1979, 1983) are shown for reference. (Reprinted with permission from *Nature*, **310**, 198–202, Copyright 1984, Macmillan Magazines Ltd.)

Further systematic model age uncertainties arise from the assumption of a bulk earth reference curve, since neither an exactly chondritic initial earth nor a subsequent undifferentiated mantle source for garnet and diamond can be presumed. The Nd isotopic character of real temporally and spatially associated mantle may usefully be represented by that of 3530-Ma-old Onverwacht Group volcanics (Hamilton *et al.* 1979, 1983) from the Barberton Mountain Land on the eastern side of the Kaapvaal craton. These komatiitic volcanics are the best preserved remnant of such magmas which appear to have been emplaced across the entire craton (Viljoen *et al.* 1982). For a mantle evolution curve passing through the Onverwacht datum (Fig. 3.5) and slightly steeper than the reference curve, the above model ages would be increased by about 100 Ma, pushing them back towards that of the komatiites.

The association between harzburgitic diamond host rocks and komatiites is not only temporal and spatial but also compositional, as first suggested by Boyd and Gurney (1982). For example, diamond inclusion olivine Mg/(Mg + Fe) ratios are similar to those for liquidus olivines in komatiites (Green *et al.* 1975). While the mineralogy of residual mantle or high-pressure cumulates remaining after komatiite generation remains controversial (cf. Arndt and Nisbet 1982), highly magnesian major element compositions are not disputed. Sm/Nd and Rb/Sr ratios respectively greater and less than bulk earth values may also be inferred for such trace-element-depleted residual mantle. Such trace element characteristics are the opposite of those deduced from measured subcalcic garnet Sm/Nd and $^{87}Sr/^{86}Sr$ ratios. Therefore, enrichment (metasomatism) of such residues is required after komatiite generation as a precursor to subcalcic garnet and diamond crystallization (see also Menzies, Chapter 4, this volume).

A suitable time interval between enrichment and diamond crystallization is also required to allow for dramatic differential growth of radiogenic ^{87}Sr in the enriched (high Rb/Sr, low Sm/Nd) diamond precursor. In contrast, differential growth of radiogenic ^{143}Nd is far less pronounced because differences in low Sm/Nd ratios are small compared with those in Rb/Sr ratio. In fact, metasomatic enrichment (diamond precursor derivation) and subsequent garnet/diamond crystallization are not distinguishable on the basis of Nd model age relationships alone, since an event (garnet/diamond crystallization) where there is little or no fractionation of the parent/daughter ratio (Sm/Nd; cf. previous section) cannot be resolved on a $^{143}Nd/^{144}Nd$ evolution diagram (Fig. 3.5). In the Rb/Sr system, however, this fractionation is dramatic during garnet crystallization with the Rb/Sr ratio decreasing to virtually zero (in garnet). Rb–Sr model age relationships can thus be used to estimate the time interval between metasomatic enrichment and garnet/diamond crystallization, as illustrated in an $^{87}Sr/^{86}Sr$ evolution diagram (Fig. 3.6). Hypothetical subcalcic garnet host (diamond precursor) evolution curves are shown diverging from the bulk earth curve at 3500 Ma (maximum Nd model age) towards the highest present-day $^{87}Sr/^{86}Sr$ ratios measured in unencapsulated concentrate garnets. Intersections of the respective diamond inclusion garnet and precursor curves yield a common Rb–Sr model age for diamond crystallization of approximately 3200 Ma, as predicated on the earliest possible time of precursor

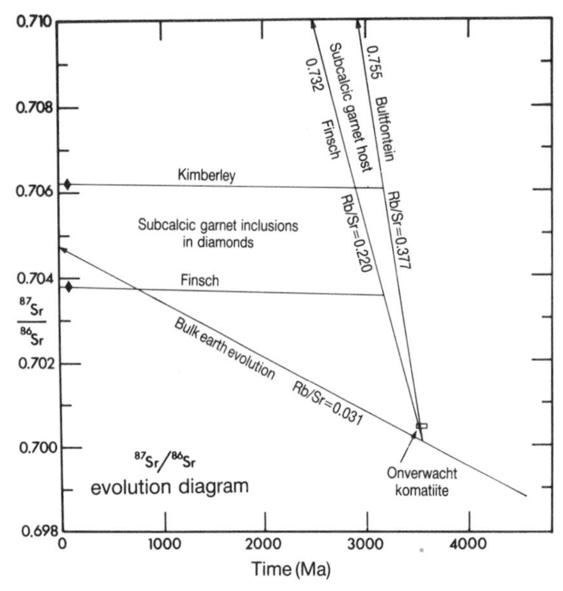

Fig. 3.6. $^{87}Sr/^{86}Sr$ evolution diagram showing evolution curves for subcalcic garnets in Kimberley and Finsch diamonds (modified from Richardson *et al.* 1984, fig. 4 and references therein). Curves are constructed from measured $^{87}Sr/^{86}Sr$ ratios, using $\lambda_{Rb} = 1.42 \times 10^{-11}$ year^{-1}. Garnet Rb/Sr ratios are so low that corresponding evolution curves are virtually flat. Diamond symbols at 90 and 118 Ma mark the approximate times of diamond transport to the surface by host kimberlites. The bulk earth evolution curve, the Onverwacht komatiite datum, and hypothetical subcalcic garnet host (diamond precursor) evolution curves, drawn between the bulk earth curve and the highest $^{87}Sr/^{86}Sr$ ratios measured in concentrate garnets (Fig. 3.4), are shown for reference. (Reprinted with permission from *Nature* **310**, 198–202, Copyright 1984, Macmillan Magazines Ltd.)

formation (enrichment of komatiite residua at 3500 Ma; for critical discussion of the above model age relationships cf. Pidgeon 1989 and Richardson 1989*a*).

As previously indicated, unencapsulated subcalcic concentrate garnets show a range of major element compositions (Fig. 3.2) and $^{87}Sr/^{86}Sr$ and $^{143}Nd/^{144}Nd$ ratios (Fig. 3.4). Unencapsulated garnet evolutionary histories involve both (1) diffusive exchange with the immediate host assemblage (re-equilibration), and (2) material exchange with the greater mantle environment (metasomatism). Process (1) accounts for the highly radiogenic Sr in essentially Rb-free garnets. Process (2) accounts for the correlated Nd and Sr isotope arrays in Fig. 3.4, which stretch from the extreme (high $^{87}Sr/^{86}Sr$, low $^{143}Nd/^{144}Nd$) compositions produced by process (1) back to values typical of more calcic garnets in lithospheric lherzolite xenoliths (Richardson *et al.* 1985). In constructing dependent Sr model age relationships, it is therefore appropriate to use the most extreme (high $^{87}Sr/^{86}Sr$) garnet compositions because they represent the least influence of process (2). In principle, more extreme concentrate garnets may remain to be analysed. Thus, precursor evolution curves may have been even steeper than indicated in Fig. 3.6, with a correspondingly smaller time interval between enrichment and diamond crystallization.

3.4. PERIDOTITIC DIAMOND THERMOBAROMETRY

Temperatures and pressures of crystallization of harzburgitic diamonds can be ascertained in cases where olivine, orthopyroxene, and garnet occur as separated inclusions in the same diamond. The assumption of chemical equilibrium at the time of diamond crystallization is supported by the observation that multiple inclusions of a particular mineral in the same diamond generally have the same composition (Gurney *et al.* 1979). The thermometer is based on Fe–Mg partitioning between olivine and garnet (O'Neill and Wood 1979) while the barometer is based on the Al content of orthopyroxene (MacGregor 1974; Perkins *et al.* 1981). Unfortunately, diamonds with inclusions of all three of the above phases are exceedingly rare and only one example from the Finsch kimberlite has been documented (Tsai *et al.* 1979). This yields an estimated crystallization temperature of 1070°C and a pressure of 56 kbar corresponding to a depth of about 175 km (Boyd and Finnerty 1980; Boyd *et al.* 1985). On a temperature–depth diagram (Fig. 3.7) this plots within the experimentally determined diamond stability field (Kennedy and Kennedy 1976) and fairly close to a continental shield geotherm calculated for a heat flow of 40 mW m^{-2} (Pollack and Chapman 1977). Diamonds with olivine–garnet inclusion pairs are more common and, since the relevant thermometer has a small pressure sensitivity of 5°C $kbar^{-1}$, temperature estimates can be obtained assuming pressures within the diamond stability field. Thus, estimates for a set of 29 representative specimens from Finsch lie in a *P–T* band approximately 250°C wide with a slope of 5°C $kbar^{-1}$ within the diamond stability field (Fig. 3.7; Boyd *et al.* 1985).

The majority of peridotitic diamonds from Finsch and elsewhere on the Kaapvaal craton thus appear to have crystallization temperatures below about 1200°C (Boyd *et al.* 1985). The solidus for peridotitic compositions in the presence of H_2O and CO_2 (Eggler and Wendtlandt 1979) is also indicated in Fig. 3.7 such that the majority of peridotitic diamonds would appear to be of subsolidus origin (Boyd and Finnerty 1980; Boyd *et al.* 1985). Finally, two diamondiferous garnet lherzolite

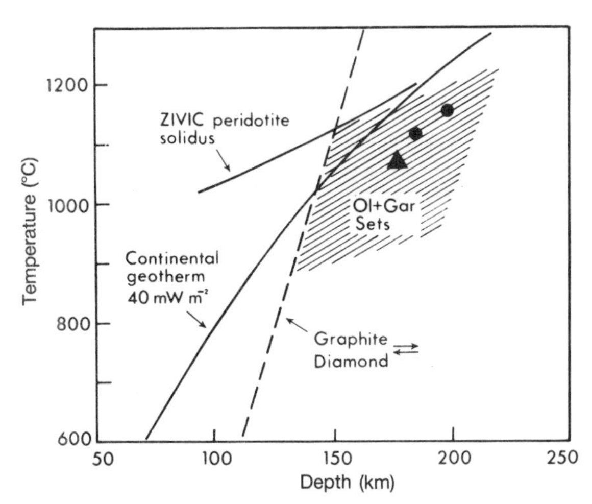

Fig. 3.7. Estimated temperature–depth equilibration conditions for peridotitic-inclusion-bearing diamonds and diamondiferous peridotite xenoliths from the Finsch kimberlite (after Boyd *et al.* 1985, fig. 1 and references therein). Symbols: black triangle, olivine–orthopyroxene–garnet inclusion assemblage in a single diamond (Tsai *et al.* 1979); black circles, diamondiferous lherzolite xenoliths (Shee *et al.* 1982); ruled lines, band of temperature estimates for 29 olivine–garnet inclusion pairs. The continental shield geotherm calculated for a heat flow of 40 mW m^{-2} (Pollack and Chapman 1977), diamond–graphite equlibrium curve (Kennedy and Kennedy 1976), and solidus for peridotite in the presence of H_2O and CO_2 (Eggler and Wendlandt 1979) are shown for reference. (Reprinted with permission from *Nature* **315**, 387–9, Copyright 1985, Macmillan Magazines Ltd.)

xenoliths from Finsch have been described by Shee *et al.* (1982). Significantly, the garnets were not, or at least are no longer, subcalcic and the xenolith minerals were subject to diffusive exchange with each other before sampling by kimberlite during the Cretaceous. Equilibration conditions for these rocks as plotted in Fig. 3.7 also lie close to the shield geotherm within the diamond stability field. Overall, the data in Fig. 3.7 suggest that *P–T* conditions close to those calculated for a present-day shield geotherm have been maintained in continental upper mantle beneath the Kaapvaal craton for more than 3 billion years.

3.5. IMPLICATIONS FOR CONTINENTAL MANTLE STABILIZATION

Since peridotitic diamonds and their inferred garnet harzburgite hosts are confined to the Kaapvaal craton, they must have formed and been stored in a continental mantle keel or tectosphere (Jordan 1978, 1988; see also Anderson, Chapter 1, this volume), which extended into the diamond stability field at depths of 150–200 km and temperatures of 900–1200°C more than 3 billion years ago, and which survived convective disruption leaving diamonds available for episodic sampling by kimberlite (Fig. 3.8). The residual (harzburgitic) yet trace-element-enriched (high Rb, LREE, C but low Ti) character of the keel may have resulted from extensive komatiite magmatism approximately 3500 Ma ago followed by asthenosphere-derived alkali carbonate metasomatism (Shimizu and Richardson 1987) and advective thickening (Jordan 1978, 1981) prior to diamond crystallization by approximately 3200 Ma ago.

In this scenario continental upper mantle was stabilized with respect to the higher interior temperature of the Archaean mantle by the greater degree of depletion in basaltic components induced by komatiite magmatism compared to modern mid-ocean ridge basalt magmatism, i.e. a chemical boundary layer (Jordan 1988) or chemical lithosphere (Richter 1988) was formed. The melts/fluids responsible for the subsequent metasomatism of this residual mantle appear to be asthenosphere-derived on the basis of the relatively narrow range of $\delta^{13}C$ values (-1 to -8) observed for peridotitic inclusion-bearing diamonds world-wide (Galimov 1985), which peaks at values similar to those for mid-ocean ridge basalts (e.g. Des Marais and Moore 1984). An alkali carbonate character for the metasomatism (see also Watson *et al.*, Chapter 6, this volume) would appear to satisfy the chemical requirements of being Rb-, LREE-, and C-rich but Ti-poor. A petrological analogue might be sought in the 2-billion-year-old

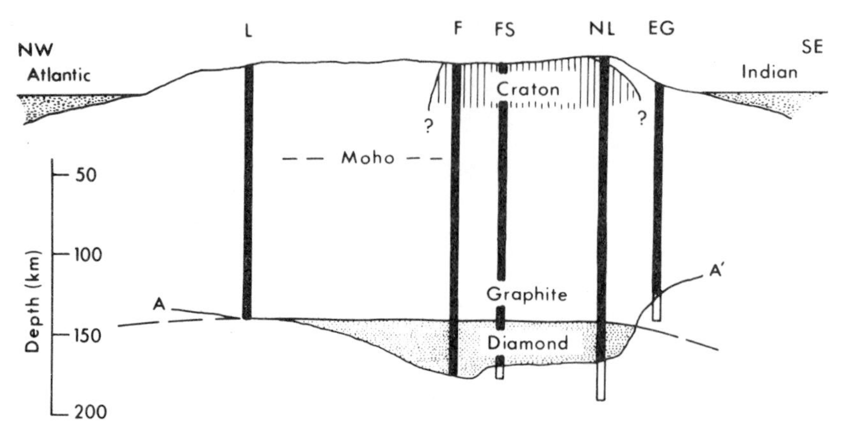

Fig. 3.8. Cross-section of the lithosphere beneath southern Africa based on thermobarometry of peridotite xenolith suites (after Boyd and Gurney 1986, fig. 6). Vertical exaggeration is 4:1 below sea level and surface relief is greatly exaggerated. Vertical bars represent xenolith suites from: L, Louwrencia; F, Finsch; FS, Frank Smith; NL, Northern Lesotho; and EG, East Griqualand. The line A–A′ represents the approximate locus of points of inflection of the xenolith geotherms. The graphite–diamond transition is represented by a line drawn through the points of intersection of xenolith geotherms with the equilibrium boundary of Kennedy and Kennedy (1976). The shaded area represents that part of the lithospheric keel in which peridotitic diamonds could have formed and been stored for more than 3 billion years. (Reprinted with permission from *Science*, **232**, 472–7, Copyright 1986, AAAS.)

phlogopite-rich Phalaborwa alkaline igneous complex (Hanekom *et al.* 1965) on the north-eastern side of the Kaapvaal craton. Finally, advective thickening is a possible mechanism for triggering diamond crystallization 200–300 Ma after initial metasomatism.

Meanwhile, secular cooling of the earth proceeded apace and komatiite magmatism waned towards the end of the Archaean. At or near the Archaean–Proterozoic boundary the interior temperature of the mantle may have reached the point where chemically distinct mantle was no longer required to ensure preservation (Richter 1988) such that continental lithosphere of Proterozoic and Phanerozoic age could be stabilized thermally as for oceanic lithosphere (see also Menzies, Chapter 4, this volume).

3.6. PERIDOTITES, ECLOGITES, AND ECLOGITIC DIAMONDS

The dominant non-diamondiferous lithospheric peridotite xenoliths from southern African kimberlites, which evidently lost a major basaltic component prior to stabilization in the lithosphere, show an extended range of isotopic signatures between those of residual yet enriched (high Rb/Sr, low Sm/Nd) peridotitic diamond host assemblages and those of more fertile yet depleted (low Rb/Sr, high Sm/Nd) asthenospheric peridotite xenoliths (Richardson *et al.* 1985). This indicates enriched component addition (metasomatism) and redistribution (by scavenging) within the lithosphere through time with re-equilibration on only a local scale.

Thermobarometry for such xenolith suites from various kimberlites on the Kaapvaal craton shows inflection points in the geotherm at depths of 150–200 km (Boyd and Gurney 1986). The inferred gradients of the high T portions of these geotherms are still controversial (Finnerty 1989), varying from steeper than the low T lithospheric portion and representing asthenospheric thermal anomalies (Boyd and Gurney 1986) to shallower than the low T conductive portion and approaching an asthenospheric adiabat (Bertrand *et al.* 1986). In any event, the inflections could still correspond to the lithosphere/asthenosphere boundary which shelves to depths < 150 km beneath the Proterozoic mobile belts peripheral to the craton (Fig. 3.8; Boyd and Gurney 1986). This would be consistent with coincidence of chemical and thermal boundary layers (Jordan 1988) in an approximately 200 km rather than 400 km thick tectosphere.

Eclogite xenoliths from southern African kimberlites appear to be more diverse in origin. Some eclogite xenoliths (type II) may represent ancient subducted oceanic crust caught up during continental lithosphere formation (Jagoutz *et al.* 1983; MacGregor 1985). Other eclogite xenoliths (type I) and eclogitic-inclusion-bearing diamonds appear to reflect entrapment, mixing, and crystallization of asthenosphere-derived magmas (which may include crustal components recycled by subduction) in the tectospheric keel during the Proterozoic (see also Kyser, Chapter 7, this volume).

In the case of eclogitic (or lherzolitic) diamonds, cogenetic garnet and clinopyroxene pairs (separated by diamond) are required to yield significant isochron age and initial ratio information. For example, at the 1180 ± 30 Ma old Premier kimberlite, eclogitic garnet and clinopyroxene inclusions define a Sm–Nd isochron age of 1150 ± 60 Ma for diamond crystallization from a source with a trace-element-depleted history akin to that of the asthenosphere (Richardson 1986). These errors allow for a relatively short mantle residence time of 1–10 Ma for these eclogitic diamonds as required by their nitrogen aggregation characteristics, prior to removal to the surface. Isotopic data (including a $^{87}Sr/^{86}Sr$ ratio of 0.72587 ± 5) for a single lherzolitic diopside inclusion from a Premier diamond are consistent with coeval crystallization of Premier lherzolitic diamonds but with a trace-element-enriched source resembling that of reworked Archaean harzburgitic diamond host rocks (Richardson 1989*b*).

At the 118-Ma-old Finsch kimberlite, a suite of eclogitic garnet and clinopyroxene inclusions define an isochron age of 1580 ± 50 Ma but with indications of involvement of both trace-element-depleted and -enriched end-members during diamond formation (Richardson 1989*b*). On the basis of available evidence, the crystallization ages (and thus mantle storage times) of harzburgitic and eclogitic diamonds brought to the surface by the Finsch kimberlite differ by more than 1.5 billion years. At the 93-Ma-old Orapa kimberlite (cf. Gurney *et al.* 1984), a similar suite of eclogitic garnet and clinopyroxene inclusions define an isochron age of 990 ± 50 Ma but with evidence of involvement of at least two trace element-enriched components during diamond formation (Richardson 1989*b*). In this case, eclogitic diamond crystallization appears to have been followed by mantle storage for some 900 Ma prior to sampling by the Orapa kimberlite in the Cretaceous.

Finally, in the case of diamonds with garnet inclusions showing pyroxene solid solution (cf. Moore and Gurney 1985) there is isotopic evidence (Richardson 1989*b*) consistent with derivation of such diamonds directly from the asthenosphere in Mesozoic time.

3.7. CONCLUSIONS

The Sm-Nd and Rb-Sr isotopic systematics of subcalcic garnet inclusions in diamonds and their macrocryst counterparts from ~100-Ma-old intracratonic kimberlites in southern Africa indicate that diamond crystallization occurred in residual yet trace-element-enriched harzburgites 3200–3300 Ma ago. Furthermore, the thermobarometry of peridotitic diamond inclusion assemblages from these localities yields diamond crystallization temperatures of 900–1200°C. Therefore, since these diamonds and their inferred garnet harzburgite hosts are confined to the Kaapvaal craton, they must have formed and been stored in a continental mantle keel or tectosphere, which extended into the diamond stability field at depths of 150–200 km and temperatures of 900–1200°C more than 3 billion years ago, and which survived convective disruption leaving harzburgitic and later eclogitic and lherzolitic diamonds available for episodic sampling by kimberlite.

ACKNOWLEDGEMENTS

The contributions of De Beers Consolidated Mines and F. R. Boyd, A. J. Erlank, J. J. Gurney, J. W. Harris, S. R. Hart, J. B. Hawthorne, and N. Shimizu to the conception, planning, and conduct of the research on which this chapter is based are gratefully acknowledged. Manuscript production credits are due to L. O'Neill and S. Wisdom. Financial support for this writing endeavour was provided from FRD funding to A. J. Erlank and friends.

REFERENCES

Arndt, N. T. and Nisbet, E. G. (ed.) (1982). *Komatiites*. Allen and Unwin, London.

Barrett, D. R. and Berg, G. W. (1975). Complementary petrographic and strontium-isotope ratio studies of South African kimberlites. *Physics and Chemistry of the Earth*, **9**, 619–35.

Bertrand, P., Sotin, C., Mercier, J.-C.C., and Takahashi, E. (1986). From the simplest chemical system to the natural one: garnet peridotite barometry. *Contributions to Mineralogy and Petrology*, **93**, 168–78.

Boyd, F. R. and Finnerty, A. A. (1980). Conditions of origin of natural diamonds of peridotitic affinity. *Journal of Geophysical Research*, **85**, 261–7.

Boyd, F. R. and Gurney, J. J. (1982). Low-calcium garnets: keys to craton structure and diamond crystallization. *Carnegie Institution of Washington, Yearbook*, **81**, 261–7.

Boyd, F. R. and Gurney, J. J. (1986). Diamonds and the African lithosphere. *Science*, **232,** 472–7.

Boyd, F. R., Gurney, J. J., and Richardson, S. H. (1985). Evidence for a 150–200 km thick Archaean lithosphere from diamond inclusion thermobarometry. *Nature*, **315,** 387–9.

Davis, G. L. (1977). The ages and uranium contents of zircons from kimberlites and associated rocks. *Carnegie Institution of Washington, Yearbook*, **76,** 631–5.

Des Marais, D. J. and Moore, J. G. (1984). Carbon and its isotopes in mid-oceanic basaltic glasses. *Earth and Planetary Science Letters*, **69,** 43–57.

Eggler, D. H. and Wendlandt, R. F. (1979). Experimental studies on the relationship between kimberlite magmas and partial melting of peridotite. In *Kimberlites, diatremes and diamonds* (ed. F. R. Boyd and H. O. A. Meyer), pp. 330–8. American Geophysical Union, Washington, DC.

Finnerty, A. A. (1989). Xenolith-derived mantle geotherms: whither the inflection? *Contributions to Mineralogy and Petrology*, **102,** 367–75.

Galimov, E. M. (1985). The relation between formation conditions and variations in isotope composition of diamonds. *Geochemistry International*, **22,** 118–42.

Green, D. H., Nicholls, I. A., Viljoen, M. J., and Viljoen, R. P. (1975). Experimental demonstration of the existence of peridotitic liquids in earliest Archaean magmatism. *Geology*, **2,** 11–14.

Griffin, W. L., Jaques, A. L., Sie, S. H., Ryan, C. G., Cousens, D. R., and Suter, G. F. (1988). Conditions of diamond growth: a proton microprobe study of inclusions in West Australian diamonds. *Contributions to Mineralogy and Petrology*, **99,** 143–58.

Gurney, J. J. and Harte, B. (1980). Chemical variations in upper mantle nodules from southern African kimberlites. *Philosophical Transactions of the Royal Society, London*, **A297,** 273–93.

Gurney, J. J. and Switzer, G. S. (1973). The discovery of garnets closely related to diamonds in the Finsch pipe, South Africa. *Contributions to Mineralogy and Petrology*, **39,** 103–16.

Gurney, J. J., Harris, J. W., and Rickard, R. S. (1979). Silicate and oxide inclusions in diamonds from the Finsch kimberlite pipe. In *Kimberlites, diatremes and diamonds* (ed. F. R. Boyd and H. O. A. Meyer), pp. 1–15. American Geophysical Union, Washington, DC.

Gurney, J. J., Harris, J. W., and Rickard, R. S. (1984). Silicate and oxide inclusions in diamonds from the Orapa mine, Botswana. In *Kimberlites II: the mantle and crust–mantle relationships* (ed. J. Kornprobst), pp. 3–10. Elsevier, Amsterdam.

Hamilton, P. J., Evenson, N. M., O'Nions, R. K., Smith, H. S., and Erlank, A. J. (1979). Sm–Nd dating of Onverwacht Group volcanics, southern Africa. *Nature*, **279,** 298–300.

Hamilton, P. J., O'Nions, R. K., Bridgwater, D., and Nutman, A. (1983). Sm–Nd studies of Archaean metasediments and metavolcanics from West Greenland and their implications for the Earth's early history. *Earth and Planetary Science Letters*, **62,** 263–72.

Hanekom, H. J., Van Staden, C. M. V. H., Smith, P. J., and Pike, D. R. (1965). The geology of the Phalaborwa Igneous Complex. *Memoirs of the Geological Survey of South Africa*, **54,** 1–185.

Harris, J. W. and Gurney, J. J. (1979). Inclusions in diamond. In *The properties of diamond* (ed. J. E. Field), pp. 555–91. Academic Press, London.

Harris, J. W., Hawthorne, J. B., and Oosterveld, M. M. (1984). A comparison of diamond characteristics from the De Beers Pool mines, Kimberley, South Africa. In *Annales Scientifiques de l'Universite de Clermont-Ferrand II* (ed. J. Kornprobst), **74,** 1–13.

Jacobsen, S. B. and Wasserburg, G. J. (1980). Sm–Nd isotopic evolution of chondrites. *Earth and Planetary Science Letters*, **50,** 139–55.

Jagoutz, E., Spettel, B., Wanke, H., and Dawson, B. (1983). Identification of early differentiation processes on the earth. *Meteoritics*, **18,** 319–20.

Jordan, T. H. (1978). Composition and development of the continental tectosphere. *Nature*, **274,** 54–8.

Jordan, T. H. (1981). Continents as a chemical boundary layer. *Philosophical Transactions of the Royal Society, London*, **A301,** 359–73.

Jordan, T. H. (1988). Structure and formation of the continental tectosphere. In *Oceanic and continental lithosphere: similarities and differences* (ed. M. A. Menzies and K. G. Cox), pp. 11–37. Oxford University Press, Oxford.

Kennedy, C. S. and Kennedy, G. C. (1976). The equilibrium boundary between graphite and diamond. *Journal of Geophysical Research*, **81,** 2467–70.

Kramers, J. D. (1979). Lead, uranium, strontium, potassium and rubidium in inclusion-bearing diamonds and mantle-derived xenoliths from southern Africa. *Earth and Planetary Science Letters*, **42**, 58–70.

Lawless, P. J. (1978). Some aspects of the mineral chemistry of the peridotite xenolith suite from the Bultfontein diamond mine, Kimberley, South Africa. Unpublished D.Phil. thesis. University of Cape Town.

MacGregor, I. D. (1974). The system $MgO–Al_2O_3–SiO_2$: solubility of Al_2O_3 in enstatite for spinel and garnet peridotite compositions. *American Mineralogist*, **59**, 110–19.

MacGregor, I. D. (1985). The Roberts Victor eclogites: ancient oceanic crust? *Geological Society of America, Abstracts with Programs*, **17**, 650.

Meyer, H. O. A. (1987). Inclusions in diamond. In *Mantle xenoliths* (ed. P. H. Nixon), pp. 501–22. Wiley, Chichester.

Meyer, H. O. A. and Tsai, H.-M. (1976). The nature and significance of mineral inclusions in natural diamond: a review. *Minerals Science and Engineering*, **8**, 242–61.

Moore, R. O. and Gurney, J. J. (1985). Pyroxene solid solution in garnets included in diamonds. *Nature*, **318**, 553–5.

O'Neill, H. St. C. and Wood, B. J. (1979). An experimental study of Fe–Mg partitioning between garnet and olivine and its calibration as a geothermometer. *Contributions to Mineralogy and Petrology*, **70**, 59–70.

Perkins III, D., Holland, T. J. B., and Newton, R. C. (1981). The Al_2O_3 contents of enstatite in equilibrium with garnet in the system $MgO–Al_2O_3–SiO_2$ at 15–40 kbar and 900–1600°C. *Contributions to Mineralogy and Petrology*, **78**, 99–109.

Pidgeon, R. T. (1989). Archaean diamond xenocrysts in kimberlites—how definitive is the evidence? In *Kimberlites and related rocks* (ed. J. Ross *et al.*), Geological Society of Australia, Special Publication No. 14, pp. 1070–4. Blackwell, Melbourne.

Pollack, H. N. and Chapman, D. S. (1977). On the regional variation of heat flow, geotherms and lithospheric thickness. *Tectonophysics*, **38**, 279–96.

Richardson, S. H. (1986). Latter-day origin of diamonds of eclogitic paragenesis. *Nature*, **322**, 623–6.

Richardson, S. H. (1989*a*). As definitive as ever: a reply to Archaean diamond xenocrysts in kimberlites—how definitive is the evidence? by R. T. Pidgeon. In *Kimberlites and related rocks* (ed. J. Ross *et al.*), Geological Society of Australia, Special Publication No. 14, pp. 1070–4. Blackwell, Melbourne.

Richardson, S. H. (1989*b*). Radiogenic isotope studies of diamond inclusions. In *Extended Abstract Volume, Workshop on Diamonds* (ed. F. R. Boyd and H. O. A. Meyer), pp. 87–90. 28th International Geological Congress, Washington, DC.

Richardson, S. H., Gurney, J. J., Erlank, A. J., and Harris, J. W. (1984). Origin of diamonds in old enriched mantle. *Nature*, **310**, 198–202.

Richardson, S. H., Erlank, A. J., and Hart, S. R. (1985). Kimberlite-borne garnet peridotite xenoliths from old enriched subcontinental lithosphere. *Earth and Planetary Science Letters*, **75**, 116–28.

Richter, F. M. (1988). A major change in the thermal state of the earth at the Archean–Proterozoic boundary: consequences for the nature and preservation of continental lithosphere. In *Oceanic and continental lithosphere: similarities and differences* (ed. M. A. Menzies and K. G. Cox), pp. 39–52. Oxford University Press, Oxford.

Shee, S. R., Gurney, J. J., and Robinson, D. N. (1982). Two diamond-bearing peridotite xenoliths from the Finsch kimberlite, South Africa. *Contributions to Mineralogy and Petrology*, **81**, 79–87.

Shimizu, N. and Richardson, S. H. (1987). Trace element abundance patterns of garnet inclusions in peridotite-suite diamonds. *Geochimica et Cosmochimica Acta*, **51**, 755–8.

Smith, C. B. (1983). Pb, Sr and Nd isotopic evidence for sources of southern African Cretaceous kimberlites. *Nature*, **304**, 51–4.

Sobolev, N. V. (1977). *Deep-seated inclusions in kimberlites and the problem of the composition of the upper mantle* [English translation, D. A. Brown]. American Geophysical Union, Washington, DC.

Sobolev, N. V., Laurentev, Y. G., Pokhilenko, N. P., and Usova, L. V. (1973). Chrome-rich garnets from the kimberlites of Yakutia and their paragenesis. *Contributions to Mineralogy and Petrology*, **40**, 39–52.

Tsai, H.-M., Meyer, H. O. A., Moreau, J., and Milledge, H. J. (1979). Mineral inclusions in diamond: Premier, Jagersfontein and Finsch kimberlites, South Africa, and Williamson Mine, Tanzania. In *Kimberlites, diatremes and diamonds* (ed. F. R. Boyd and H. O. A. Meyer), pp. 16–26. American Geophysical Union, Washington, DC.

Viljoen, M. J., Viljoen, R. P., and Pearton, T. N. (1982). The nature and distribution of Archaean komatiite volcanics in South Africa. In *Komatiites* (ed. N. T. Arndt and E. G. Nisbet). 53–79. Allen and Unwin, London.

4

Archaean, Proterozoic, and Phanerozoic lithospheres

Martin A. Menzies

4.1. INTRODUCTION

Many crust–mantle evolutionary models assume that the continents are underlain by chemically homogeneous mantle with an isotopic composition identical to the source of mid-ocean ridge basalts. However, the vast data base for oceanic and continental volcanic rocks indicates that this is not the case and that the upper mantle comprises several discrete chemical reservoirs. One of the most significant reservoirs is the continental lithospheric mantle, the rigid part of the upper mantle that immediately underlies the continental crust and has cooled to below 1280°C. In turn the lithosphere rests on a substratum of normal temperature asthenosphere which is that part of the upper mantle that flows in response to stress and has temperatures above 1280°C. Theoretical considerations (McKenzie and Bickle 1988; McKenzie 1989) and the analysis of micro-inclusions in diamonds (Navon *et al.* 1988) indicate the presence, in the asthenosphere, of small volume melt fractions rich in potassium, water, and carbonates. Since the asthenosphere has formed a hot, fertile substratum for the lithospheric plates since the Archaean, that part of the lithosphere immediately juxtaposed to the asthenosphere has continually equilibrated with the adiabatic interior. Consequently, any part of the lithosphere that is K–Fe–Ti rich is believed to have been modified by the upward passage of asthenospheric melts or fluids (Menzies and Hawkesworth 1987*b*) including those which are carbonate-rich (Watson *et al.*, Chapter 6, this volume). As a result of this the thermal boundary layer at the base of the lithosphere has maintained a chemical identity close to that of the underlying asthenosphere whilst isotopic heterogeneities have developed in the overlying more isolated mechanical boundary layer.

Samples of the thermal and mechanical boundary layers of the lithosphere are entrained as xenoliths (e.g. spinel and garnet peridotites) and megacrysts (e.g. diamonds) in alkaline magmas erupted through Phanerozoic, Proterozoic, and Archaean crust and, as such, provide a vital monitor of the temporal evolution of the Earth's upper mantle (e.g. Nixon 1987). Kimberlite-borne xenoliths offer a unique insight into the nature of the Archaean lithospheric mantle since they are essentially restricted to Archaean terrains with only minor occurrences in Proterozoic to Phanerozoic circum-cratonic mobile belts. In contrast, basalt-borne xenoliths constrain the nature of the post-Archaean lithospheric mantle. Not all xenoliths, however, are derived from the lithosphere and one has to carefully consider all aspects of their petrology and chemistry to ascertain their exact origin. Metamorphic and igneous rocks coexist in many mantle xenolith suites occasionally associated with a variety of single crystals or megacrysts. Basalt-borne and kimberlite-borne peridotite xenoliths have considerable textural variety from coarse granular to sheared porphyroclastic types. It has been demonstrated that *sheared or porphyroclastic* textures (higher temperature) are short-lived and constitute a disequilibrium texture and their presence in xenolith suites has been taken as an indication of instability or weakness due to the existence of shear zones or deformation around mantle plumes. In contrast, the presence of a *granuloblastic* texture (lower temperature) in xenoliths has been taken as an indication of stability or rigidity of the lithosphere. Whilst granuloblastic peridotites can be shown to be the oldest component in the mantle, igneous-textured pyroxenites and amphibolites (higher temperature) are a more recent derivative of magmatic processes that upwell from the asthenosphere. In many instances the transfer of magma through lithospheric conduits or in upwelling diapirs is thought to be responsible for the sheared xenoliths. The interaction of lithosphere with underlying asthenosphere can be further evaluated by consideration of the geothermometry of xenoliths. Geophysical considerations have defined a continental geotherm of 40 mW m^{-2} (i.e. 1 HFU) valid for conductive heat transport. Equilibration conditions of mantle xenolith suites either conform to this geotherm or record higher temperatures for a specific depth. The higher temperature of sheared xenoliths is consistent with diapiric heat transport and/or magmatic activity in the upper mantle. In contrast, the lower temperature of the granular xenoliths and their coincidence with the continental geotherm (Finnerty and Boyd 1987) are compatible with derivation from the mechanical boundary layer. Interestingly, the intersection of high- and low-temperature xenolith suites in *P–T* space occurs at around 1000–1200°C, very close to the temperature of the thermal boundary layer and the asthenosphere (>1280°C). The thermobarometric and textural differences in mantle xenolith suites are matched by compositional differences. Although asthenosphere could convert to lithosphere simply by isochemical cooling, extraction of highly magnesian melts from the asthenosphere in the Archaean has produced a compositional difference between the asthenosphere (>1280°C) and the Archaean lithospheric mantle (<1280°C) and, as such, the mechanical boundary layer beneath Archaean crust is in part a chemical boundary layer.

In very general terms the lithospheric mantle is believed to be cold, coarse, and rigid while the asthenosphere is 'hot' (>1280°C), sheared, and flows in response to stress. Interaction between the lithosphere and the asthenosphere is apparent in xenolith suites. Indeed, the relationships between different rock types (e.g. mica–amphibole–rutile–ilmenite–diopside veins (i.e. MARID) in garnet peridotites) or the juxtaposition of textural types (e.g. sheared Fe–Ti rich lherzolites and Mg-rich lherzolites) can help constrain the temporal relationship between chemically distinct mantle domains in the lithosphere and asthenosphere. Chemically depleted spinel and garnet peridotites are frequently cross-cut by igneous textured amphibolites and pyroxenites that are rich in large-ion lithophile (LIL) and light rare earth elements (LREE). The depleted component is believed to represent residual lithospheric mantle wall-rock and the igneous material is believed to represent polybaric derivatives of alkaline magmas from the asthenosphere or lithosphere. Studies of basalt-borne xenoliths indicate that in the spinel stability field of the lithosphere silicate melts are transported by crack propagation with localized development of 'mantle breccia pipes'. In contrast, potassic small volume melt fractions are more common in the deeper garnet stability field according to studies of kimberlite-borne xenoliths. For example, metamorphic garnet peridotites and pyroxenites are occasionally cross-cut by igneous-textured MARID assemblages that are rich in LREE. Such igneous-textured rocks may have been produced by veining of mantle wall-rock with small volume melts (e.g. kimberlites and lamproites) that could have originated within the asthenosphere (if their isotopic composition is similar to oceanic basalts) or the lithospheric mantle (if they have a more heterogeneous isotopic character more akin to that of micaceous kimberlites or lamproites). Megacryst suites (e.g. olivine, clinopyroxene, amphibole, apatite, feldspar) are believed to have been produced by: (1) disruption of lithospheric mantle wall rock (Mg-rich); or (2) precipitation from upwelling asthenospheric melts (Fe–Ti rich). The most important megacrysts are undoubtedly diamonds which retain evidence of deep-seated mantle processes (Richardson, Chapter 3, this volume).

4.2. CONTINENTAL MANTLE DOMAINS

In the following section the geochemistry of continental lithospheric xenoliths entrained from beneath discrete crustal age provinces (i.e. Archaean, Proterozoic, and Phanerozoic) will be compared with actual oceanic mantle material and with oceanic and continental mantle end-member compositions inferred from studies of volcanic rocks. It is assumed throughout that the asthenosphere has a relatively homogeneous isotopic composition similar to that of mid-ocean ridge basalts. While much of the isotopic heterogeneity in the upper mantle is assumed to be within the continental lithosphere, it is appreciated that elemental heterogeneity also occurs within the oceanic lithosphere but that insufficient time has elapsed to allow for the 'ageing' of such heterogeneities.

4.2.1. Domain chemistry

A first order subdivision of lithospheric mantle types can be made on the basis of major and minor element chemistry of basalt-borne and kimberlite-borne xenoliths (Maaloe and Aoki 1977). When compared with basalt-borne spinel lherzolites, kimberlite-borne garnet lherzolites are richer in silica and magnesium and poorer in calcium and aluminium. Although spinel lherzolites are stable at a lower presssure than garnet lherzolites these chemical differences were not interpreted by Maaloe and Aoki (1977) as an indication of chemical stratification of the lithosphere. Instead it was taken to mean that kimberlite-borne xenoliths were more depleted than basalt-borne xenoliths. This was a fundamentally significant observation that illustrated a major difference between sub-Proterozoic lithosphere (lherzolite) and sub-Archaean lithosphere (harzburgite). The modal and compositional variation of olivines in lithospheric peridotites similar to those studied by Maaloe and Aoki (1977) can be used as an indication of first-order differences between lithospheric mantle types (Fig. 4.1). In Fig. 4.1 the composition of Archaean lithospheric mantle is constrained with the use of data from kimberlite-borne peridotites, Proterozoic lithospheric mantle with the use of basalt-borne xenoliths, and Phanerozoic lithospheric mantle with the use of basalt-borne xenoliths and ophiolitic, orogenic, and abyssal peridotites. Archaean granular, cold peridotites are compositionally different from post-Archaean granular peridotites and Proterozoic–Phanerozoic continental lithospheres are compositionally similar to Phanerozoic oceanic lithosphere (Fig. 4.1). Phanerozoic oceanic peridotites from pre-oceanic (Zabargad island, Red Sea), passive margin (Spitzbergen and Iberia), and supra-subduction (Tonga and Mariana) environments are chemically distinct (Bonatti and Michael 1989), their chemical variability (Fig. 4.1) having been explained in terms of variable degrees of dry and wet melting. Supra-subduction peridotites are highly refractory perhaps due to hydrous melting or higher temperatures of melting, although the latter hypothesis is inconsistent with the low equilibration temperatures reported in supra-subduction peridotites relative to mid-ocean ridge peridotites (Bonatti and Michael 1989). Collectively the data in Fig. 4.1 indicate that low-temperature peridotites from beneath the Archaean crust of Africa are significantly more depleted in basaltic elements than post-Archaean peridotites and estimates of undepleted fertile peridotite (e.g. pyrolite). The low-temperature granular peridotites are compositionally unique with little or no petrological or geochemical similarity to continental or oceanic mantle beneath Proterozoic or Phanerozoic crust. The characteristic highly magnesian nature of these granular peridotites may point to an origin as a high-pressure residue left after extraction of komatiite (high geothermal gradient) in the production of early oceanic crust (Boyd 1987, 1989; Boyd and Mertzman 1987; Chase and Patchett 1988). Alternatively several billion years of 'exposure' of Archaean lithosphere to active margin processes during accretion of Proterozoic and Phanerozoic lithosphere around the margins of cratonic nucleii may account for its refractory 'highly processed' nature (Ashwal and Burke 1989). Subduction may initiate hydrous melting of the mantle wedge at the craton margins, a process that would tend to deplete the stabilized residual lithosphere in a manner not too dissimilar to what is happening at the Pacific margins today. The west Pacific, in particular, is noted for highly refractory peridotites that differ only slightly in composition to the Archaean lithospheric mantle (Fig. 4.1). In contrast to the low-temperature peridotites, high-temperature peridotites from sub-Archaean lithosphere are compositionally similar to peridotites from mobile belts (i.e. Proterozoic mantle) and oceanic regions (i.e. Phanerozoic mantle). This led Boyd and Mertzmann (1987) to suggest that these high-temperature peridotites were low-pressure residua produced by the extraction of tholeiitic melts. Their present location beneath the Archaean crust of South Africa was thought to be due to subduction and subsequent underplating of residual oceanic lithosphere. Recent geochemical data does indicate, however, that *both* the low- and high-temperature peridotites were isolated from the asthenosphere for several billion years and, as such, must both reside within the mechanical boundary layer of the subcrustal lithosphere (Walker *et al.* 1989) rather than comprise an old lithospheric component (low-temperature) and a recently added asthenospheric or recycled lithospheric component (high-temperature).

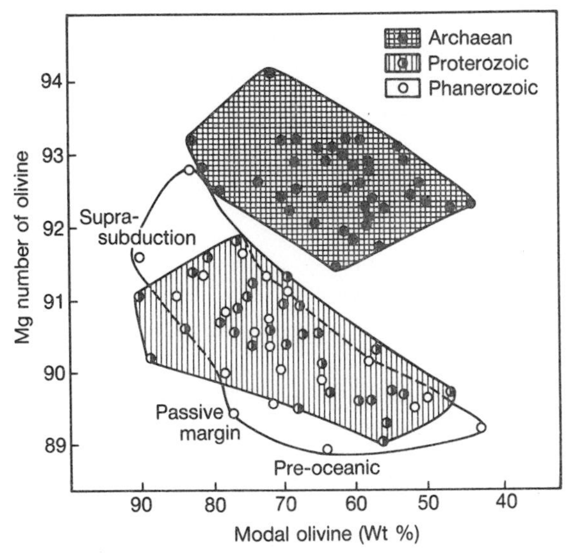

Fig. 4.1. Magnesium number of olivine versus modal amount of olivine in lithospheric mantle peridotites from areas overlain by Archaean, Proterozoic, and Phanerozoic crust (after Boyd 1989; Menzies 1991 and references therein). Note that the Archaean lithospheric mantle is different from post-Archaean lithospheres. The highly magnesian character of the low-temperature granular peridotites that constitute a major part of the Archaean lithosphere may be due to extraction of komatiitic melts (or hydrous melting above subduction zones). Proterozoic and Phanerozoic continental lithospheres are similar to Phanerozoic oceanic lithospheres which range in composition from pre-oceanic and passive margin 'undepleted' compositions to depleted compositions in supra-subduction or active margin environments like the west Pacific.

It appears from these major and minor element considerations that the lithospheric mantle found beneath crust of all ages is 'residual', meaning that a partial melt has been extracted leaving a peridotite depleted in incompatible elements (e.g. Rb, Nd, and U). Residua generated by melt extraction or depletion processes are characterized by low Rb/Sr, low Nd/Sm, and low U/Pb ratios. A time-integrated response to such elemental abundances generates low $^{87}Sr/^{86}Sr$, high $^{143}Nd/^{144}Nd$, low $^{206}Pb/^{204}Pb$, and low $^{207}Pb/^{204}Pb$ ratios. If we accept the conclusions based on the data in Fig. 4.1 then it follows that most of the subcrustal lithosphere, regardless of age, should have an isotopic composition similar to depleted mantle or perhaps the residue left after extraction of mid-ocean ridge basalt.

Mantle xenoliths entrained from beneath crust of Phanerozoic, Proterozoic, and Archaean age display a considerable range in neodymium isotopic ratio (expressed as ε_{Nd}) (Fig. 4.2). Whilst the range of ε_{Nd} in Phanerozoic and Proterozoic lithospheric mantle is, to some degree, consistent with an origin as a partial melting residue (i.e. $\varepsilon_{Nd} > 0$), other processes must have affected the lithosphere giving rise to ε_{Nd} values significantly different to MORB (i.e. $\varepsilon_{Nd} < 0$). Furthermore, many of the samples of Phanerozoic and Proterozoic lithosphere with $\varepsilon_{Nd} > 0$ are enriched in the light rare earth elements (LREE) like Nd, but insufficient time has passed for such elemental heterogeneities to register as isotopic heterogeneities. It is apparent (Fig. 4.2) that the range in neodymium isotopic composition (i.e. ε_{Nd}) observed in mantle xenoliths is very different from that observed in mid-ocean ridge basalts (MORB) and is therefore inconsistent with an origin as a simple melt residue. The lithosphere appears to have been modified since formation and stabilization by chemical enrichment that has not affected the major and minor element chemistry. Such enrichments can be achieved by migration of small volume partial melts from the asthenosphere that may not mix with melts from the lithosphere. The considerable range in ε_{Nd} for the Archaean lithospheric *mantle* is similar to the *lower crust* but differs significantly from the Phanerozic and Proterozoic lithosphere and MORB (i.e., depleted MORB mantle reservoir). This isotopic heterogeneity is believed to be a by-product of age and mixing processes. It should be noted that the most non-radiogenic neodymium isotopic compositions (negative ε_{Nd} values) have been recorded in xenoliths from the Archaean lithospheric mantle (Basu and Tatsumoto 1980; Menzies and Murthy 1980*a*; Richardson *et al.* 1984; Menzies *et al.* 1987; Menzies and Halliday 1988) indicating either significant 'ageing' of trace element enrichment processes or recycling of material (Kyser, Chapter 7, this volume) with an old provenance age. This contrasts with the petrology of kimberlite-borne xenoliths which points to a depletion in those elements concentrated in

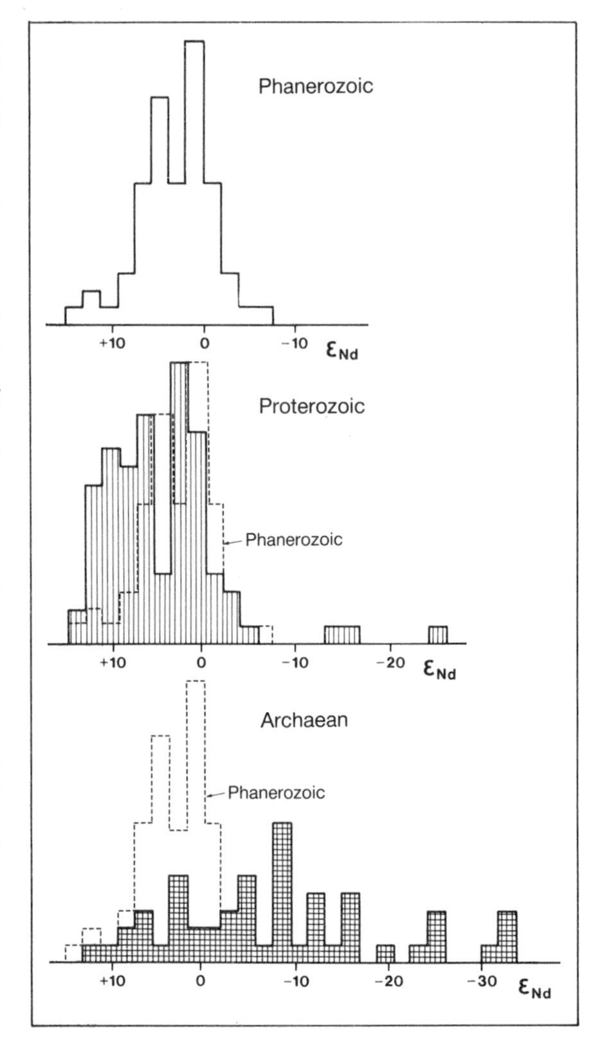

Fig. 4.2. Histograms of neodymium isotopic composition in lithospheric peridotites (amphiboles and clinopyroxenes) from regions overlain by crust of different age (after Menzies and Hawkesworth 1987*a*,*b* and references therein; Menzies and Halliday 1988; Stolz and Davies 1988). Note that, as in Fig. 4.1, Archaean and post-Archaean lithospheres are very different, a difference that cannot be wholly explained by all lithospheres being partial melting residua (i.e. positive ε_{Nd}). Phanerozoic and Proterozoic lithospheric mantle are isotopically similar to 'oceanic mantle' and may be partial melting residua, but the Archaean lithospheric mantle has a much greater range in isotopic composition than can be allowed for if they are simple residua. ε_{Nd} ranges from values close to MORB to values close to the 'lower crust'. This may be due to plume–lithosphere hybridization involving a depleted Archaean protolith and hot spots encountered during post-Archaean times, or recycling of oceanic lithosphere.

basalt. Such material would have evolved to give radiogenic $^{143}Nd/^{144}Nd$ ratios ($\varepsilon_{Nd} > 0$) had the depletion 'aged' since the Archaean. The absence of such depleted isotopic compositions indicates that, if they existed, their presence has been very effectively obliterated by Archaean or post-Archaean processes. The range in neodymium isotopic composition in Archaean lithospheric peridotites allows one to speculate that: (1) a peridotitic residue was initially produced in the Archaean, thus explaining the mineralogy and the major and minor element chemistry of the xenoliths (Fig. 4.1); and (2) that later subsolidus processes (Cox *et al.* 1987), or the migration of small volume melts, led to an elemental redistribution within this residue. It is the time-integrated response to this elemental redistribution and enrichment that has produced the present-day heterogeneity in ε_{Nd}. Overall the Archaean lithosphere data base (Fig. 4.2) points to the passage of a considerable amount of time since its formation producing isotopic heterogeneity from initial elemental heterogeneity and possible mixing of an originally depleted protolith with small volume melt fractions such that a spread of isotopic data results. Archaean lithospheric mantle is perhaps more enriched in incompatible elements (Nd) due to prolonged absorption of small volume melts from the underlying asthenosphere (Kyser, Chapter 7, this volume; Watson *et al.*, Chapter 6, this volume).

In summary, whilst the major element data indicate a residual origin for Archaean, Proterozoic–Phanerozoic lithosphere by extraction of komatiitic and basaltic melts, respectively, the range in Nd isotopic composition points to subsequent metasomatism or enrichment of these residua by upward migration of small volume melts enriched in the LREE generated either within the lithosphere or the underlying asthenosphere.

4.2.2. Domain type

Major elements and neodymium isotopes indicate a first-order difference between lithospheric mantle produced in the Archaean to that produced in the Proterozoic and Phanerozoic. A more precise idea of the origin of the lithospheric mantle can only be uncovered with the use of several isotopic systems. The four mantle components already defined for the source region of oceanic basalts (Zindler and Hart 1986) will be used as a basis for interpretation of available mantle xenolith data. Zindler and Hart (1986) used several acronyms to describe isotopically discrete mantle domains or reservoirs. These acronyms will be used in this chapter and are defined as follows: DMM (depleted MORB mantle); HIMU (high U/Pb or μ); EM1 (enriched mantle 1 characterized by low $^{143}Nd/^{144}Nd$ and $^{87}Sr/^{86}Sr$ ratios), and EM2 (enriched mantle 2 characterized by high $^{87}Sr/^{86}Sr$ and intermediate $^{143}Nd/^{144}Nd$). Only combined Sr–Nd–Pb isotopic data sets for the constituent minerals (i.e. amphibole and clinopyroxene) from mantle xenoliths have been plotted in Figs 4.3 to 4.6 (Kramers *et al.* 1983; Roden *et al.* 1984; Menzies and Halliday 1988; Stolz and Davies 1988; Galer and O'Nions 1989; BenOthman *et al.* 1990) but reference will be made to other Sr, Nd, or Pb isotopic data sets. It is assumed in this chapter that Archaean, Proterozoic, and Phanerozoic lithosphere have always been floored by asthenosphere—a source for small melt fractions. Whilst all continental lithospheres were exposed to the same type of elemental enrichments, the cumulative amount of enrichment may depend on the length of exposure to the hot asthenosphere.

The lithospheric mantle is heterogeneous for Sr and Nd isotopes (Fig. 4.3) and the compositional variability extends well beyond the limit set by oceanic and continental volcanic rocks to values more commonly associated with crustal rocks. Had the lithospheric mantle been a simple residua produced by partial melting processes, all the available isotopic data should have plotted close to the DMM end-member in Fig. 4.3. This is not the case and the Sr and Nd isotopic data for mantle xenoliths are bounded by several depleted (DMM) and enriched (EM1 and EM2) domains. The *Archaean lithospheric mantle*, as represented by granuloblastic peridotites and igneous-textured pyroxenites, is confined within the area bounded by the HIMU, EM1, and EM2 end-members. Since many of these xenoliths are from a restricted area within South Africa these data indicate considerable local heterogeneity over a vertical distance of 50 km in the lithospheric mantle (Waters and Erlank 1988). Other Sr–Nd isotopic data for mantle xenoliths or megacrystic inclusions entrained from beneath Archaean crust similarly indicate: (1) an EM1 or EM2 domain under the Archaean of South Africa preserved as silicate inclusions in kimberlite-borne megacrystic diamonds (Richardson *et al.* 1984, 1985; Richardson, Chapter 3, this volume); (2) an EMI domain under the Archaean of Greenland preserved in kimberlite-borne peridotite xenoliths (Scott-Smith 1987); (3) an EM1 domain under Wyoming preserved in basalt-borne pyroxenite xenoliths (Dudas *et al.* 1987); (4) an EM2 domain under the Archaean of South Africa preserved as kimberlite-borne garnet peridotites (Menzies and Murthy 1980*a*; Richardson *et al.* 1985); and (5) an EM2 domain beneath the Archaean of Montana preserved in glimmerite xenoliths (Irving *et al.* 1989). There is little or no evidence for a dominant depleted domain (e.g. DMM) in the Archaean lithospheric mantle, a surprising result when one considers that major and minor element data provide unequivocal

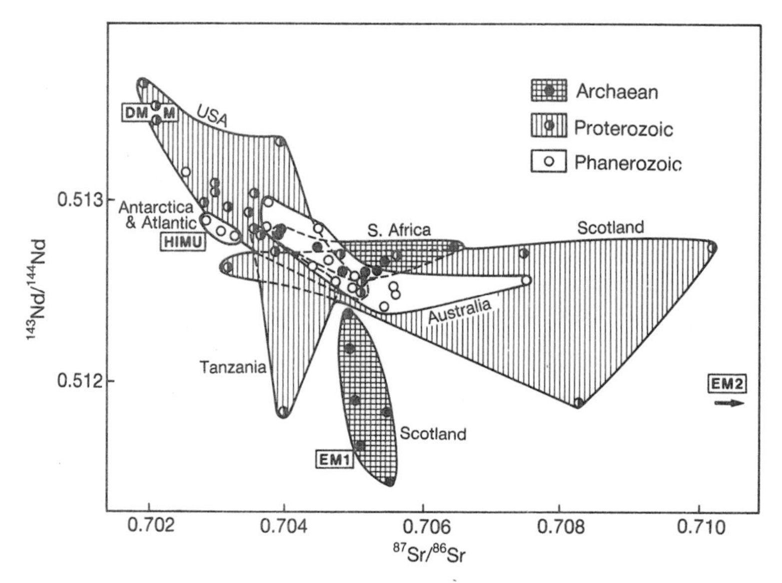

Fig. 4.3. Strontium versus neodymium isotopes in lithospheric peridotites and megacrysts entrained or emplaced through crust of variable age (Kramers *et al.* 1983; Roden *et al.* 1984; Menzies and Halliday 1988; Stolz and Davies 1988; Galer and O'Nions 1989; BenOthman *et al.* 1990). Note that: (1) if all lithospheres were simple ancient partial melting residua they would plot in the vicinity of the DMM domain; (2) the variation in isotopic composition extends beyond the region occupied by oceanic basalts and points to heterogeneous lithospheres; (3) the Sr and Nd data are enclosed within a region bounded by the DMM, HIMU, EM1, and EM2 mantle end-members defined on the basis of oceanic and continental basalts; (4) certain data sets approximate to 'pure' end-member compositions (e.g. EMI—Scotland; DMM—USA); (5) the Archaean data trend from HIMU toward the EM1 and EM2 domains with little or no evidence for a depleted domain although major and minor elements and osmium isotopes argue for a depleted protolith different in origin to DMM; (6) the Proterozoic data cover much of the diagram in particular from DMM to EM2; and (7) the Phanerozoic data stretch from HIMU toward EM2.

evidence for a depleted protolith (Fig. 4.1). This may indicate that the major and minor element chemistry is controlled by minerals like olivine and orthopyroxene that are relatively poor in Sr and LREE but constitute up to 95 per cent of the rock. In contrast, the Sr and Nd isotopic characteristics are determined by minor phases (<5 per cent) like clinopyroxene and amphibole which are rich in Sr and LREE. *Proterozoic lithospheric mantle*, as represented by spinel and garnet peridotites and pyroxenites, exhibits a considerable range in Sr and Nd isotopes bounded by the DMM, HIMU, and EM2 domains with a possible representative of the EM1 domain in spinel and garnet lherzolites from Tanzania. A vast Sr–Nd data base exists for Proterozoic lithospheric mantle and most of these data are enclosed within the area bounded by the DMM, HIMU, and EM2 domains (e.g. Menzies *et al.* 1985; Stosch and Lugmair 1986; Downes and Dupuy 1987; Roden *et al.* 1988). *Phanerozoic lithospheric mantle*, as represented by spinel peridotites, is similarly bounded by the DMM, HIMU, and EM2 domains. This is substantiated by consideration of other Sr–Nd data from the Pacific Ocean and the Red Sea (e.g. Menzies and Hawkesworth 1987*a,b*; Broxel *et al.* 1988; Brueckner *et al.* 1988).

The range in Sr and Pb isotopes (Fig. 4.4) in mantle xenoliths is similar to that observed in oceanic volcanic rocks and certain crustal rocks. *Archaean lithospheric mantle*, in South Africa and Scotland, is dominated by the HIMU, EM1, and EM2 domains as was the case for Sr–Nd isotopes (Fig. 4.3). The Scottish basalt-borne xenoliths have a Sr, Nd, and Pb isotopic composition precisely that of the EM1 domain. Moreover, some of the xenoliths from Tanzania and South Africa have a more marked trend towards HIMU than was apparent from the Sr–Nd data, as it was difficult in Fig. 4.3 to distinguish between a trend towards DMM or HIMU. *Proterozoic lithospheric mantle* is again contained within the area of DMM, HIMU, and EM2 but the close match between the sub-Proterozoic lithosphere and the DMM domain beneath the western USA (Fig. 4.3) is not substantiated in Sr–Pb space due to high ^{206}Pb/^{204}Pb ratios. The *Phanerozoic lithospheric mantle* is contained within the area bounded by the DMM, HIMU, and EM2 domains as was the case for Sr–Nd isotopes

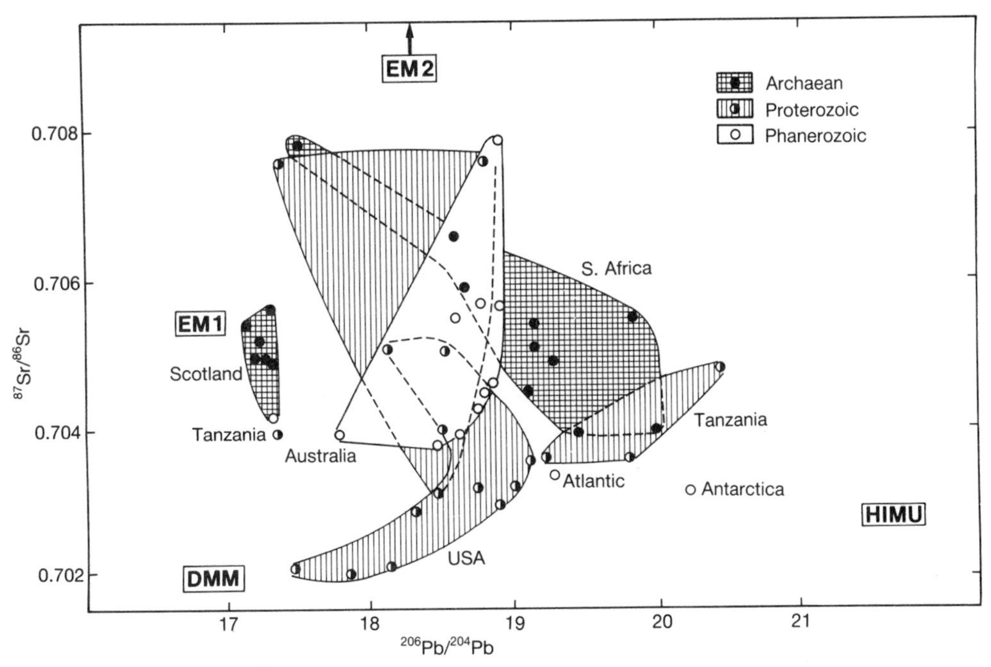

Fig. 4.4. Strontium versus lead isotopes in lithospheric mantle peridotites and megacrysts (Kramers *et al.* 1983; Roden *et al.* 1984; Menzies and Halliday 1988; Stolz and Davies 1988; Galer and O'Nions 1989; BenOthman *et al.* 1990). The range in $^{206}Pb/^{204}Pb$ in the lithospheric mantle covers much of the range recorded in ocean island basalts from the source region of mid-ocean ridge basalts (DMM) toward the source of ocean islands like St. Helena (HIMU) and Samoa (EM2). Note: (1) the location of ancient partial melting residua (DMM) in relation to the bulk of the xenoliths plotted; (2) the proximity of specific data sets to mantle end-member compositions, in particular the EM1 domain; (3) the relative positions of the Archaean, Proterozoic, and Phanerozoic datasets; (4) the lack of a depleted domain (e.g. DMM) in the Archaean and the lack of the EM1 domain in much of the Proterozoic and Phanerozoic.

(Fig. 4.3). Whilst the Antarctic–Atlantic data overlap exactly with the HIMU domain in Sr–Nd space (Fig. 4.3) this is not the case in Sr–Pb space because of low $^{206}Pb/^{204}Pb$ ratios.

Nd and Pb isotopes (Fig. 4.5) indicate that the *Archaean lithospheric mantle* is again confined within the area occupied by the HIMU and EM1 domains with less of a bias towards the EM2 domain. The Scottish basalt-borne xenoliths are the closest approximation to the EM1 domain and thus corroborate what was observed in Sr–Nd and Sr–Pb space. *Proterozoic lithospheric mantle* is within the area defined by the DMM, HIMU, and EM2 domains with the western USA data being the closest approximation to the DMM domain. The tendency of the Tanzania data to plot toward the HIMU domain is again apparent. *Phanerozoic lithospheric mantle* is contained within the region occupied by the DMM, HIMU, and EM2 domains, although the data do trend more toward the HIMU and EM2 end-members. The Antarctic data is once more the closest to the HIMU domain as it was on the two previous diagrams.

In a Pb–Pb isotope diagram (Fig. 4.6) *Archaean lithospheric mantle* varies in composition between the HIMU and EM1 domains with not such a marked bias toward the EM2 domain as was seen in previous figures. The Scottish data again represent the closest approximation to the pure EM1 domain. *Proterozoic lithospheric mantle* departs from the area bounded by the DMM, HIMU, and EM2 domains and appears to plot toward the HIMU and EM1 end-members. Moreover, none of the western USA xenoliths have low enough $^{207}Pb/^{204}Pb$ ratios to be representative of the pure DMM domain although in previous diagrams they closely approached a pure DMM domain. *Phanerozoic lithospheric mantle* plots between the DMM and EM2 domains with a bias toward the HIMU end-member as seen in previous figures. No sample, however, has a high enough $^{207}Pb/^{204}Pb$ to overlap with the pure EM2 domain. The Antarctic samples are again closest to the HIMU domain.

The variation in C and Sr isotopes (Fig. 4.7) in the lithospheric mantle (Mattey *et al.* 1989) defines two distinct trends: (1) towards $\delta^{13}C = -30$ and low

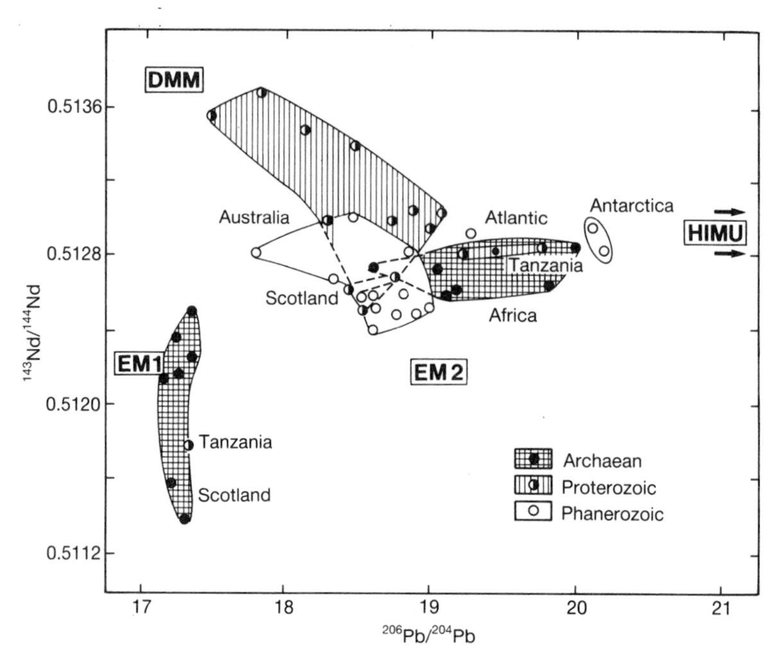

Fig. 4.5. Neodymium versus lead isotopes in lithospheric mantle peridotites and megacrysts (Kramers *et al.* 1983; Roden *et al.* 1984; Menzies and Halliday 1988; Stolz and Davies 1988; Galer and O'Nions 1989; BenOthman *et al.* 1990). The EM1 domain and the Scottish Archaean data set stand apart from the rest of the data which are adequately contained within the area bounded by the DMM, EM2, and HIMU domains. The lack of a depleted domain (e.g. DMM) in the Archaean and the lack of the EM1 domain in much of the Proterozoic and Phanerozoic are again apparent.

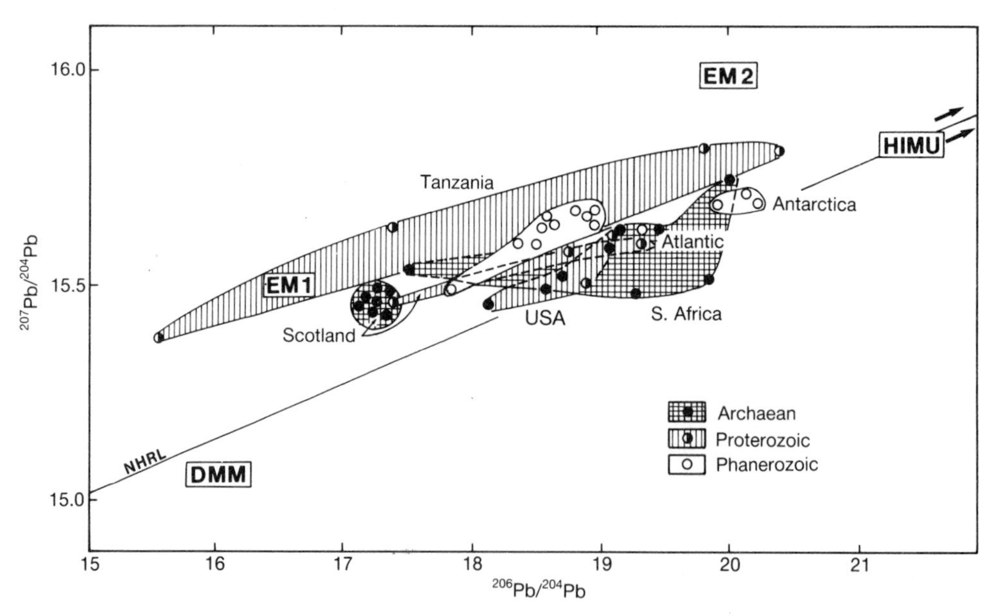

Fig. 4.6. ^{206}Pb/^{204}Pb versus ^{207}Pb/^{204}Pb diagram for lithospheric peridotites and megacrysts (Kramers *et al.* 1983; Roden *et al.* 1984; Menzies and Halliday 1988; Stolz and Davies 1988; Galer and O'Nions 1989; BenOthman *et al.* 1990). Note that all the data, except for a few samples from South Africa, plot above the northern hemisphere reference line (NHRL). Note the high ^{207}Pb/^{204}Pb and ^{206}Pb/^{204}Pb ratios of the Proterozoic lithosphere—a feature that does not compare favourably with the DMM domain as it did in previous diagrams.

$^{87}Sr/^{86}Sr$ ratios; and (2) towards $\delta^{13}C = -10$ and high $^{87}Sr/^{86}Sr$ ratios. The DMM domain can be fixed in Sr–C space at a $\delta^{13}C = -5$. The two trends emanating from the DMM domain are defined by: (1) samples of Archaean, Proterozoic, and Phanerozoic lithospheric mantle from South Africa, Antarctica, and Australia, respectively, that show only minor changes in $\delta^{13}C$ with increasing $^{87}Sr/^{86}Sr$ ratio; and (2) samples of Proterozoic and Phanerozoic lithospheric mantle from the western USA, France, and Japan, respectively, that show major changes in $\delta^{13}C$ with small changes in $^{87}Sr/^{86}Sr$. The only domain common to Archaean, Proterozoic, and Phanerozoic lithospheric mantle with that variation in $^{87}Sr/^{86}Sr$ is the EM2 domain and consequently it is speculated that $\delta^{13}C$ adds a new dimension to our understanding of the evolution of the EM2 domain. It has been proposed that the C–Sr trends result from recycling of the two main crustal carbon reservoirs—organic matter and sedimentary carbonate (Mattey *et al.* 1989). The shallow trend may result from mixing of a DMM reservoir with carbonate carbon found in carbonate-rich sediments whereas the steep trend may result from mixing DMM and pelagic carbon in organic sediments ($\delta^{13}C = -20$ to -35). It is of interest that the only Archaean data on the Sr–C diagram defines a mixing trend involving carbonate carbon and depleted mantle, but the Proterozoic–Phanerozoic data define a mixing trend involving organic carbon and depleted mantle. This may be explained by the lack of organic carbon in the Archaean due to an absence of abundant life forms and the vast amount of organic carbon in Proterozoic–Phanerozoic times. Alternatively, the shallow trend may indicate mixing of a DMM protolith with small volume carbonate melts from sublithospheric sources.

4.3. TEMPORAL AND SPATIAL ASPECTS OF MANTLE DOMAIN EVOLUTION

As discussed previously the lithospheric mantle may have been significantly modified since it stabilized and will thus retain evidence of partial melting events (depletion) and magmatic processes (enrichment), the latter being due to plume–lithosphere interaction, asthenosphere–lithosphere interaction (e.g. Menzies *et al.* 1985; Storey *et al.* 1988, 1989), or supra-subduction/recycling processes. One fundamentally important feature of mantle xenoliths is that they record the temporal interplay between different mantle domains. For ex-

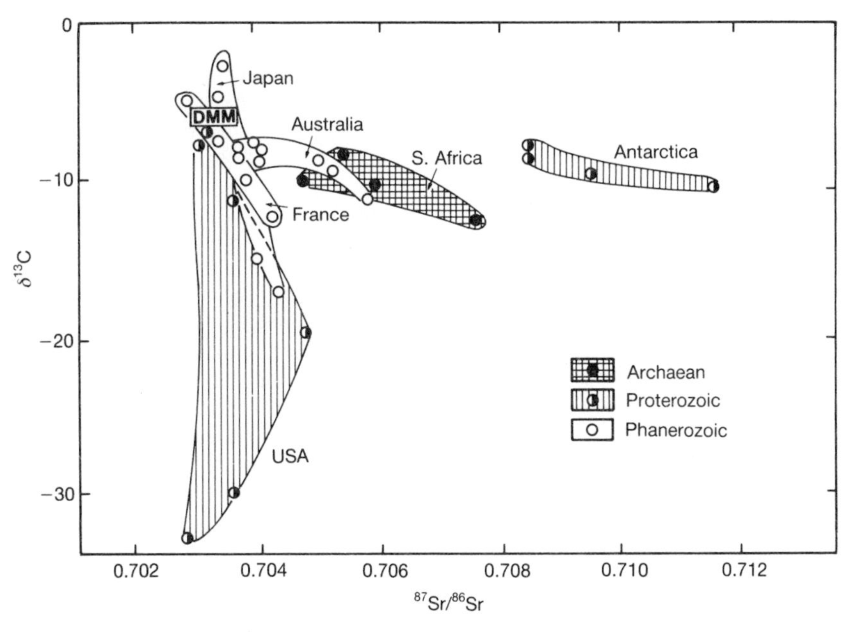

Fig. 4.7. Carbon versus strontium isotopes for lithospheric peridotites (Mattey *et al.* 1989). The DMM domain is located in the top left-hand corner and the data define two trends toward high and low $^{87}Sr/^{86}Sr$ ratios. This is interpreted as different forms of recycling relevant to the origin of the EM2 domain. Involvement of carbonate carbon (Archaean to Phanerozoic) and organic carbon (Proterozoic to Phanerozoic) in the recycling process may adequately explain the differences between the two trends.

ample, the Archaean lithospheric mantle may have experienced a plethora of Proterozoic and/or Phanerozoic magmatic episodes that may have imposed distinct chemical characteristics on the Archaean lithosphere.

4.3.1. Archaean lithospheric mantle—protolith composition, geochemical stratification, and domain chronology

The Archaean lithospheric mantle beneath South Africa contains representatives of the HIMU, EM1, and EM2 domains and the lithospheric mantle beneath the Archaean of Scotland contains the closest approximation to the pure EM1 domain. With the available data one can produce a strong case for locating the EM1 domain in the Archaean lithosphere. This is further substantiated by considerations of the regional variation in the source regions of continental volcanic rocks. Volcanic rocks tapping the EM1 reservoir appear to be restricted to the Archaean of North America (Menzies 1989). Hart *et al.* (1986) noted the close association between EM1 and HIMU domains (i.e. LoNd array) and suggested that they both formed within the subcontinental lithosphere by interaction between the lithosphere and silicate melts. This is compatible with the data for Archaean lithospheric mantle but in the case of South Africa and Scotland the HIMU and EM2 domains appear to have overprinted any prior existence of EM1 or the depleted protolith (DM = depleted mantle) inferred from Fig. 4.1. On the basis of the petrology and the major and minor element geochemistry of cratonic peridotites a very strong case was made earlier for a depleted reservoir (i.e. melt residue) beneath the Archaean crust and, consequently, one has to query whether or not the enriched domains (i.e. EM1, EM2, and HIMU) are part of the original protolith beneath the Archaean crust or due to post-Archaean processes. Support for a depleted protolith beneath the cratons is forthcoming from recent isotopic studies. Osmium isotopic ratios and the low Re/Os ratios of kimberlite-borne xenoliths from the Kaapvaal craton, South Africa indicate that the Archaean lithospheric mantle is residual in character and different in origin from MORB (Walker *et al.* 1989) and, as such, cannot be related to the genesis of modern oceanic mantle. Xenolith suites entrained from beneath the Archaean of the Kaapvaal craton South Africa and the Wyoming craton USA (e.g. Kramers 1987; Nixon 1987 and references therein; Hearn *et al.* 1989; Irving *et al.* 1989) are dominated by harzburgites and dunites, refractory rocks whose origin clearly involved melt extraction. Moreover, neodymium isotopic data (Collerson and Campbell 1989) for ultramafic units (lherzolites and websterites) within the Saglek–Hebron area of northern Labrador point to the existence of a depleted lithospheric mantle almost 4 billion years ago. In general, the Canadian shield is believed to have been an area of positive relief during Phanerozoic time due to the buoyancy of garnet-free harzburgite relative to garnet lherzolite (Hoffmann 1988). In addition the widespread occurrence of carbonatites of variable age with 'depleted' isotopic ratios (Bell and Blenkinsop 1987) erupted through the Archaean of Canada may point to the presence of a widespread depleted mantle source beneath some Archaean cratons. We must, however, evaluate to what extent carbonatites can be thought of as melts of lithospheric mantle if we intend to use these data to constrain the nature of the subcrustal mantle. To some extent recent experimental work supports a lithospheric origin for carbonatites because the carbonatite melt field lies between 21 and 31 kbar (<100 km) and 930 and 1075°C (Wallace 1989) indicating that carbonatites may form at relatively shallow pressures and temperatures. On the basis of this one could argue that the Canadian carbonatite data support the assertion that the Archaean lithospheric mantle beneath Canada was depleted in the Archaean and remained so throughout post-Archaean times. More evidence for a depleted protolith in the Archaean lithospheric mantle beneath Canada is forthcoming from the initial neodymium isotopic composition of the bulk of Archaean mantle-derived rocks. Archaean tholeiitic alkaline, and komatiitic rocks indicate that their mantle source regions were depleted in incompatible elements very early in Earth's history (e.g. Hart and Brooks 1977; Zindler 1982; Basu *et al.* 1984; Shirley and Hanson 1986) and that in many instances the source can be thought of as 'oceanic'. If we assume that the depletion in the source region was related to partial melting, then the residue that underplated Archaean crust must have had a similarly depleted isotopic composition.

Perhaps on a world-wide scale one can infer that the pristine condition of the sub-Archaean lithosphere was that of a reduced, refractory residue (Anderson, Chapter 1, this volume; Haggerty, Chapter 5, this volume) produced as a result of ancient partial melting processes. However, the continued upward passage of small volume melts from the asthenosphere for the last three or more billion years has transformed this protolith producing EM1, HIMU, and EM2 components. The lithospheric mantle beneath the Archaean of Canada may well have escaped subsequent enrichment in that its plate tectonic history may not have involved interception of upwelling mantle plumes, as is the case in Gondwanaland a supercontinent that may well have collided with the Dupal hot spots. The considerable range in ε_{Nd} values for the Archaean lithospheric mantle (Fig. 4.2) relative to post-Archaean lithosphere may in some way betray this enrichment or mixing process in that the spread of data could be interpreted as a by-product of mixing of a depleted domain and small volume melts from sublithospheric reservoirs. The marked enrichment in incompatible elements in small volume melts would enrich the depleted protolith and the passage of time would eventually produce a domain with ε_{Nd} similar to EM1 or EM2 dependent on the elemental ratios. The complex evolution of the Archaean lithospheric mantle can best be explored on an isotope–isotope diagram where some indication of the age of individual suites of xenoliths is apparent (Fig. 4.8).

Enriched domains appear to dominate the xenolith suites entrained from within the Archaean lithosphere (i.e. EM1 or EM2) as represented by garnet peridotites and silicate inclusions in diamond megacrysts with an age of 3.2–3.3 billion years (Richardson, Chapter 3, this volume). However, the compositional similarity between minerals found as inclusions in diamonds and the constituent minerals in refractory garnet harzburtites indicates that the Archaean lithosphere was composed of a depleted protolith that underwent Archaean enrichment processes. Furthermore, post-Archaean enrichment processes may explain the Sm–Nd model ages reported for metasomatized peridotites. Garnet peridotites and garnet phlogopite peridotites (EM2 hybrid) have Proterozoic Sm–Nd model ages (1.0–1.4 billion years) and phlogopite–richterite peridotites, and cross-cutting MARID rocks (EM2–HIMU hybrid) have Phanerozoic Sm–Nd model ages (<0.2 billion years) as do the megacryst suites (HIMU). Clearly, the lithospheric keel beneath the Archaean crust of South Africa records a series of Archaean and post-Archaean depletion and enrichment episodes that have produced a chemically complex reservoir. As mentioned in the 'Introduction', crucial petrographic details, like the relationship between the peridotites and MARID veins, can constrain vital time relationships between these episodes. Melt migration and concomitant enrichment of garnet peridotites (GP) occurred in: (1) the Proterozoic producing metasomatized wall rock (i.e. garnet phlogopite peridotites (GPP) and phlogopite peridotites (PP)) due to the upwelling of alkaline melts/fluids of possible asthenospheric origin (Richardson *et al.* 1985; Menzies *et al.* 1987); and (2) the Phanerozoic producing metasomatized wall rock (i.e. phlogopite K-ricchterite peridotites (PKP)) around mantle pegmatite veins (i.e. MARID) due to the migration of an asthenospheric-derived kimberlite (Kramers *et al.* 1983; Erlank *et al.* 1987; Waters 1987). In the Phanerozoic derivatives of upwelling alkaline melts may also have given rise to megacrystalline suites and pyroxenite xenoliths.

The lithospheric mantle beneath South Africa was some 200 km thick in the Archaean (Boyd *et al.* 1985; Anderson, Chapter 1, this volume; Richardson, Chapter 3, this volume) and the thermal boundary layer at the base of this keel may have been successively eroded and perturbed by upwelling deep mantle hot spots or may have interacted with the hot asthenospheric substratum (i.e. plume–lithosphere and asthenosphere–lithosphere interaction). These processes may account for the domain chronology (i.e. (DM)–EM1–EM2–HIMU) apparent in the kimberlite-borne xenoliths (Fig. 4.8) and the geochemical affinities of volcanic rocks whose source is believed to have included the lithospheric mantle. Perhaps in the late Proterozoic–Phanerozoic the lithospheric plate carrying South Africa encountered a Dupal type hot spot (i.e. EM2 domain) and in Phanerozoic times a HIMU hot spot thus accounting for the Dupal (Erlank *et al.* 1987) and HIMU xenoliths (Kramers 1977). The overall effectiveness of this prolonged hybridization process is apparent in the lack of many domains with the isotopic character of the inferred depleted protolith DM, and the ubiquity of domains with: (1) evidence for metasomatic conversion of a granular peridotite protolith by later 'magmatic' episodes; (2) the chemical characteristics of EM2 and HIMU domains; (3) Proterozoic or Phanerozoic model ages.

If continual interaction has occurred between the Archaean lithosphere and upwelling magmas from the lower mantle and the asthenosphere, then a vertical chemical variation may be apparent in the lithosphere. Such chemical stratification can be assessed, to some extent, by consideration of isotopic variability in conjunction with thermobaromatric data (Waters and Erlank 1988). A gradual decrease in $^{87}Sr/^{86}Sr$ and a gradual increase in $^{143}Nd/^{144}Nd$ appears to occur with increasing depth in the subcrustal lithosphere of South Africa (Fig. 4.9). This has important implications for the formation of the lithosphere as it could indicate a temporal change in the isotopic composition of the lithosphere. Since the keel was 200 km thick in the Archaean (Boyd *et al.* 1985; Anderson, Chapter 2, this volume) and was initially composed of a depleted protolith, the observed isotopic heterogeneity cannot be primary or due to underplating, *sensu stricto* (as no increase in thickness has been documented), but to modification of the pre-existent lithosphere due to thermal and chemical disturbances from sublithospheric sources. Whilst the mechanical boundary layer (MBL) remains essentially unaffected by any reaction with

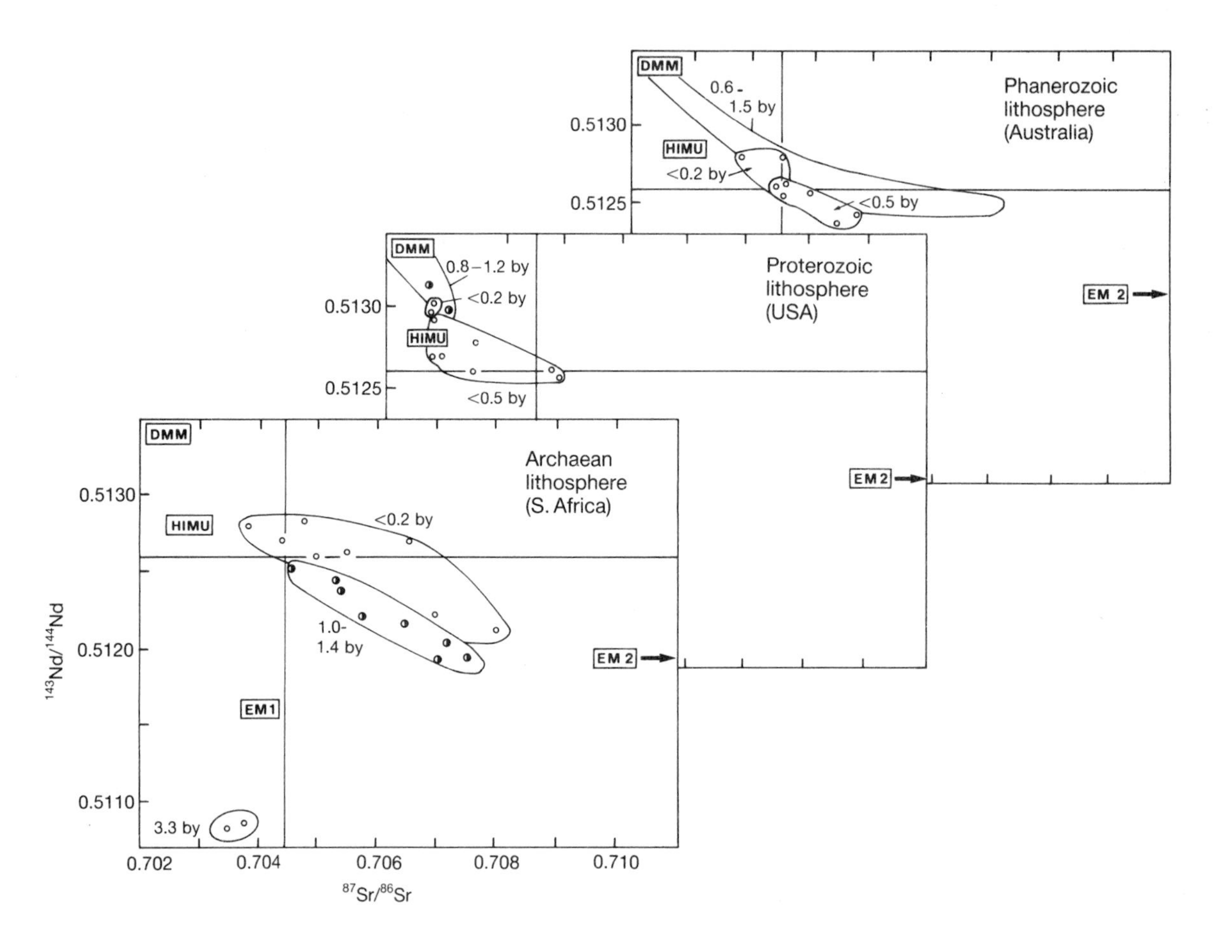

Fig. 4.8. Temporal evolution of Archaean, Proterozoic, and Phanerozoic lithosphere. Archaean lithosphere (South Africa): DM is inferred to be the depleted mantle Archaean protolith that was partially transformed to EM1 and EM2 domains. Collision of the African plate with hot spots (i.e. EM2 domain and HIMU domain) in the Proterozoic (1.0–1.4 billion years) and Phanerozoic (< 0.2 billion years), respectively, may explain the domain chronology found today beneath the Archaean of South Africa. Data are taken from several sources (Kramers *et al.* 1983; Menzies and Murthy 1980*a*; Richardson *et al.* 1984, 1985; Erlank *et al.* 1987; Waters 1987). Proterozoic lithosphere (USA): The Proterozoic protolith DMM is successively overprinted by the EM2 and HIMU domains in response to accretion and/or subduction in the Proterozoic and Phanerozoic and eventually by the DMM domain as a result of upwelling of asthenospheric plumes in the late Phanerozoic. Data are taken from several sources (BenOthman *et al.* 1990; Galer and O'Nions 1989; Menzies *et al.* 1985; Roden *et al.* 1988). Phanerozoic lithosphere (Australia): As in the Proterozoic, DMM is believed to be the oldest domain formed as a residue after partial melting in the Proterozoic. This protolith was subsequently overprinted by the EM2 domain related to Palaeozoic subduction or more likely the influence of the DUPAL hot spots. More recent interplay (< 0.2 billion years) may have involved the HIMU and DMM domains. Overall the Phanerozoic lithosphere has a simpler domain history than the Archaean lithosphere because exposure to sublithospheric contamination has been on the order of hundreds of millions of years in contrast to billions of years for the Archaean lithosphere. Data are taken from Griffin *et al.* (1988) and Stolz and Davies (1988).

sublithospheric processes, the thermal boundary layer (TBL) is continually in a state of flux as it represents a buffer between the MBL and the asthenosphere. Small volume melts within the asthenosphere or adiabatic interior (AI) are known to be rich in potassium, carbonate, and water and these represent formidable metasomatic agents (Navon *et al.* 1988; McKenzie 1989; Watson *et al.*, Chapter 6, this volume). The diamond inclusion data not only indicated that an EM1–EM2 domain underlaid the Archaean crust of South Africa to a depth of 180–200 km *in the Archaean*, but that fluids rich in LREE had permeated the base of the lithosphere (Richardson, Chapter 3, this volume). The lack of the EM1 domain other than in diamond inclusions indicates that post-Archaean magmatic activity may have been responsible for the gradual modification of the lithosphere in Proterozoic and Phanerozoic times (Figs 4.8 and 4.9). The near total obliteration of the depleted domain DM and the enriched domain EM1 could have resulted from continual exposure of the base of the lithosphere to the infiltration of asthenospheric melts. If this is the case the survival of the EM1 domain at 180–200 km might appear to be paradoxical but it is largely due to the coating of diamond. Indeed one could argue that the most likely place for the lithospheric protolith (i.e. DM) to survive is immediately beneath the crust well within the mechanical/chemical boundary layer (MBL/CBL) because in this location it had the protection of several hundred kilometers of lithospheric keel. The influence of the thermal boundary layer (TBL) and ultimately the asthenosphere is thus minimized and the MBL/CBL is shielded from any major re-equilibration due to infiltration of small volume melts.

In summary the Archaean domain chronology beneath South Africa appears to record a change from a predominantly depleted domain (DM) to enriched domains EM1–EM2–HIMU, the result of:

(1) crustal extraction in Archaean times—depleted residue (DM);

(2) Archaean asthenosphere–lithosphere interaction involving carbonate melts (low Rb/Sr ratios)—enriched domain (EM1);

(3) post-Archaean interaction with the deep mantle DUPAL anomaly—enriched domain (EM2);

(4) Phanerozoic interaction with other hot spots—enriched domains (EM2 and HIMU).

4.3.2. Proterozoic lithosphere—subduction modified MORB mantle

The Proterozoic lithospheric mantle contains representatives of the DMM and HIMU domains (USA) with possible evidence for an EM2 domain (Fig. 4.8). The western USA contains the closest approximation to the pure DMM domain except for Pb isotopes. Moreover, the highest $^{87}Sr/^{86}Sr$ ratios are to be found in the sub-Proterozoic lithosphere (Tanzania) and, although characteristic of the EM2 reservoir, the lack of enrichment in $^{207}Pb/^{204}Pb$ presents a problem and indicates that perhaps Pb is decoupled from the Sr and Nd.

The variability in Sr and Nd isotopes (Fig. 4.8) in lithospheric xenoliths from the western USA in association with 'age' information can help constrain the temporal evolution of the Proterozoic lithosphere (Menzies *et al.* 1985; Roden *et al.* 1990). The oldest component (0.8–1.2 billion years) within the Proterozoic lithosphere appears to be granuloblastic olivine-rich peridotites interpreted as the closest approximation of the DMM domain (i.e. a Proterozoic reduced protolith?). This residue has been transformed by a sequence of events. In Proterozoic–Phanerozoic times (*c.* 0.5 billion years) HIMU trace element characteristics were superimposed on the DMM residue. In the Phanerozoic (<0.2 billion years) accretion or recycling of crustal material (Kyser, Chapter 7, this volume) produced the EM2 domain which overprints the DMM and HIMU domains. Finally, the HIMU–EM2 domains were themselves overprinted (<0.2 billion years) by the influx of small volume melts from the asthenosphere (i.e. DMM domain). These small volume melts enriched and metasomatized the HIMU–EM2 protolith.

Vertical stratification of the Proterozoic lithosphere in the western USA is apparent from the chemical variability of spinel- and garnet-bearing xenoliths entrained by volcanic rocks erupted through the Colorado Plateau (Roden *et al.* 1990). Shallow spinel peridotites are chemically depleted and are underlain by garnet peridotites enriched in incompatible elements, an enrichment caused by asthenosphere–lithosphere interaction. Elsewhere the spatial geometry of mantle domains beneath Proterozoic crust (Hart *et al.* 1989) has been partially constrained through the study of volcanic rocks from Ethiopia and Afar. Hart *et al.* (1989) proposed that the Proterozoic lithosphere is chemically stratified from a shallow EM2 domain (produced by subduction or recycling processes during pan-African accretion) to a deeper HIMU or LoNd domain.

In summary the Proterozoic domain chronology beneath the western USA and eastern Africa appears to be DMM–EM2–HIMU–DMM, essentially producing continental lithospheric mantle with 'oceanic' characteristics. This differs from the domain chronology beneath the Archaean lithospheric mantle (i.e. DMM–EM1–EM2–HIMU) only in that the Archaean depleted protolith had to be inferred. The Proterozoic domain

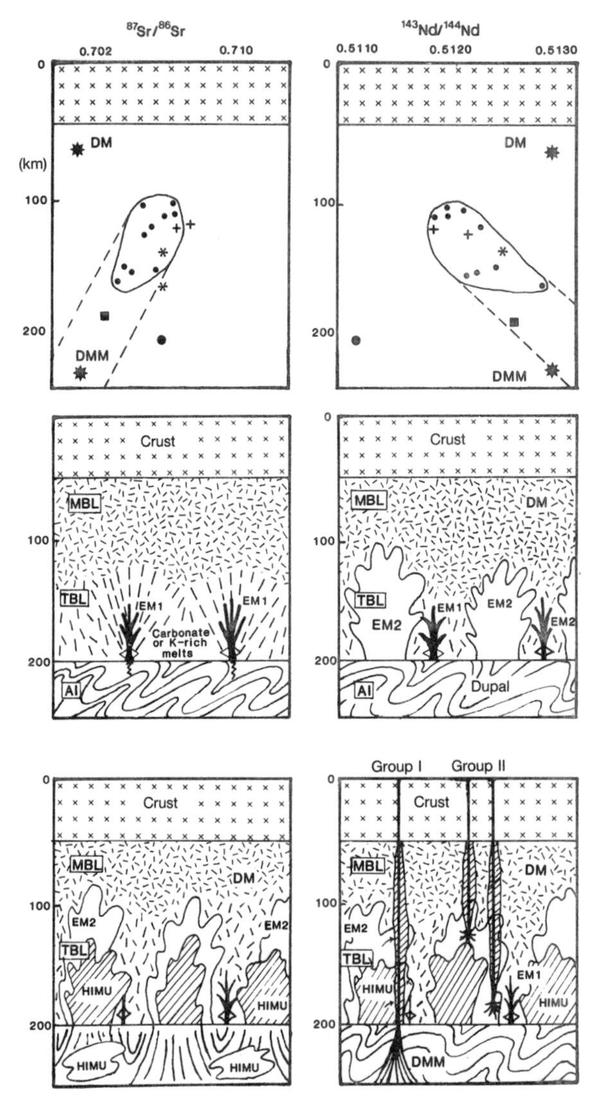

Fig. 4.9. (Top two figures) Vertical variation in Sr and Nd isotopes in the Archaean lithosphere of South Africa may be the result of interaction of a depleted protolith with upwelling sublithospheric melts and fluids. In a rather general sense the data trend from Dupal type signatures around 100 km depth through a region of HIMU signatures near the base of the lithosphere at 150–200 km and, if projected, intersect the composition of DMM domain in the asthenosphere. Note that: (1) the peridotite xenoliths shown by crosses do not have *P–T* data but are petrographically similar to rocks with known *P–T* conditions; (2) the sample shown by an asterisk is a whole-rock value as the clinopyroxene ratio appears to be anomalously high; (3) the square is the range of isotopic compositions for clinopyroxenites assumed to have formed at the base of the lithosphere; (4) the black dot is the isotopic composition of garnet inclusions in diamonds; (5) the DMM star is the approximate composition of the sublithospheric mantle; and (6) the DM star symbol close to crust–mantle boundary represents the approximate composition of the Archaean depleted mantle protolith had it not been affected by interaction with small volume melts and had remained a pristine residue of partial melting processes. Data are taken from several sources (Menzies and Murthy 1980*a*; Richardson *et al.* 1984, 1985; Erlank *et al.* 1987; Waters and Erlank 1988). (Lower four figures—top left to bottom right) A protolith composition (depleted mantle = DM) occurring within the mechanical (MBL) and the thermal boundary layer (TBL) of the Archaean lithosphere can be petrologically and chemically converted dependent on the composition of upwelling magmas from the adiabatic interior (AI) or the lower mantle. Transformation of the Archaean protolith lithosphere is believed to have

chronology is believed to have resulted from the following processes:

(1) Archaean?–Proterozoic partial melting—melt residue (DMM);

(2) Proterozoic–Phanerozoic supra-subduction processes associated with terrain accretion and recycling—enriched domains (EM2, HIMU);

(3) Phanerozoic plume–lithosphere interaction—depleted domain (DMM).

4.3.3. Phanerozoic lithospheric mantle—Dupal modified MORB mantle

The Phanerozoic lithospheric mantle contains representatives of the DMM and EM2 domains with some evidence for the existence of HIMU domains in Antarctic samples. These samples, however, are megacrystic and possible derivatives of sublithospheric melts and could, as such, point to the presence of an HIMU domain beneath the lithosphere.

The variation in Sr and Nd isotopes in the Phanerozoic lithospheric mantle of eastern Australia (McDonough and McCulloch 1987; Griffin *et al.* 1988; Stolz and Davies 1988) provides vital information about domain interplay (Fig. 4.8). The oldest component in the lithosphere must be the protolith that existed prior to the oldest identifiable enrichment or metasomatic events. McDonough and McCulloch (1987) report the existence of a Proterozoic protolith (0.6–1.5 billion years) in the form of a depleted lherzolite similar to the DMM domain (see also O'Reilly, unpublished data). This domain may have been emplaced as a result of intraplate processes associated with a continental rift system. We can therefore speculate that the sub-Phanerozoic lithospheric mantle was initially composed of a DMM domain that was overprinted in the early Phanerozoic (<0.5 billion years) by an EM2 domain related either to Palaeozoic subduction processes (Stolz and Davies 1988) or the interception of the Australian plate with the DUPAL 'plumes' (i.e. EM2) as suggested by Ewart *et al.* (1988). During the remainder of the Phanerozoic (<0.2 billion years) the EM2 domain was gradually modified by upwelling of a sublithospheric depleted domain (e.g. DMM).

Both sub-Archaean and sub-Proterozoic lithosphere are believed to be chemically stratified and there is no reason to suppose that sub-Phanerozoic lithosphere evolved in a different fashion. In the case of eastern Australia a depleted protolith (DMM) underplates the Phanerozoic crust as it is ubiquitous amongst spinel peridotite suites. Phanerozoic overprinting with the EM2 domain is apparent and one can speculate that the lower garnet-bearing lithosphere is predominantly the EM2 domain. This can be verified to some degree by consideration of the identity of lithospheric mantle sources for volcanic rocks. Hawaiites and leucitites are of particular interest in that they frequently contain spinel lherzolite xenoliths indicative of a mantle origin at greater depths. Indeed leucitites are believed to be derived from garnet–phlogopite sources at depths of 100 km or more, a factor that would limit their genesis to within the lower Phanerozoic lithosphere. Regional considerations (Ewart *et al.* 1988) indicate that the arc signature (i.e. EM2) in many of these volcanic rocks is due to the participation of the lithosphere in their petrogenesis. While leucitites may be lithospheric in origin, the alkaline volcanic rocks of Tasmania point to a DMM asthenosphere underlying the lithosphere. Taken together with the xenolith data we can construct a chemically stratified lithosphere that varies from DMM (spinel-facies xenoliths) to EM2 (leucitite garnet-facies source) underlain by DMM asthenosphere (Tas-

involved migration of small volume melts (potassic and carbonate rich) with variable Rb/Sr ratios in the Archaean–Proterozoic resulting in EM1 and EM2 domains now trapped within diamond inclusions (top left). Interaction between the lithosphere and recycled material (e.g. Dupal) in the late Proterozoic–Phanerozoic may have led to the formation of EM2 domains (top right). Recycling of oceanic lithosphere in the Phanerozoic may have helped generate HIMU domains (bottom left) and upwelling of the Bouvet or Marion hot spots may have produced a DMM domain in the late Phanerozoic (bottom right) during eruption of basaltic kimberlites (Group I). Furthermore reactivation of the EM2 domain may have been responsible for eruption of micaceous kimberlites (Group II). Much of the metasomatic conversion of the lithosphere is apparent between 100 and 150 km (Waters and Erlank 1988), but the transfer of magmas to the surface can presumably extend the zone of metasomatism to shallower levels (50–100 km) where carbon dioxide rich sodic metasomatism may be as widespread as potassic metasomatism between 100 and 150 km. Basaltic kimberlites are believed to be sublithospheric in origin and transport HIMU and EM2 xenoliths to the surface whereas micaceous kimberlites are believed to be derived from Dupal-type (EM2) lithosphere (LeRoex 1986). It is implied in this figure that the DM protolith may only survive outwith areas of active volcanism well within the mechanical boundary layer. Proximity to the thermal boundary layer or any volcanic conduit essentially means that enrichment or metasomatic processes could obliterate any evidence of a DM protolith.

manian alkali basalts). This produces a domain chronology of DMM–EM2–HIMU/DMM consistent with:

(1) Proterozoic partial melting—depleted domain (DMM);

(2) Phanerozoic subduction or DUPAL–lithosphere interaction—enriched domain (EM2);

(3) Phanerozoic upwelling of asthenosphere (DMM).

4.3.4. Continental lithospheres

The peridotite protoliths that initially underplated the continental crust have all been designated as depleted domains because of an origin as partial melting residua. Whilst the chemistry of sub-Archaean lithosphere can be linked to the removal of komatiitic melts from an undepleted Archaean precursor (Boyd 1989) oceanic lithospheres beneath Proterozoic and Phanerozoic crust are more fertile perhaps indicating removal of less magnesian (tholeiitic?) melts. Although a case has been made for a major difference between the Archaean and post-Archaean lithospheres on the basis of major elements and isotopes, the depletion apparent in the MORB lithosphere, so common beneath post-Archaean crust, may have been produced in the early Archaean. Spinel and garnet peridotites entrained from beneath post-Archaean crust have Pb, Sr, and Nd model ages in the range 1.0–4.0 billion years indicating that some basalt-borne xenoliths may represent very old residua formed early in earth's history. Other more convincing lines of evidence point to an Archaean depletion in the MORB source. Rare gas studies reveal that the MORB source underwent major degassing within the first billion years of Earth's history (Allegre *et al.* 1983) and, as mentioned earlier, neodymium isotopic data for Archaean volcanic rocks point to a source region depleted in the light rare earth elements in the early Archaean (e.g. Zindler 1982).

Why does DMM lithosphere underplate post-Archaean and not Archaean crust? Perhaps it did underplate the Archaean crust (e.g. Ashwal and Burke 1989) but the protolith has been modified beyond recognition? The petrological and chemical integrity of the depleted protolith beneath the Archaean cratons presumably survives within the mechanical boundary layer in areas not contaminated by asthenosphere–lithosphere and plume–lithosphere interaction. For the last four billion years the continual exposure of the Archaean keel to lower mantle plumes and the hot asthenosphere has perhaps modified it beyond recognition particularly with respect to incompatible element content and radiogenic isotope geochemistry. In addition, subduction processes around the cratonic nucleii may have further processed the residual protolith thus enhancing the depletion in incompatible elements and producing a highly refractory residue. Osmium isotope data (Walker *et al.* 1989), however, argue against any link between the petrogenetic processes responsible for modern ocean basins and those responsible for the Archaean lithosphere. It may be that the Archaean lithospheric mantle was the residue remaining after extraction of the crust and that the MORB depletion was a much earlier event. Indeed, the widespread belief that the depletion in the MORB source is balanced by the enrichment in the continental crust has recently been questioned by Silver *et al.* (1988) mainly because the depletion in the MORB source was apparently established in the early Archaean before extraction of the bulk of the continental crust from the mantle (Pollack 1986 and references therein).

Finally, one might speculate that the domain chronologies outlined in this chapter record the propensity of continental plates to move away from regions of hot mantle to regions of cold mantle (Anderson, Chapter 1, this volume). All the lithospheres studied comprise an initial depleted protolith that appears to be a residue of intraplate processes. Consequently, one could argue that such protolith material accreted to the crust during deep mantle upwelling and that it was this hot upwelling that caused the migration of the continental plate to a region of cold downwelling thus accounting for the subsequent superposition of a subduction signature (i.e. EM2) on many of these depleted lithospheres. Cessation of subduction or the appearance of a slab-window may account for the resurgence of mantle upwelling and the renewed influx of asthenospheric material (i.e. DMM). As such the geochemical (and redox) stratification of the lithosphere may record the movement of continental plates.

4.4. SUMMARY

The subcrustal lithosphere has inherited much of its petrological and geochemical diversity from partial melting or depletion processes related to crust–mantle differentiation. This accounts for the widespread existence of depleted protoliths in subcrustal lithospheres of all ages (Fig. 4.10). However, interaction of lithosphere with deep mantle hot spots, the asthenosphere, or recycled material has significantly altered this residue and resulted in a range in Sr, Nd, and Pb isotopes in Archaean, Proterozoic, and Phanerozoic lithospheric mantle that encompasses the fields occupied by oceanic

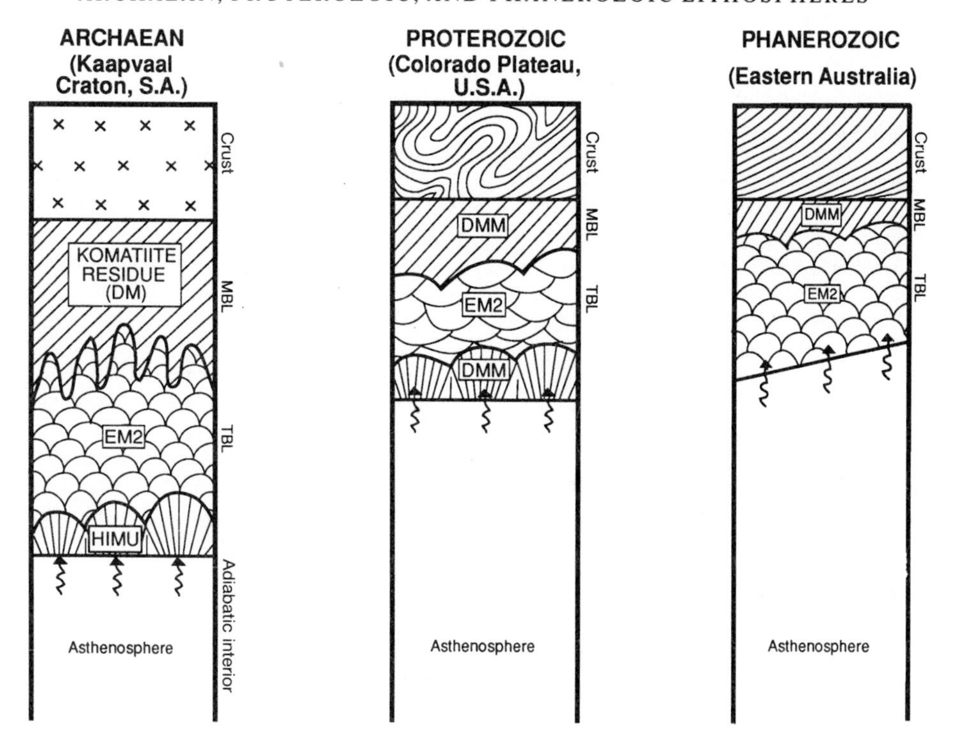

Fig. 4.10. Continental lithospheres. The depleted protolith, that presumably underplated continental crust of all ages, remains essentially intact within the mechanical boundary layer (MBL) whereas within the thermal boundary layer (TBL) the depleted protolith is transformed by ingress of melts from the asthenosphere or lower mantle (after Menzies 1990). See text for discussion.

volcanic rocks, uncontaminated continental volcanic rocks, and a significant number of lower and upper crustal rocks. Cratonic lithosphere, beneath Archaean crust, is a highly magnesian residue left after extraction of komatiitic magmas or the result of billions of years of 'processing' by active margin processes. Whilst the character of this depleted protolith can be constrained by mineralogical, major and minor elements and osmium isotopes, the Sr, Nd, and Pb isotope geochemistry has been extensively modified by post-Archaean processes. These include local migration of small-degree partial melts resulting in enriched domains (EM1) in the Archaean and superposition of other chemically discrete domains (EM2 and HIMU) in the Proterozoic and Phanerozoic due to plume–lithosphere interaction (e.g. Dupal and Gough). Thus the depleted protolith has been very effectively obliterated by the superposition of enrichment processes throughout geological time (Fig. 4.10). Oceanic lithosphere occurs beneath Proterozoic and Phanerozoic crust in continental environments. It contrasts markedly with cratonic lithosphere in that it comprises a less refractory lherzolitic protolith whose major, minor, and trace element and isotopic characteristics are compatible with an origin as a partial melt residue (DMM). This protolith has been overprinted by the EM2 and HIMU domains generated by more recent enrichment processes related to accretion (EM2), crustal recycling (EM2), plume–lithosphere interaction (HIMU/EM2), and upwelling of asthenosphere (DMM).

ACKNOWLEDGEMENTS

Steve Richardson is thanked for his comments and Joe Boyd for preprints of unpublished work.

REFERENCES

Allegre, C. J., Staudacher, T., Sarda, P., and Kurz, M. (1983). Constraints on evolution of Earth's mantle from rare gas systematics. *Nature*, **303,** 762–6.

Ashwal, L. D. and Burke, K. (1989). African lithosphere structure, volcanism and topography. *Earth and Planetary Science Letters*, **96,** 8–14.

Basu, A. R. and Tatsumoto, M. (1980). Nd isotopes in selected mantle-derived rocks and minerals and their implications for mantle evolution. *Contributions to Mineralogy and Petrology*, **75,** 43–54.

Basu, A. R., Goodwin, A. M., and Tatsumoto, M. (1984). Sm–Nd study of Archaean alkalic rocks from the Superior Province of the Canadian Shield. *Earth and Planetary Science Letters*, **70,** 40–6.

Bell, K. and Blenkinsop, J. (1987). Archaean depleted mantle: Evidence from Nd and Sr initial isotopic ratios of carbonatites. *Geochimica et Cosmochimica Acta*, **51,** 291–8.

BenOthman, D., Tilton, G., and Menzies, M.A. (1990). Pb, Nd and Sr isotopic investigations of kaersutite and clinopyroxene from ultramafic nodules and their host basalts: the nature of the subcontinental mantle. *Geochimica et Cosmochimica Acta* (in press).

Bonatti, E. and Michael, P. J. (1989). Mantle peridotites from continental rifts to ocean basins to subduction zones. *Earth and Planetary Science Letters*, **91,** 297–311.

Boyd, F. R. (1987). High and low temperature garnet peridotite xenoliths and their possible relation to the lithosphere–asthenosphere boundary beneath southern Africa. In *Mantle xenoliths* (ed. P. Nixon), pp. 403–12. John Wiley and Sons, Chichester.

Boyd, F. R. (1989). Compositional distinction between oceanic and cratonic lithosphere. *Earth and Planetary Science Letters*, **96,** 15–26.

Boyd, F. R. and Mertzman, S. A. (1987). Composition and structure of the Kaapvaal lithosphere, southern Africa. In *Magmatic processes and physicochemical principles* (ed. B. Mysen), pp. 13–24. Geochemical Society, Pennsylvania.

Boyd, F. R., Gurney, J. J., and Richardson, S. H. (1985). Evidence for a 150–200 km. thick Archaean lithosphere from diamond inclusion thermobarometry. *Nature*, **315,** 387–9.

Broxel, M., Tatsumoto, M. and Clague, D. A. (1988). Sr, Nd and Pb isotopes of spinel lherzolite xenoliths Koloa volcanics, Kauai, Hawaii. *Transactions of the American Geophysical Union*, **69,** 1517.

Brueckner, H. K., Zindler, A., Seyler, M., and Bonatti, E. (1988). Zabargad and the isotopic evolution of the sub-Red Sea mantle and crust. *Tectonophysics*, **150,** 163–76.

Chase, C. G. and Patchett, P. J. (1988). Stored mafic/ultramafic crust and early Archaean mantle depletion. *Earth and Planetary Science Letters*, **91,** 66–72.

Collerson, K. D. S., and Campbell, L. M. (1989). Sm–Nd isotope systematics of early Archaean ultramafic rocks in the North Atlantic craton: a sample of ancient depleted mantle? *Transactions of the American Geophysical Union*, **70,** 1390 (abstract).

Cox, K., Smith, M. R., and Beswetherick, S. (1987). Textural studies of garnet lherzolites: evidence of exsolution origin from high-temperature harzburgites. In *Mantle xenoliths* (ed. P. H. Nixon), pp. 537–50. John Wiley and Sons, Chichester.

Downes, H. and Dupuy, C. (1987). Textural, isotopic and REE variations in spinel peridotite xenoliths, Massif Central, France. *Earth and Planetary Science Letters*, **82,** 121–35.

Dudas, F. O., Carlson, R. W., and Eggler, D. H. (1987). Regional mid-Proterozoic enrichment of the sub-continental mantle source of igneous rocks from central Montana. *Geology*, **15,** 22–5.

Erlank, A. J., Waters, F. G., Hawkesworth, C. J., Haggerty, S. E., Allsopp, H. L., Rickard, R. S., and Menzies, M. A. (1987). Evidence for mantle metasomatism in peridotite nodules from the Kimberley Pipes, South Africa. In *Mantle metasomatism* (ed. M. A. Menzies and C. J. Hawkesworth), pp. 221–311. Academic Press, London.

Ewart, A., Chappell, B. W., and Menzies, M. A. (1988). An overview of the geochemical and isotopic characteristics of the Eastern Australian Cainozoic volcanic provinces. In *Oceanic and continental lithosphere: similarities and differences* (ed. M. A. Menzies and K. G. Cox), pp. 225–75. Oxford University Press, Oxford.

Finnerty, A. A. and Boyd, F. R. (1987). Thermobarometry for garnet peridotites: basis for the determination of thermal and compositional structure of the upper mantle. In *Mantle xenoliths* (ed. P. Nixon), pp. 381–402. J. Wiley and Sons, Chichester.

Galer, S. J. G. and O'Nions, R. K. (1989). Chemical and isotopic studies of ultramafic inclusions from the San Carlos volcanic field, Arizona. Bearing on their petrogenesis. *Journal of Petrology*, **30,** 1033–64.

Griffin, W. L., O'Reilly, S. Y., and Stabel, A. (1988). Mantle metasomatism beneath western Victoria, Australia: II Isotopic geochemistry of Cr–diopside lherzolites and Al-augite pyroxenites. *Geochimica et Cosmochimica Acta*, **52**, 449–59.

Hart, S. R. and Brooks, C. (1977). The geochemistry and evolution of the Early Precambrian mantle. *Contributions to Mineralogy and Petrology*, **61**, 109–28.

Hart, S. R., Gerlach, D. C., and White, W. M. (1986). A possible new Sr–Nd–Pb mantle array and consequences for mantle mixing. *Geochimica et Cosmochimica Acta*, **50**, 1551–7.

Hart, W. K., WoldeGabriel, G., Walter, R. C., and Mertzman, S. A. (1989). Basaltic volcanism in Ethiopia: constraints on continental rifting and mantle interactions. *Journal of Geophysical Research*, **94**, 7731–49.

Hearn, B. C. Jr, Collerson, K. J. D., Macdonald, R. A., and Upton, B. G. J. (1989). Mantle–crust lithosphere of north-central Montana, U.S.A.: evidence from xenoliths. *New Mexico Bureau of Mines and Mineral Resources, Bulletin*, **131**, 125 (abstract).

Hoffmann, P. F. (1988). United Plates of America, the birth of a craton: early Proterozoic assembly and growth of Laurentia. *Annual Reviews of Earth and Planetary Science*, **16**, 543–663.

Irving, A. J., O'Brien, H. E., and McCallum, I. S. (1989). Montana potassic volcanism: geochemical evidence for interaction of asthenospheric melts and metasomatically veined PreCambrian subcontinental mantle lithosphere. *New Mexico Bureau of Mines and Mineral Resources, Bulletin*, **131**, 140 (abstract).

Kramers, J. D. (1977). Lead and strontium isotopes in Cretaceous kimberlites and mantle-derived xenoliths from southern Africa. *Earth and Planetary Science Letters*, **34**, 419–31.

Kramers, J. D. (1987). Link between Archaean continent formation and anomalous sub-continental mantle. *Nature*, **325**, 47–50.

Kramers, J. D., Roddick, J., and Dawson, J. (1983). Trace element and isotope studies on veined, metasomatic and MARID xenoliths from Bultfontein South Africa. *Earth and Planetary Science Letters*, **65**, 90–106.

LeRoex, A. P. (1986). Geochemical correlation between southern African kimberlites and South Atlantic hotspots. *Nature*, **324**, 243–5.

Maaloe, S. and Aoki, A. (1977). The major element composition of the upper mantle estimated from the composition of lherzolites. *Contributions to Meneralogy and Petrology*, **63**, 161–73.

Mattey, D., Exley, R. A., Pillinger, C. T., Menzies, M. A., Porcelli, D. R., Galer, S., and O'Nions, R. K. (1989). Relationships between C, He, Sr, and Nd isotopes in mantle diopsides. In *Kimberlites and related rocks: their mantle/crust setting, diamonds and diamond exploration*. Geological Society of Australia Special Publication No. 14, pp. 913–21. Geological Society of Australia, Carlton, Victoria.

McDonough, W. E. and McCulloch, M. C. (1987). Isotopic heterogeneity in the south east Australian subcontinental lithospheric mantle. *Earth and Planetary Science Letters*, **86**, 327–40.

McKenzie, D. P. (1989). Some remarks on the movement of small melt fractions in the mantle. *Earth and Planetary Science Letters*, **95**, 53–72.

McKenzie, D. P. and Bickle, M. J. (1988). The volume and composition of melt generated by extension of the lithosphere. *Journal of Petrology*, **29**, 625–79.

Menzies, M. A. (1989). Cratonic, circum-cratonic and oceanic mantle domains beneath the western U.S.A. *Journal of Geophysical Research*, **94**, 7899–915.

Menzies, M. A. (1990). Effects of small volume melts. *Nature*, **343**, 312–13.

Menzies, M. A. (1991). Oceanic peridotites: MORB lithosphere modified by within-plate and active margin processes. In *Oceanic basalts* (ed. P. Floyd). Blackie Publishing, Glasgow (in press).

Menzies, M. A. and Halliday, A. J. (1988). Lithospheric mantle domains beneath the Archaean and Poterozoic crust of Scotland. In *Oceanic and continental lithosphere: similarities and differences* (ed. M. A. Menzies and K. G. Cox), pp. 275–302. Oxford University Press, Oxford.

Menzies, M. A. and Hawkesworth, C. J. (1987*a*). Upper mantle processes and composition. In *Mantle xenoliths* (ed. P. Nixon), pp. 725–38. John Wiley and Sons, Chichester.

Menzies, M. A. and Hawkesworth, C. J. (eds) (1987*b*). *Mantle metasomatism*. Academic Press, London.

Menzies, M. A. and Murthy, V. R. (1980*a*). Enriched mantle: Nd and Sr isotopes in diopsides from kimberlite nodules. *Nature*, **283**, 634–6.

Menzies, M. A. and Murthy, V. R. (1980*b*). Nd and Sr isotope geochemistry of hydrous mantle nodules and their host alkali basalts: implications for local heterogeneities in metasomatically veined mantle. *Earth and Planetary Science Letters*, **46**, 323–34.

Menzies, M. A., Kempton, P. D., and Dungan, M. (1985). Interaction of continental lithosphere and asthenospheric melts below the Geronimo volcanic field Arizona, U.S.A. *Journal of Petrology*, **26**, 663–93.

Menzies, M. A., Halliday, A. N., Palacz, Z., Hunter, R., Upton, B. G. J., Aspen, P., and Hawkesworth, C. J. (1987). Evidence from mantle xenoliths for an enriched lithospheric keel under the Outer Hebrides. *Nature*, **325**, 44–7.

Navon, O., Hutcheon, I. D., Rossman, G. R., and Wasserburg, G. J. (1988). Mantle-derived fluids in diamond micro-inclusions. *Nature*, **335**, 784–9.

Nixon, P. N. (ed.) (1987). *Mantle xenoliths*. John Wiley and Sons, Chichester.

Pollack, H. N. (1986). Cratonization and the thermal evolution of the mantle. *Earth and Planetary Science Letters*, **80**, 175–82.

Richardson, S. H., Gurney, J. J., Erlank, A. J., and Harris, J. W. (1984). Origin of diamonds in old enriched mantle. *Nature*, **310**, 198–202.

Richardson, S. H., Erlank, A. J., and Hart, S. R. (1985). Kimberlite-borne garnet peridotite xenoliths from old enriched subcontinental lithosphere. *Earth and Planetary Science Letters*, **75**, 116–28.

Roden, M., Irving, A. J., and Murthy, V. R. (1988). Isotopic and trace element composition of the upper mantle beneath a young continental rift: results from Kilbourne Hole, New Mexico. *Geochimica et Cosmochimica Acta*, **52**, 461–74.

Roden, M., Smith, D., and Murthy, V. R. (1990). Geochemical constraints on lithosphere composition and evolution beneath the Colorado Plateau. *Journal of Geophysical Research* (in press).

Roden, M. K., Hart, S. R., Frey, F. A., and Melson, W. G. (1984). Sr, Nd and Pb isotopic and REE geochemistry of St. Paul's Rocks: the metamorphic and metasomatic development of an alkali basalt mantle source. *Contributions to Mineralogy and Petrology*, **85**, 379–400.

Scott-Smith, B. H. (1987). Greenland. In *Mantle xenoliths* (ed. P. H. Nixon), pp. 23–32. John Wiley and Sons, Chichester.

Shirley, S. B. and Hanson, G. N. (1986). Mantle heterogeneity and crustal recycling in Archaean granite–greenstone belts—evidence from Nd isotopes and trace elements in the Rainy Lake area Ontario. *Geochimica et Cosmochimica Acta*, **50**, 2631–51.

Silver, P. G., Carlson, R. W., and Olson, P. (1988). Deep slabs, geochemical heterogeneity, and the large-scale structure of mantle convection: investigation of an enduring paradox. *Annual Reviews of Earth and Planetary Sciences*, **16**, 477–541.

Stolz, A. J. and Davies, G. R. (1988). Chemical and isotopic evidence from spinel lherzolite xenoliths for episodic metasomatism of the upper mantle beneath southeastern Australia. In *Oceanic and continental lithosphere: similarities and differences* (ed. M. A. Menzies and K. G. Cox), pp. 303–30. Oxford University Press, Oxford.

Storey, M., Saunders, A. D., Tarney, J., Leat, P., Thirwall, M. F., Thompson, R. N., Menzies, M. A., and Marriner, G. F. (1988). Geochemical evidence for plume–mantle interactions beneath Kerguelen and Heard Islands Indian Ocean. *Nature*, **336**, 371–4.

Storey, M., Saunders, A. D., Tarney, J., Gibson, I. L., Norry, M. J., Leat, P., Thirlwall, M. F., Thompson, R. N., and Menzies, M. A. (1989). Geochemical consequences of plume–lithosphere and plume–asthenosphere interactions: Indian Ocean. *Nature*, **338**, 574–6.

Stosch, H-G. and Lugmair, G. W. (1986). Trace element and Sr and Nd isotope geochemistry of peridotite xenoliths from the Eifel (West Germany) and their bearing on the evolution of the sub-continental lithosphere. *Earth and Planetary Science Letters*, **90**, 281–98.

Walker, R. J., Carlson, R. W., Shirey, S. B., and Boyd, F. R. (1989). Os, Sr, Nd and Pb isotope systematics of southern African peridotite xenoliths: implications for the chemical evolution of sub-continental mantle. *Geochimica et Cosmochimica Acta*, **53**, 1583–95.

Wallace, M. E. (1989). Stability of amphibole and carbonate in an oxidised upper mantle and the genesis of carbonatites. Unpublished PhD thesis, University of Tasmania.

Waters, F. (1987) A suggested origin of MARID xenoliths in kimberlites by high pressure crystallisation of an ultrapotassic rock such as lamproite. *Contributions to Mineralogy and Petrology*, **95**, 523–33.

Waters, F. and Erlank, A. J. (1988). Assessment of the vertical extent and distribution of mantle metasomatism below Kimberley, South Africa. In *Oceanic and continental lithosphere: similarities and differences* (ed. M. A. Menzies and K. G. Cox), pp. 185–204. Oxford University Press, Oxford.

Zindler, A. (1982). Nd and Sr isotopic studies of komatiites and related rocks. In *Komatiites* (ed. N. T. Arndt and E. G. Nisbet), pp. 399–420. Allen and Unwin, London.

Zindler, A. and Hart, S. R. (1986). Chemical geodynamics. *Annual Reviews of Earth and Planetary Science*, **14**, 493–571.

5

Redox state of the continental lithosphere

Stephen E. Haggerty

5.1. INTRODUCTION

Reduction–oxidation (redox) equilibrium studies have been focal points of attention for at least the past two decades: on the grand scale of solar nebula dynamics and condensation, in planet-wide bulk differentiation models rationalizing the existence of cores, mantles, crusts, hydrospheres, and atmospheres, and in the genesis of planetary magmas (Basaltic Volcanism Study Project 1981). Many of these studies remain inconclusive. The singular exception is a consensus that a *bulk* or *average* condition, whether redox or any other intrinsic parameter, cannot be reasonably assigned to any planetary body that continues to undergo mass volatile differentiation through melt extraction and recycling. The range of redox conditions from iron meteorites (and enstatite chondrites) to carbonaceous chondrites is very nearly equivalent to the simplistic redox view of the Earth as a two-component coupling having a highly reduced metallic core and an oxidized hydrated crust.

The bulk of the Earth, with its lower and upper mantle, remains enigmatic in physical properties, chemical composition, and redox state. Even the precise natures of the core–mantle and mantle–crust boundaries are obscure, and the obvious questions, in the context of redox equilibria, are: Can one predict the redox state of the mantle based on homogeneous or heterogeneous accretion models of the Earth? Is the mantle a perched redox zone resembling neither the core nor the crust? Is the mantle gradational in redox state? To what extent do redox interactions exist between the lower mantle and the core and the upper mantle and the crust? Is the upper mantle at a constant redox condition, now and in the past, or do tectonic settings and evolution play central roles?

While these questions may never be fully answered, it is relevant to note that the density diluent required in the Earth's core includes such elements as oxygen, carbon, hydrogen, and sulphur (and potassium or silicon), species that have a demonstrable effect in buffering the redox condition of any portion of the Earth, or any planet. These are, furthermore, species that abound in the Earth's crust. It is an improbable scenario to expect that the mantle is an isolated entity in redox state, but one question persists: Is the mantle more core-like or does the mantle have closer affinities to the crust? A seemingly obvious first estimate is that the lower mantle is in equilibrium with a metallic phase, whereas the upper mantle is uniformly more oxidized and akin to the crust. Little if any attention has been given to the former, in contrast to the redox state of the upper mantle which is a current and popular forum of active debate. The issues are: whether homogeneous or heterogeneous redox states pertain; whether extremely reducing conditions exist; or whether data in support of extreme reduction are an artefact of experimental technique?

Several excellent reviews have recently discussed the dichotomy of upper mantle redox states. This chapter covers some of the territory included in the review by Arculus (1985), but it expands neither on the comprehensive thermochemical review in Woermann and Rosenhauer (1985) nor the siderophile treatment in Arculus and Delano (1987).

The major aim of this chapter is a synthesis of upper mantle redox data that include several important studies since 1985, one of which is the formulation of a new oxygen geobarometer. Unlike other reviews, the data are treated in a geological context with the goal of formulating a model that is internally consistent–mineralogically, petrologically, and tectonically. A major conclusion from this review is that the upper mantle is heterogeneous in redox state. A second conclusion is that the upper mantle, having evolved geochemically, has also evolved in redox state. Both conclusions were reached earlier (Haggerty and Tompkins 1983), but the evidence is now much stronger. This should not be misconstrued as a statement that the problem is solved. Much has been accomplished but refinements to existing techniques and thermochemical data and new approaches are required, as are suites of upper mantle samples fully characterized to include the broad spectrum of mineral, bulk chemical, and isotopes to establish more precisely whether a redox measurement is primitive and pristine or superimposed and secondary.

5.2. REDOX STANDARDS

Discussions in the following sections are referenced to standard oxygen buffers, selected to allow for high-pressure corrections. Revised experimental data for these buffers (e.g. Chou 1978) show only slight variations and do not materially affect the overall conclusions, the comparison among data sets, or the oxygen fugacity (fO_2) of determinative techniques. Some buffers are given at 1 bar total pressure, others at 30 kbar, and still others are corrected to a cratonic average geothermal gradient corresponding to a surface heat flow of 40 mW m^{-2}.

The buffers are all based on multiple valence states, with the lower state (e.g. Fe^{2+}) stable at lower oxidation values and the higher state (e.g. Fe^{3+}) stable at correspondingly higher oxidation values. The buffers involve oxides, silicates, and elements, and an intrinsic property is that fO_2 increases with increasing temperature (*T*) and total pressure (*P*). The buffers employed, with increasing fO_2 at constant *P* and *T*, are:

IW	iron–wüstite (Eugster and Wones 1962)
WM	wüstite–magnetite (Eugster and Wones 1962)
FMQ	fayalite–magnetite–quartz (Wones and Gilbert 1969)
NNO	nickel–nickel oxide (Huebner and Sato 1970)
$MnO–Mn_3O_4$ Mn–Ha	manganosite–hausmanite (Huebner and Sato 1970).

5.3. REDOX STATE OF THE UPPER MANTLE

Why is there an interest in the redox condition of the upper mantle and why should it be so hotly disputed? The answer in part lies in the following.

5.3.1. Influence of redox state on the physicochemistry of magmas and the mantle

A large proportion of magmatic material reaching the Earth's surface originates in the upper mantle, as attested to by voluminous flood basalts on the continents (see also Menzies and Kyle, Chapter 8, this volume) and larger volumes that cover over 80 per cent of the Earth's surface on the ocean floors. At magmatic temperatures, oxygen fugacity (fO_2) is a major controlling factor in elemental partitioning. At high values of fO_2, iron is partially transformed into Fe^{3+} and oxide minerals crystallize in preference to Fe^{2+}-bearing silicates. At lower values of fO_2, the available iron is competitively partitioned between and among silicate and oxide phases. The higher fO_2 condition leads to the familiar *Bowen* trend, whereas lower fO_2 results in the *Fenner* crystallization trend, with each trend having a contrasting mineralogy and, hence, differences in rock type designation (Osborne 1962).

Redox equilibrium studies in basalt-type systems of $CaO–MgO–Al_2O_3–SiO_2–Fe–O$ have shown that Fe^{2+} is a network modifier, that Fe^{2+}/Fe^{3+} is linearly correlated with polymerization (expressed as non-bridging oxygens per tetrahedrally coordinated cations, NBO/T) and with Al/Al+Si of the melt, and that fO_2 affects fractional crystallization trends (Mysen *et al.* 1985). With increasing pressure, from 1 bar to 10 kbar, $Fe^{3+}/\Sigma Fe$ decreases, and ferric iron undergoes a coordination transformation from a network-former to a network-modifier resulting in depolymerization (Mysen and Virgo 1985). Polymerization affects viscosity, and viscosity increases with polymerization and $Fe^{3+}/\Sigma Fe$ (Dingwell and Virgo 1987), thus influencing magma extraction and magmatic flow in conduits from the upper mantle.

From an equally broad geophysical perspective, such fundamental mantle properties as electrical conductivity (Duba and Nicholls 1973; Duba *et al.* 1974) and creep rate in olivine (Ryerson *et al.* 1989) are shown experimentally to be affected by oxygen fugacity.

5.3.2. Redox buffering in basaltic magmas and lavas

The major gas species in volcanic eruptions is water, followed by CO_2 (Gerlach and Nordlie 1975*a,b*; Gerlach 1979, 1982; Basaltic Volcanism Study Project 1981), which are both capable of exercising a significant to overriding influence on the redox states of erupting magmas, and on the redox states of melt source regions. Data summarized by Haggerty (1976) suggest that melts and volcanic gases are approximately in equilibrium and that basalts, for example, are redox probes to source regions in the upper mantle, not only of the Earth but also of other planetary and asteroidal bodies (Haggerty 1978; Basaltic Volcanism Study Project 1981). That view has been challenged by Sato (1978) on the basis that ubiquitous but minor contents of elemental carbon in the upper mantle have a high buffering capacity. Moderately oxidized basalts are interpreted as having undergone degassing and wall-rock interaction, but the inferred source region is reduced. Mathez (1984) has followed this line of reasoning with the postulate that upper mantle-derived magmas are oxidized on ascent. A plausible scenario is based on the abundance of gas species as a function of oxygen fugacity, carbon gas species saturation as graphite, and oxidized and reduced iron species in the melt. Rapid loss of hydrogen by diffusion, with an attendant increase in H_2O/H_2, increases fO_2 and has been modelled as the mechanism to account for oxidized crusts in ponded lava lakes (Sato and Wright 1966) and oxidation in the interiors of some plateau basalts (Watkins and Haggerty 1968).

Sulphur also has been invoked as a self-regulating internal redox buffer that maintains a relatively reduced state in rapidly quenched glasses (see also Kyser, Chapter 7, this volume), but, on SO_2-degassing, equivalent crystalline lavas are oxidized (Anderson and Wright 1972; Carmichael and Ghiorso 1986). Alkali-rich, silica-poor basic lavas may contain sulphate-bearing minerals (such as anhydrite, nosean, or haüyne) and are, therefore, considerably more oxidized but undegassed in comparison, for example, to tholeiites (Carmichael and Ghiorso 1986).

From data summarized earlier (Haggerty 1976) and updated in this chapter, there is a remarkable adherence to the solid-state buffer fayalite–magnetite–quartz (FMQ) for basalt-type lavas, whether continental or oceanic. The envelope is around FMQ+1 to FMQ−1 (i.e. FMQ±1 order of magnitude of $\log_{10}fO_2$) and applies equally to Archaean lavas as it does to recent eruptives. This averaging around FMQ has not been satisfactorily answered or adequately addressed.

5.3.3. Gas solubilities and upper mantle metasomatism

In their study of volatile solubilities in mafic magmas, Holloway and Jakobsson (1986) conclude that, if melt volatiles are high in H_2O+CO, the equilibrium fluid is enriched in H_2+CO_2. This fractionation is interpreted to imply that a reduced upper mantle will yield magmas and volcanic gases that are more oxidized than the source and, inversely, that oxidized upper mantle will give rise to more reduced magmas and more reduced volcanic gases. The gases are predominantly H_2O with minor $CO_2+CO+H_2+CH_4$. If Holloway and Jakobsson's (1986) proposition is correct, it is difficult to conceive how their notion of a 'neutral' redox state at FMQ is achieved.

There is uniform consensus that the solubility of H_2O exceeds that of CO_2 in mafic melts (Myssen and Boettcher 1975; Brey and Green 1977; Eggler 1987; Wyllie 1987), and there is also the proposition, based on experimental work, that H_2 is more soluble than CO_2 (Luth and Boettcher 1986). In the continental upper mantle, free CO_2 is possible only at depths shallower than ~80 km (Eggler 1987; Wyllie 1987; Woermann and Rosenhauer 1985 and references therein), and free H_2O is possible at depths down to ~100 km. The gas species of significance at greater depths are modelled as C–O–H–S (see also Kyser, Chapter 7, this volume). Fractionation may occur, and partial melting may be induced, but H_2O, CO_2, and H_2 will rapidly dissolve into the melt (Wyllie 1987), making the melt a potent modifier to redox state along the volcanic conduit as the melt rises through the upper mantle in response to density contrasts. Modification is expressed as metasomatism, with deposition at shallower levels of hydrous minerals such as phlogopite and amphibole; metasomatism increases the overall state of oxidation.

5.3.4. Redox-melting and upper mantle redox stratigraphy

In a detailed series of experiments on an iron-free system at fO_2 of three to four orders of magnitude of $\log_{10}fO_2$ below IW (iron–wüstite), Taylor and Green (1987) laid the foundation for their 'exploration' of redox melting in the upper mantle (Green *et al.* 1987). They propose that $CH_4>H_2O>H_2$, and that CO_2–H_2O–H_2 fluids may give rise to melting by the decomposition of methane ($CH_4+O_2 \rightarrow 2H_2O+C$) which yields water to lower the liquidus and precipitates carbon (in some instances as diamond). As envisaged, the redox melting model requires a reduced, CH_4-dominated fluid and an oxidized segment to the upper mantle. Foley (1988) has drawn on redox data largely from oceanic regions and has modelled the range from rifts to cratons for the genesis of continental basic alkaline magmas by a mechanism of redox melting.

Reduced volatiles in the system C–O–H were studied by Eggler and Baker (1982) who concluded that CH_4 is the dominant volatile under reduced conditions in the upper mantle. Their formulation of EMOG (enstatite, magnesite, olivine, and graphite) and EMOD (D = diamond) implies upper mantle redox conditions between FMQ and WM (wüstite–magnetite). The question of methane is controversial, but Eggler (1987) states emphatically, based on experiments of fluid and gas species in equilibrium with upper mantle peridotites, that the dominant species are H_2O+CO_2 and not $CO+CH_4$. Gas release studies on diamond (Melton *et al.* 1972; Giardini and Melton 1975) show that CH_4 is present as an impurity, and Deines (1980) has used methane to rationalize the carbon isotopic signature of diamonds; the very existence of fluid inclusions, however, was questioned by Meyer (1985), but confirmed by Navon *et al.* (1988).

Implicit in the redox-melting models is that the upper mantle is redox-stratified. A stratified model proposed by Haggerty and Tompkins (1983) contrasts with the version by Green *et al.* (1987), in so far as the lower reaches of the continental lithosphere on the one hand are oxidized (Green *et al.* 1987) but, on the other hand, it is modelled as being reduced (Haggerty and Tompkins 1983).

Redox melting of the upper mantle (Eggler and Baker 1982; Green *et al.* 1987; Taylor and Green 1987; Foley 1988) is an attractive hypothesis, but free CH_4 is questionable. More relevant is whether the upper mantle is oxidized or reduced, and whether redox stratigraphy is one in which the asthenosphere is reduced and the lithosphere oxidized (Green *et al.* 1987) or vice versa (Haggerty and Tompkins 1983).

5.3.5. Analytical techniques and controversial very low redox states

Among the variety of techniques (mineral chemistry, experimental, thermodynamic) employed to determine the redox state of the upper mantle, the most direct and yet the most controversial is the electrochemical intrinsic oxygen fugacity method proposed by Sato (1970, 1972). This technique has been widely employed for the determination of the redox states of lunar basalts (Sato 1976), meteorites (Brett *et al.* 1977), and terrestrial rocks ranging from layered intrusions (Sato and Valenza 1980; Elliott *et al.* 1982), to kimberlites (Ulmer *et al.* 1976), carbonatites (Friel and Ulmer 1974), upper mantle-derived lherzolites (Arculus and Delano 1980, 1981), dunites in alkali basalts (Sato 1978), and discrete ilmenite megacrysts from the upper mantle entrained in kimberlites (Arculus *et al.* 1984; Ulmer *et al.* 1987; Loureiro *et al.* 1988; Virgo *et al.* 1988).

Much of the debate surrounding the uncertainty of the redox state of the upper mantle, but also much of the incentive to achieve a better understanding of this intrinsic property, stems from data on lherzolites by Arculus and Delano (1980, 1981). They show that some Type 1 (also Type A) spinel lherzolites (chromian diopside-bearing and geochemically depleted) are highly reduced and lie both below and just above the IW buffer. These data contrast with Type 2 (also Type B) spinel lherzolites (aluminium augite-bearing and geochemically fertile) that have a redox state in the region of FMQ–NNO (nickel–nickel oxide). The difference between Type 1 and Type 2 is approximately five orders of magnitude in $\log_{10} fO_2$, and is equivalent to the differences in redox states of magnetite solid solution-bearing terrestrial basalts and metallic iron-bearing lunar basalts. That such diverse redox conditions are present in the upper mantle has prompted considerable criticism of the electrochemical technique, and a division into two redox camps: those who consider the upper mantle to be uniformly oxidized, and those who entertain the possibility that multiple redox states may be present.

Both auto-oxidation and auto-reduction are demonstrated in the electrochemical method (Ulmer *et al.* 1987). This irreversible behaviour results in part from contaminating carbonate, or decrepitating and oxidizing fluid inclusions, or from finely dispersed, and reducing, discrete carbon (Sato 1978; Virgo *et al.* 1988). Other possible sources for irreversibility include crystal defects (Green 1985), electrochemical-cell memory (Loureiro *et al.* 1988), and the reheating at 1 bar of high-pressure phases to $T > 1200°C$ (Haggerty and Tompkins 1983).

Some samples respond better than others to the electrochemical method; among the former are discrete ilmenite megacrysts from kimberlites. Ilmenites free of carbon respond coherently, and interlaboratory comparisons confirm that the redox states recorded are equivalent to a range from WM to NNO, a bracket equivalent to approximately two orders of magnitude in $\log_{10} fO_2$ (Arculus *et al.* 1984; Loureiro *et al.* 1988; Virgo *et al.* 1988). From other electrochemical data on ilmenite and associated xenoliths, the range of redox conditions is from below IW to approximately WM (Ulmer *et al.* 1987). The latter data have a 'high confidence' level and extreme measures were exercised in establishing criteria for acceptance or rejection of data.

Thermochemical methods on ilmenite-bearing upper mantle-derived silicate xenoliths yield a range of redox states from WM to FMQ (Eggler 1983), consistent with the EMOG–EMOD buffer formulated by Eggler and Baker (1982).

Experimental calibration of the activity of the synthetic magnetite (Fe_3O_4) component in spinel ($MgAl_2O_4$) has been determined by Mattioli and Wood (1986) and Mattioli *et al.* (1987), and applied to Type 1 and Type 2 spinel lherzolites in alkali basalts. Lower-temperature (800–1000°C) equilibrated Type 1 xenoliths are close to FMQ, whereas the higher-temperature (1100–1200°C) equilibrated Type 2 spinel lherzolites are approximately FMQ + 3 (i.e. three orders of magnitude of fO_2 above FMQ).

A thermodynamically formulated olivine–orthopyroxene–spinel oxygen geobarometer has been used to determine the fO_2 of an extensive suite of spinel-bearing peridotites (O'Neill and Wall 1987). The results show, with few exceptions, that fO_2 is bracketed between FMQ and WM. This bracket includes samples from which the controversial lower fO_2 values of Type 1 spinel lherzolites were obtained by the electrochemical method (Arculus and Delano 1981).

Co-equilibrated ilmenite solid solution ($FeTiO_3$–Fe_2O_3) and ulvöspinel solid solution (Fe_2TiO_4–Fe_3O_4) members yield unique values of temperature and fO_2 (Buddington and Lindsley 1964; Spencer and Lindsley 1981; Andersen and Lindsley 1988). The method has been widely employed in igneous and metamorphic rocks (Haggerty 1976), and has been applied to subsolidus reduced kimberlitic ilmenites that contain crystallographically coherent lamellar spinel (Haggerty and Tompkins 1984). Results show that ilmenites record a redox state between WM and FMQ, with some data at approximately FMQ+2. These data are confirmed by the electrochemical method (Loureiro *et al.* 1988; Virgo *et al.* 1988) and, in the sense of Ulmer *et al.* (1987), have a 'high confidence' level.

Other determinations of upper mantle fO_2 states are by Ryabchikov *et al.* (1981) and Saxena (1989) who base their estimate of FMQ to FMQ+3 on thermodynamic and experimental data in the system C–CO–H_2O. In an exhaustive thermodynamic and Schreinemaker's treatment of the systems CaO–MgO–SiO_2–C–H–O and MgO–SiO_2–C–H–O, and in an extensive review of available, albeit diverse, data on upper mantle redox states, Woermann and Rosenhauer (1985) conclude that the overwhelming body of evidence favours a heterogeneous oxidized and reduced upper mantle with fO_2 spanning a range from IW+1 to FMQ+3.

5.3.6. Limits of upper mantle redox states

The redox state of the upper mantle may be broadly constrained between an upper fO_2 limit defined by the oxidation of elemental carbon species (graphite or diamond) to CO or CO_2 and a lower fO_2 limit defined by the precipitation of metallic Ni–Fe alloy from olivine (Woermann and Rosenhauer 1985; Arculus and Delano 1987; O'Neill and Wall 1987).

The stability of diamond as defined by C–CO ranges from FMQ to well below IW (Eggler 1987). The lower fO_2 value is supported by the presence of metallic iron (Ni-free) inclusions in diamond (Sobolev *et al.* 1981; Meyer and McCallum 1986). Carbonates such as brucite, magnesite, and dolomite that are postulated thermodynamically and demonstrated experimentally to be present in upper mantle peridotites (e.g. Wyllie 1989; see also Watson *et al.*, Chapter 6, this volume), but shown to be extremely rare in xenolith suites (Smith 1987), range from estimates of WM (Eggler 1987) to approximately FMQ+3 (Ryabchikov *et al.* 1981; Ryabchikov 1983).

Nickel concentrations in upper mantle-derived olivines vary from 2000 to 3300 p.p.m. (e.g. Boyd and Finnerty 1980; Hervig *et al.* 1980). Based on Fo_{90} olivine and 2000 p.p.m. Ni, Woermann and Rosenhauer (1985) calculate that metal precipitation at 70 kbar and 1000°C would take place at about one order of magnitude of fO_2 below the IW buffer. Using Fo_{90} and 3500 p.p.m. Ni, the Ni precipitation curve determined by O'Neill and Wall (1987) is virtually coincident with the IW buffer, being slightly below IW at $T>1200$°C and slightly above IW at T between 800 and 1200°C.

O'Neill and Wall (1987) dismiss the significance of metallic iron inclusions in diamond as a redox mineral indicator but have omitted to consider the ubiquitous presence of Ni–Fe sulphides in diamonds (which are orders of magnitude more abundant than any other diamond inclusion) and in xenoliths, and the possibility that sulphidation of free metal may have occurred. In addition, there is sufficient uncertainty in the thermodynamic data, as cautioned by Woermann and Rosenhauer (1985, p. 315), to entertain the possibility that conditions in the upper mantle may *locally* achieve redox states in proximity to IW or slightly below IW as shown by the 'high confidence' data set of Ulmer *et al.* (1987). Notwithstanding these uncertainties, a realistic lower fO_2 limit in the upper mantle, imposed by Ni metal precipitation and as presently constituted and recorded in xenolith suites, is at about IW.

5.3.7. Fluid and solid-state inclusions in upper mantle materials

Fluid and solid-state inclusions are potential redox barometers that have guided experimental studies in the judicious selection of volatile species deemed appropriate for the stability of peridotites in the upper mantle. The seminal studies of Roedder (1965, 1984) demonstrate that CO_2 fluid inclusions are ubiquitous in upper mantle-derived olivines. This is confirmed by CO_2 inclusions in dunites, peridotites, and pyroxenites (Murck *et al.* 1978), amphiboles (Matson *et al.* 1984), clinopyroxene (Andersen *et al.* 1984; Bergman and Dubessy 1984), and feldspar (Bergmann and Dubessy 1984), all from xenoliths in alkali basalts. Data for xenoliths of deeper origin in kimberlites confirm that CO_2 is the dominant gas species (Pasteris 1987).

The widespread occurrence of CO_2 fluid inclusions has resulted in an upper mantle paradox, in which carbonated peridotites are postulated as primary sources for kimberlite (see review by Wyllie 1980), and yet carbonate species are extremely rare in xenoliths on a global scale. Carbonated conditions are equivelent to relatively oxidized states (FMQ+3), but the upper mantle cannot be buffered by C–O alone, given the presence of hydrous minerals such as phlogopite and amphibole and eruptive gas compositions that are overwhelmingly dominated by water.

In an extensive review of the literature and also using evidence based on personal observations, Pasteris (1987) makes an excellent case for the late-stage secondary trapping of fluid inclusions in upper mantle xenoliths. She concludes that the primary encapsulation of fluids or gases of deep-seated origin is most unlikely, given that free CO_2 and H_2O do not exist at depths of >80 km and >100 km, respectively (Eggler 1987; Wyllie 1987). Most fluid inclusions show evidence of reaction with the mineral host and many have glass-lined walls (e.g. Murck *et al.* 1978); others contain daughter minerals of graphite and sulphide (Mathez and Delaney 1981), or amphibole (Bergman and Dubessy 1984), or amphibole accompanied by mica and apatite (Andersen *et al.* 1984). The glass-lined (melt) walls to inclusions will readily remove water because of the high solubility of H_2O in melts, and hydrogen may diffuse from the system. Equally, the presence of hydrous minerals indicates a precursor fluid composition possibly saturated in H_2O. Mathez and Delaney (1981) make the interesting proposition that graphite forms from the disproportionation of CO, catalysed by sulphide–oxide surfaces on the walls of vesicles. Catalysed diamond on sulphides was proposed by Marx (1972), and may account for the widespread occurrence of sulphides occluded in diamonds (Haggerty 1986).

Although estimates of fO_2 have been made on fluid inclusions, with conditions of <FMQ (Mathez and Delaney 1981) and of FMQ to FMQ+2 (Bergman and Dubessy 1984), it is clear from the review by Pasteris (1987) and later work on the thermal homogenization of fluid inclusions (Pasteris and Wanamaker 1988) that fO_2 estimates and any judgement of fluid composition must account both for the solid components present and the fractional solubility of fluids into melts and minerals. The compositions of trapped fluids dominated by CO_2 are most reasonably understood from the relatively insoluble and incompatible nature of carbon dioxide; and the presence of glass wall linings (to inclusions) by decompression melting and the lowering of host mineral liquidi by H_2O.

The solubility of carbon in olivine (Freund *et al.* 1980) may provide another estimate of upper mantle fO_2 states, but the considerable uncertainty surrounding these data (e.g. Tingle and Green 1987) makes the method, at present, unreliable.

5.3.8. Summary

In summary, knowledge of the redox state of the upper mantle will increase our understanding of the Fe^{2+}/Fe^{3+} effects on: physical properties; chemical partitioning and fractionation of elements among minerals; and the nature and compositions of gaseous fluids. Current topics of debate include: homogeneous–heterogeneous redox states; redox stratigraphy in the asthenosphere and lithosphere; the effects of metasomatism on redox equilibria; reduction-induced partial melting; and the existence of highly reduced states.

5.4. REDOX DATA SETS FOR THE UPPER MANTLE

Temperature and oxygen fugacity data sets in this section are updated from earlier presentations (Haggerty 1976; Basaltic Volcanism Study Project 1981). In some cases there are negligible changes (e.g. continental basalts), but in others (e.g. upper mantle suites), the data base has dramatically increased. The emphasis is on upper mantle-derived xenoliths, but melt fO_2 data on glasses and crystalline rocks (basalts, kimberlites, and carbonatites) are included. Discussion is biased toward the continental lithosphere in keeping with the theme of this book. Data from the oceans are presented to demonstrate a remarkable degree of uniformity in redox state between oceanic basalts and broadly similar compositions from the continents; substantially different redox conditions are, however, inferred for deeper lithospheric environments.

5.4.1. Basalts

Oceanic island and ocean floor basalts below ~950°C are tightly clustered around FMQ but become more oxidized and more reduced at higher temperatures, reaching a maximum at ~1200°C of FMQ+1.5 and a minimum at ≃1000°C of FMQ−1 (Fig. 5.1). These data are obtained exclusively by the Buddington and Lindsley (1964) and Spencer and Lindsley (1981) method of co-equilibrated Fe–Ti oxides.

Oxygen fugacity estimates for basaltic glass compositions from a variety of oceanic settings are reported, based on precise determinations of Fe^{2+}/Fe^{3+} (Carmichael and Ghiorso 1986; Christie *et al.* 1986), and using previously established ratios in experimental glasses (Sack *et al.* 1980; Kilinc *et al.* 1983). Temperatures are fixed at 1200°C and the range in fO_2 is from FMQ+1.5 to FMQ−2.5 (Fig. 5.1). Higher fO_2 values from glasses in, or associated with, crystalline basalt are consistent with the basalt suite. The lower fO_2 values are for glasses only and the interpretation is that glasses have remained reduced because of the retention of sulphur species and

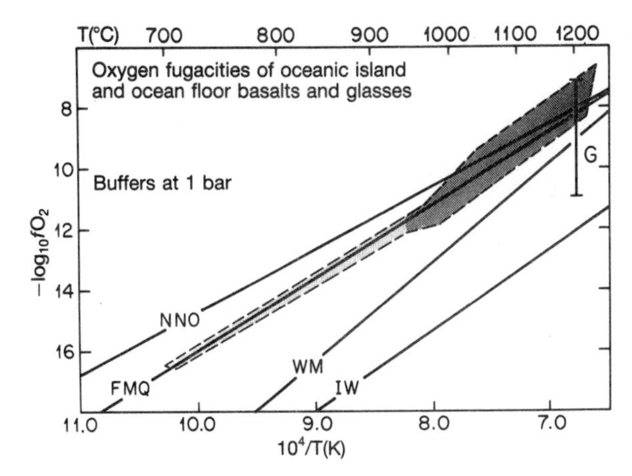

Fig. 5.1. Temperature–oxygen fugacity data envelope for oceanic island and ocean floor basalts and glasses. Sources are largely from Haggerty (1976), updated at lower temperatures for basalts and at higher temperatures to include glass (G) and other determinations in the region of 1200°C. Darker shading represents more abundant data. IW, MW, FMQ, and NNO are standard solid buffers defined in the text.

an absence of disproportionation of H_2O into hydrogen and oxygen (Carmichael and Ghiorso 1986).

Continental basalts (Fig. 5.2) in the range 1050–1200°C are in the fO_2 bracket of WM to FMQ. Equilibration toward lower temperatures, for which there are few data, show that mild reduction is in effect with minimum values at FMQ−1 and $T=700°C$. The

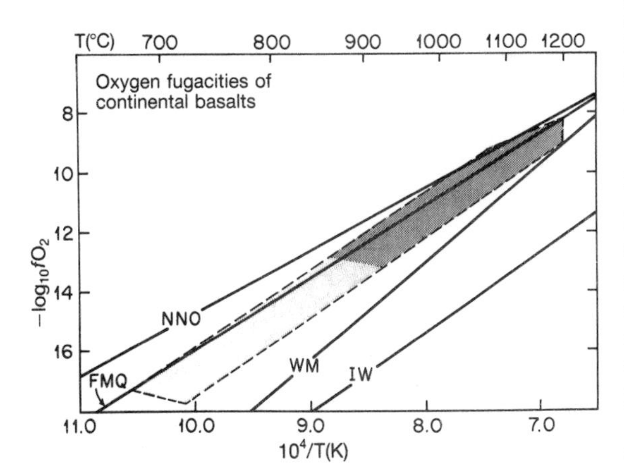

Fig. 5.2. Temperature–oxygen fugacity data envelope for continental basalts, updated from Haggerty (1976). Shading as in Fig. 5.1. IW to NNO are standard solid buffers (see text).

method of analysis is the Buddington and Lindsley (1964) technique.

5.4.2. Kimberlites

Intrinsic fO_2 data for kimberlites from the Premier Mine (Ulmer *et al.* 1976), compare favourably with the oxide geothermometer technique applied in the Benfontein Kimberlite Sills (McMahon and Haggerty 1984). A cross-over at $T \sim 1000°C$ results in FMQ+1 at higher temperatures and FMQ−1 at lower temperatures (Fig. 5.3(a)).

5.4.3. Lamprophyres and lamproites

Using a combination of Fe^{2+}/Fe^{3+} in bulk compositions and oxidation ratios of iron in chromian spinel, Foley (1985) and Venturelli *et al.* (1988) report equilibration curves of T and fO_2 for lamproites from Gaussberg, Antarctica (~NNO), the Leucite Hills, Wyoming (NNO+2), West Kimberley, Australia (WM), and Spain (WM−1 to NNO), as illustrated in Fig. 5.3(b).

5.4.4. Carbonatites

Data for carbonatites are restricted to the Oka complex in Canada. An intrinsic fO_2 determination yields FMQ−0.5 (Friel and Ulmer 1974), in close agreement with the carbonated kimberlites of Premier and Benfontein (Fig. 5.3). A second determination by Treiman and Essene (1984), based on the eutectic assemblage periclase–dolomite–calcite, in association with apatite, forsterite, magnesioferrite, pyrrhotite, and alabandite, lies very close to FMQ at $T=640°C$ and $fO_2=10^{-18.6}$; this is in reasonable agreement with the intrinsic fO_2 determination at FMQ−0.5 (Fig. 5.3(a)).

5.4.5. Silicate and silicate–oxide xenoliths

Figure 5.3(a) incorporates the controversial electrochemical data by Arculus and Delano (1981) on Type 1 depleted spinel lherzolites, which bracket the IW buffer at $T=1100$–1200°C. An example of irreversibility for olivine in a dunite (Sato 1978) is shown in Fig. 5.3(b), with the initial heating curve *above* IW, followed by reduction and cooling *below* IW.

Fluid inclusion data for a veined wehrlite (Lunar Crater Volcanic Field, Nevada) plot close to FMQ for the host, and at FMQ+2.6 for the vein (Bergman and Dubessy 1984). Pressure and temperature are, respectively, 10.3 kbar and 1200°C (Fig. 5.3(b)).

An unusual garnet-bearing clinopyroxenite from

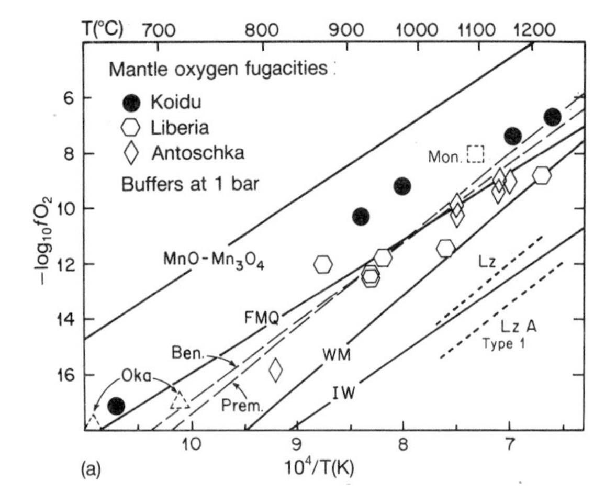

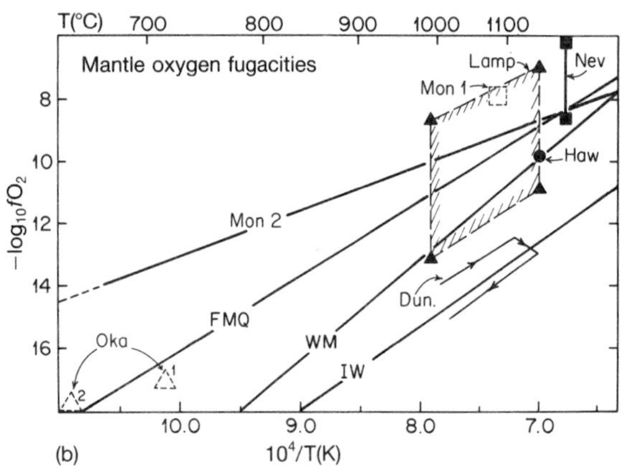

Fig. 5.3. (a and b) Upper mantle-derived oxygen fugacity–temperature data points and reactions with reference to standard buffers defined in the text. Koidu, Liberia, and Antoschka refer to xenocrystic ilmenite–spinel assemblages from kimberlites in west Africa (Haggerty and Tompkins 1983). Prem = Premier kimberlite (Ulmer *et al.* 1976). Ben = Benfontein (McMahon and Haggerty 1984). Mon- = Monastery ilmenite (Loureiro *et al.* 1988). Mon 1 and Mon 2 = Monastery ilmenites (McMahon 1984). Nev = Nevada for fluid inclusions (Bergman and Dubessy 1984). Lamp = lamprophyres and lamproites (Foley 1985; Venturelli *et al.* 1988). Haw = Hawaii for coexisting oxides in clinopyroxene (Sen and Jones 1988). Dun = dunite in Hawaiian alkali basalt, tracking the effect of electrochemical cell auto-reduction (Sato 1978). Lz A is the controversial low fO_2 data by Arculus and Delano (1981) for spinels in depleted Type 1 (also Type A) spinel lherzolites. Oka = Oka carbonatite with one (Friel and Ulmer 1974) and two (Treiman and Essene 1984) determinations.

Hawaii containing lamellar garnet, ilmenite, and a spinel solid-solution member shows a last upper mantle equilibration at $P = 19$ kbar, $T = 1153°C$, and $fO_2 = 10^{-9.7}$ bar, which corresponds to WM at 1 bar (Fig. 5.3(b)) but, corrected to 15 kbar, is close to FMQ (Sen and Jones 1988).

The T and fO_2 field for ilmenite-bearing xenoliths is illustrated in Fig. 5.4, with a large population of data clustering between WM and FMQ and a small number of xenoliths at FMQ + 2 at 700°C (Eggler 1983).

Data by Mattioli and Wood (1986) for Type 1 and Type 2 spinel lherzolites are close to FMQ and at FMQ + 2, respectively (Fig. 5.4). Other spinel lherzolite data by O'Neill and Wall (1987) form an envelope about WM, ranging from WM − 0.5 to WM + 1.8. Relative to FMQ, the range is FMQ − 0.5 to FMQ − 1 (Fig. 5.4). Some spinel data from the literature treated by O'Neill and Wall (1987) have very low or no Fe^{3+} present. These may represent much lower values of fO_2, to at least IW (shown in Fig. 5.4 by T1–Type 1 arrows), but may equally be due to analytical error.

A comprehensive data set by Mattioli *et al.* (1989) is summarized in Fig. 5.5 for xenoliths in alkali basalts from a variety of oceanic and continental rift settings. The range of redox conditions recorded is WM + 0.5 to

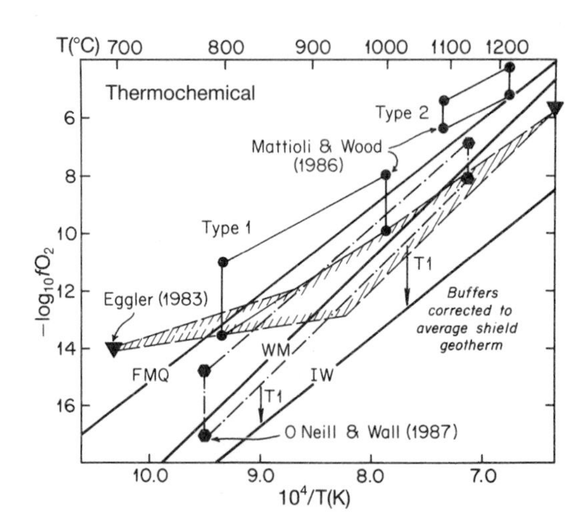

Fig. 5.4. Temperature–oxygen fugacity data for upper mantle xenoliths determined thermochemically from the activities of magnetite in spinel (Mattioli and Wood 1986; O'Neill and Wall 1987) and haematite in ilmenite (Eggler 1983) relative to standard but *P*-corrected buffers. Type 1 and 2 refer to depleted and fertile spinel lherzolites, respectively. The T1 arrows refer to Fe^{3+}-absent spinels in Type 1 spinel lherzolites that may be more reduced than the envelope defined by O'Neill and Wall (1987).

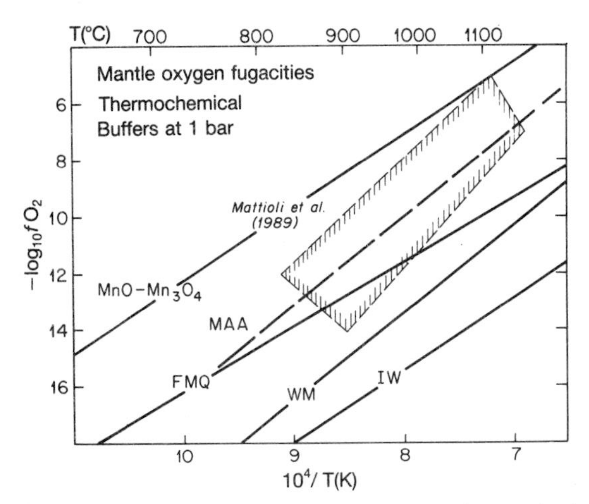

Fig. 5.5. Oxygen fugacity envelope relative to the standard buffers for a screened set of xenolith data, some of which are considered to have undergone metasomatism. MAA refers to an averaging of the data and is defined as the Mantle Average Array (Mattioli *et al.* 1989). Method employed is the Mattioli and Wood (1986) spinel oxygen geobarometer.

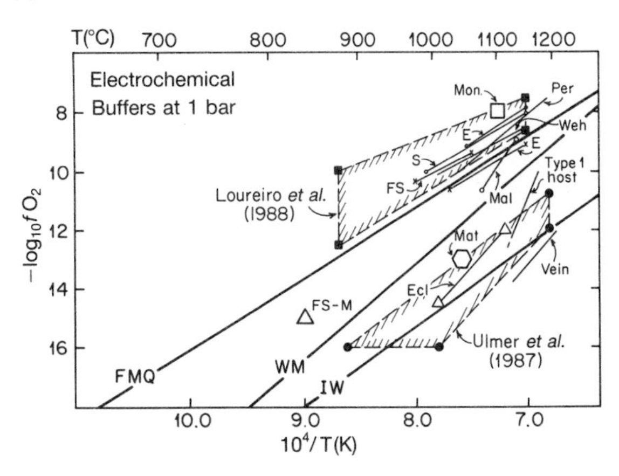

Fig. 5.6. Oxygen-fugacity data for upper mantle xenoliths determined by the electrochemical intrinsic oxygen fugacity method, in relation to the standard buffers. Ecl = a graphite- and diamond-bearing eclogite from Roberts Victor, and Mat = a garnet–orthopyroxene–olivine xenolith from Matsoku, Lesotho (Ulmer *et al.* 1987). FS–M = an olivine–garnet–clinopyroxene xenolith from Frank Smith (McMahon 1984). Vein and host refer to a Type 1 spinel lherzolite, Weh = wehrlite, Per = peridotite, and E, S, FS, and Mal, respectively, refer to ilmenite megacrysts from the Excelsior, Sekameng, and Frank Smith kimberlites and Malaita alnoite (Arculus *et al.* 1984). The data envelope by Ulmer *et al.* 1987) is for ilmenites from kimberlites. Mon = a Monastery ilmenite electrochemical data point emphasized within the ilmenite data envelope by Loureiro *et al.* (1988) to illustrate the reliability of the method for ilmenites; the value is virtually coincident with the Monastery ilmenite determination by McMahon (1984) who used the thermal gas equilibration method (Mon 1 on Fig. 5.3(b)).

FMQ+4; this includes samples interpreted to have undergone metasomatism by subduction. A *Mantle Array Average* (MAA) is proposed that approaches FMQ at ~750°C and is at FMQ+3 at ~1100°C (Fig. 5.5). While the MAA is appropriate to the data set, it is not relevant to the upper mantle at large because the Mattioli and Wood (1986) technique cannot be applied to high Cr content spinels, and, therefore, to highly depleted and, by implication, reduced peridotites.

A variety of electrochemical data are shown in Fig. 5.6 for xenoliths from kimberlites and alkali basalts. The xenoliths from kimberlites include a diamond- and graphite-bearing eclogite from Roberts Victor at IW to IW+1, and a garnet–orthopyroxene–olivine xenolith from Matsoku at IW+1.2, equivalent to WM−1.4 (Ulmer *et al.* 1987); also included is an olivine–garnet–clinopyroxene xenolith from the Frank Smith kimberlite pipe at WM+1.4 (McMahon 1984). The xenoliths from alkali basalts include a peridotite–wehrlite pair at ~FMQ+1, and a Type 1 spinel lherzolite that lies between IW and WM with an associated diopside-rich vein at IW−0.2 to IW−1 (Arculus *et al.* 1984).

5.4.6. Ilmenite megacrysts and xenocrysts

Ilmenite is ideally suited to redox determinations because the mineral is sensitive to minor fluctuations in fO_2, expressed as variations in Fe_2O_3 (haematite solid solution) content, or by characteristic decomposition assemblages that include spinel, and hence variations in magnetite ($FeO \cdot Fe_2O_3$) activity.

Haggerty and Tompkins (1983) reported on ilmenite–spinel pairs from kimberlites in west Africa and showed that equilibration was at either WM–FMQ or at FMQ+1 to FMQ+2 (Fig. 5.3(a)). Correcting the respective buffers to 30 kbar, and to the temperature of an average cratonic geothermal gradient at a depth equivalent to 30 kbar (Haggerty and Tompkins 1984), all data are bracketed between IW and FMQ (Fig. 5.7).

Megacryst ilmenite data from kimberlite pipes in South Africa, Lesotho, and Namibia, and from an alnoite in Malaita, determined by the electrochemical method (Arculus *et al.* 1984), lie close to or above FMQ, at FMQ+1 to FMQ+1.5 (Fig. 5.5). A second set of

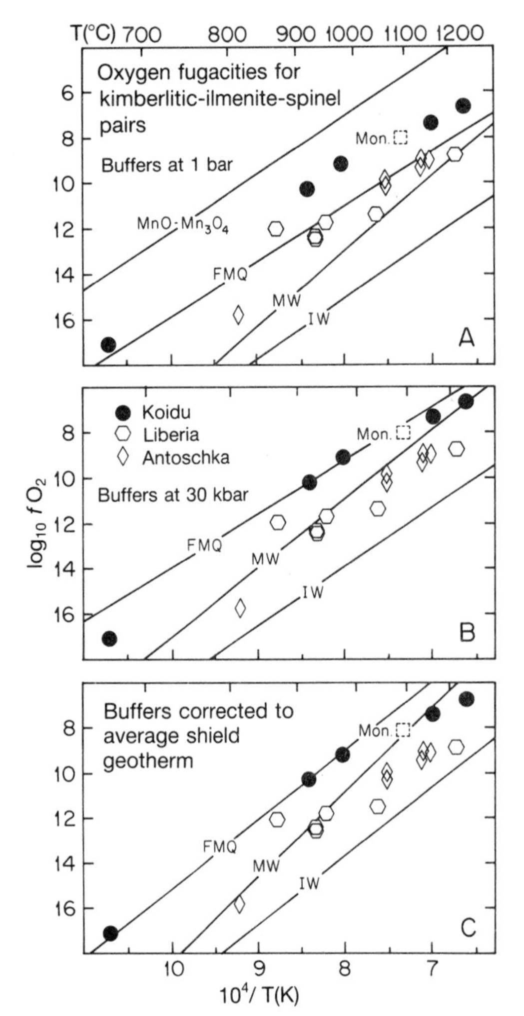

Fig. 5.7. Selected data from Fig. 5.3(a) plotted to illustrate the effect of pressure and temperature on the standard buffers at 30 kbar assuming a typical geothermal gradient for a shield with surface heat flow of 40 mW m^{-2}. Increasing P results in increased fO_2.

electrochemical data by Loureiro *et al.* (1988) includes ilmenites from the Monastery kimberlite, an alnoite in Malaita, and ilmenites from basanites in Algeria and Niger. The data encompass those determined by Arculus *et al.* (1984) but extend to approximately FMQ+2.5 at $T<900°C$, relative to the 1 bar buffer. The two data sets are in very close agreement. The electrochemical data for Monastery (Fig. 5.6) agree well with data by McMahon (1984), shown in Fig. 5.3(b), where the method was by gas-mixing equilibration. In contrast, a third, 'high confidence', electrochemical data set on ilmenites (Ulmer *et al.* 1987) lies at IW±1 (Fig. 5.6). A close correspondence to the first two of these data sets is shown by the data from Algeria as reported by Haggerty *et al.* (1985) using co-equilibrated oxide pairs that yield FMQ−0.25 (open circles in Fig. 5.7).

Temperature and fO_2 determinations for ilmenite megacrysts in melilitites that have reacted to form diopside+perovskite are at about FMQ−1 (Haggerty *et al.* 1985) and are very similar to megacrysts in basanites from Algeria (Fig. 5.8). Note that these data correspond closely to the carbonated kimberlites at Premier and Benfontein (Fig. 5.8). Also shown in Fig. 5.8 are two perovskite-forming reactions (Carmichael and Nicholls 1967):

Perv 1 is diopside + forsterite + ulvöspinel + O_2 = magnetite + perovskite + enstatite;

Perv 2 is akermanite + ulvöspinel + O_2 = magnetite + perovskite + diopside.

Perv 2 is similar to WM and is possibly the lower fO_2 limit for perovskite-forming reactions involving titaniferous spinels in melilitites.

Figure 5.9 summarizes the result of 1140 electron microbeam analyses of upper mantle-derived ilmenites (Haggerty 1989*a*) plotted in the ternary $FeTiO_3$–$MgTiO_3$–Fe_2O_3, and superimposed on the phase and fO_2 grid of Woermann *et al.* (1970) at 1300°C and 1 bar

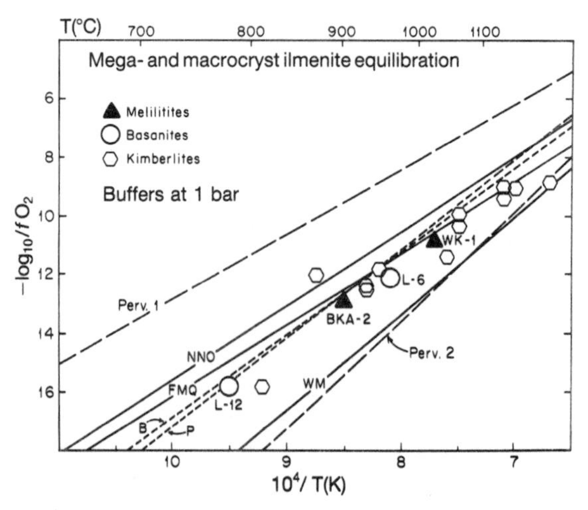

Fig. 5.8. Temperature–oxygen fugacity data for megacrystic ilmenite in melilitites and basanites (Haggerty *et al.* 1985), with reference to the standard buffers, to data for the carbonated Premier (P) and Benfontein (B) kimberlites, and ilmenite macrocysts from west African Kimberlites (see Fig. 5.3).

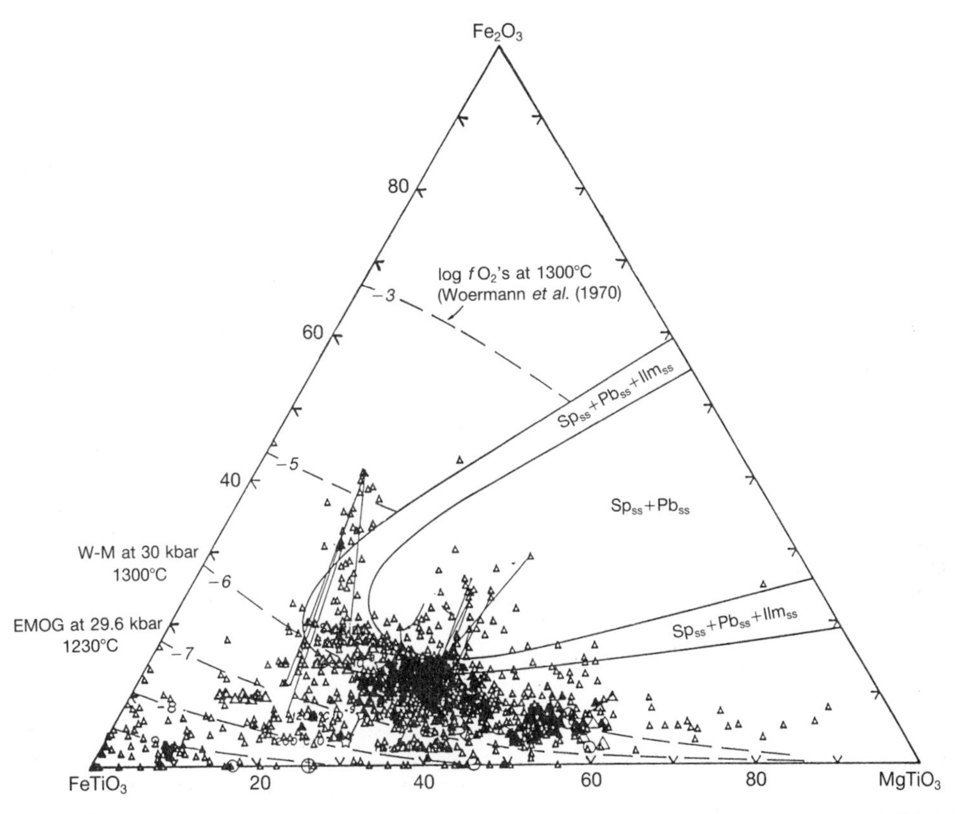

Fig. 5.9. Upper mantle-derived ilmenite compositions (1140 analyses) in the ternary ilmenite ($FeTiO_3$)–haematite (Fe_2O_3)–geikielite ($MgTiO_3$), illustrating a clustering of the data at fO_2 values of 10^{-6} and 10^{-7} bar; these fO_2 values are equivalent to the WM buffer at 30 kbar and 1300°C and to the EMOG buffer at 1230°C and 29.6kbar. Oxygen fugacity contours and the decomposition loops are from Woermann *et al.* (1970). Sp = spinel, Pb = pseudobrookite, and Ilm = ilmenite.

total pressure. The greatest clustering of data lie along the fO_2 contours of 10^{-6} and 10^{-7} bar, which is equivalent to WM at 30 kbar and 1300°C and to EMOG at 29.6 kbar and 1230°C.

These data are in accord with the ilmenite megacryst data in Figs 5.3, 5.4, and 5.6–5.8 and demonstrate unequivocally that a restricted range of fO_2 conditions are recorded in these samples on a globe-wide basis.

5.4.7. Diamond inclusions

Metallic iron and moissanite (SiC), inferred Cr^{2+} in olivine, spinel, and garnet, coupled with zero to extremely low contents of Fe^{3+} in oxides and silicates, and the abundance of sulphides in diamond inclusions are testament to the reduced state of ultramafic suite diamonds originating in subcratonic lithospheres (Haggerty 1986). This view is substantiated by data on spinel inclusions in diamonds (Fig. 5.10) from southern Africa (Daniels and Gurney 1990) and Australia (Jaques *et al.* 1990) using the O'Neill and Wall (1987) spinel oxygen geobarometer. With a single exception, occluded spinels are highly reduced and are bracketed by IW and WM (Fig. 5.10).

5.4.8. Diamond oxidation

Resorption and dissolution of diamond surfaces are recorded in all diamond-bearing kimberlites and lamproites, which fact has been interpreted as being due to oxidation on ascent from source regions in the upper mantle to emplacement in the crust (e.g. Robinson 1978). An alternative view is that resorption takes place within the source regions, and that local highly oxidized conditions result in diamond combustion (Haggerty 1986).

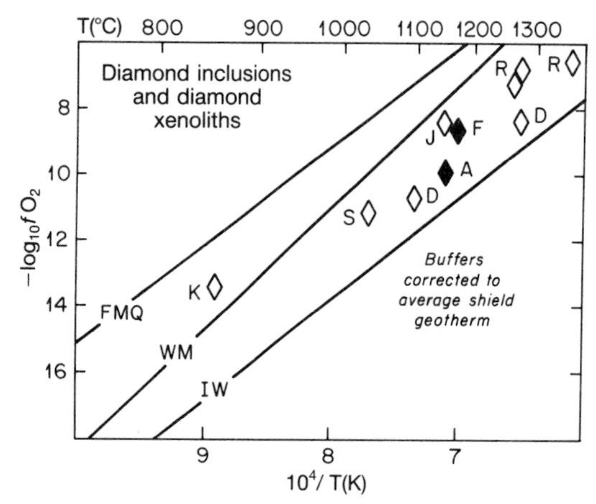

Fig. 5.10. Temperature and oxygen fugacity estimates for spinels included in diamond (open symbols) and for diamond-bearing xenoliths (closed symbols), determined by the O'Neill and Wall (1987) method at 40 kbar. Data for Argyle (A) in Australia are from Jaques *et al.* (1990), and values for Dokolowayo (D), Finsch (F), Jwaneng (J), Koffiefontein (K), Roberts Victor (R), and Star (S), in southern Africa are from Daniels and Gurney (1990).

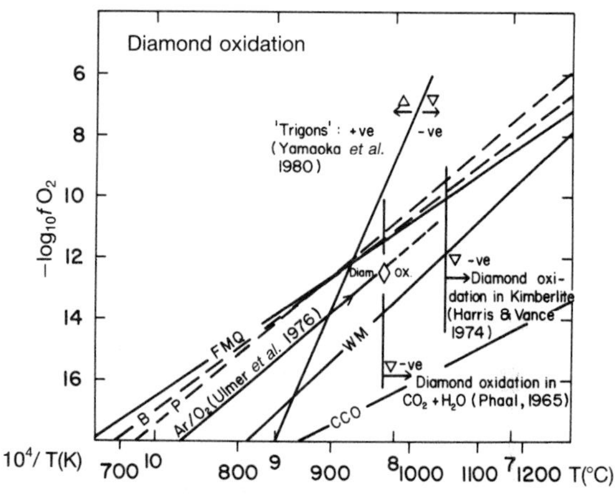

Fig. 5.11. Temperature–oxygen fugacity data for diamond oxidation experiments at 1 bar, with reference to the standard buffers and the Premier (P) and Benfontein (B) carbonated kimberlites. Trigons are etch pits, defined as positive and negative respectively for apex-up and apex-down geometries.

Experimental data on diamond oxidation at 1 bar total pressure are summarized in Fig. 5.11, along with CCO and FMQ reference buffers and equilibrium T–fO_2 curves for the carbonated Premier and Benfontein kimberlites. The range of experiments include gas mixtures of $Ar+O_2$ (Ulmer *et al.* 1976), CO_2+H_2O (Phaal 1965), CO_2+H (Yamaoka *et al.* 1980), and diamond oxidation in a kimberlite (Harris and Vance 1974). The data selected do not include amorphous carbon (or graphite) studies, but instead are restricted, with the exception of the result by Ulmer *et al.* (1976), to investigations in which trigon (triangular etch pit) formation is documented. Trigons are described as either positive (apex up) or negative (apex down).

Diamond oxidation and negative trigon formation are recorded at $T>950$°C, but positive and negative trigons are T and fO_2 dependent (Yamaoka *et al.* 1980) with a slope much steeper than the standard buffers. These data indeed demonstrate that diamond trigons may form in kimberlite, and that diamonds would be oxidized in the Premier and Benfontein intrusions. That oxidation is initiated at about FMQ − 1 (Ulmer *et al.* 1976) and at conditions more reducing than the Premier and Benfontein kimberlites (Fig. 5.9), which are both highly carbonated and, therefore, highly oxidized, means that the source region in which diamonds formed was at least FMQ − 1. This value of fO_2 is bracketed by FMQ and WM, a range recorded by most ilmenite megacrysts (Figs 5.3, 5.4, and 5.6–5.8). In other words, diamonds would be oxidized and would not survive in source regions of ilmenite megacryst formation. This is consistent with the extremely rare occurrence of ilmenite as a primary diamond inclusion (Meyer 1985), and with the conclusion that ultramafic suite diamonds crystallize in the deeper regions of depleted and reduced lithosphere rather than in oxidized and fertile asthenosphere (Haggerty 1986).

5.4.9. Summary

Five methods have been applied to obtain the fO_2 data sets presented. Glass and ilmenite megacryst fO_2 determinations are based on Fe^{2+}/Fe^{3+}, and this ratio is used also in the thermodynamic determinations of fO_2 in ilmenite- and spinel-bearing xenoliths. The remaining three methods are electrochemical, equilibration of coexisting mineral oxides, and thermal gas equilibration.

Oceanic settings record crystalline basalt fO_2 values that are, in general, more oxidized than those recorded by basaltic glasses. Continental basalts are bracketed by WM and NNO. Kimberlites and carbonatites are close to FMQ, whereas lamprophyres are more oxidized and lamproites are more reduced. Type 1 depleted spinel lherzolites are more reduced than Type 2 geochemically fertile spinel lherzolites. Ilmenite-bearing xenoliths and ilmenite megacrysts from kimberlites are WM–FMQ. Ilmenite megacrysts in basanites and melilitites are FMQ. Some silicate xenoliths in kimberlites and alkali basalts, determined exclusively by the electrochemical intrinsic oxygen fugacity method, are consistently WM–IW, or record values more reducing at IW − 1. The results from diamond inclusions and of diamond oxidation experiments imply a source region more reducing than that recorded by ilmenite megacrysts or by ilmenite-silicate xenoliths, consistent with the low fO_2 range (IW to IW + 1) recorded for diamond–graphite–bearing eclogites.

5.5. CONTINENTAL UPPER MANTLE REDOX MODEL

A starting point in developing a redox model for the upper mantle is to consider how the upper mantle reached its present configuration, particularly in relation to the crust, and specifically in the context of the differentiation of the upper mantle into lithosphere and asthenosphere. Of particular interest is the process of craton formation (e.g. Pollack 1986; see also Anderson, Chapter 1, this volume), and the constraints placed on craton evolution by the ages of diamonds and diamond inclusions (Richardson *et al.* 1984; Richardson, Chapter 3, this volume). Rapid growth of continental crust in the early evolution of the Earth, and the existence of stabilized cratonic lithospheres to depths of ~200 km is becoming an increasingly popular model. Some estimates for continental lithospheric thicknesses are considerably greater than 200 km (Jordan 1988), but the evidence from xenolith suites in kimberlites (Finnerty and Boyd 1987) places maximum P and T no higher than ~50 kbar and ~1000°C, respectively—equivalent to ~180 km.

Continental, subcratonic lithosphere is geochemically depleted (see reviews in Nixon *et al.* 1981; Nixon 1987). This has resulted in large measure from extraction of basalt and related rocks and formation of the crust. Geochemical depletion is coupled with devolatilization and iron extraction. The resulting upper mantle lithosphere, therefore, has a lower density, a lower overall temperature, and is anhydrous and brittle, relative to the deeper asthenosphere. Anhydrous conditions and a decrease in the activity of Fe translate into a lower overall oxidation state for the lithosphere relative to the asthenosphere.

This overall condition is, however, an end-member and ideal case. There is substantial evidence for early lateral accretion of continental crust (e.g. Hoffman 1988), and possibly associated foundering, subduction, and recycling. Under these conditions, the mobile asthenosphere is unlikely to maintain its pristine character; the more rigid lithosphere is likely to be less affected, but is equally unlikely to escape the effects of metasomatism (Menzies and Hawkesworth 1987*a,b*; Haggerty 1989*b*). Depleted and reduced lithospheric rocks become highly enriched in large-ion lithophile (LIL) and high field strength (HFS) elements and enrichment is accompanied by the formation of hydrous phlogopite and amphibole and by enhanced oxidation.

Metasomatism in the deeper portions (> 100 km depth) of the lithosphere is restricted to the formation of phlogopite and ilmenite, but at shallower depths (< 100 km) the assemblage is phlogopite + amphibole + clinopyroxene + Ba–K–Sr–Cr–Zr–Nb titanates (Haggerty 1983, 1987, 1989*a*; Erlank *et al.* 1987). Intense metasomatism is restricted to depths of about 75 to 100 km, and may be understood in terms of rising melts that freeze on intersection with the C–H–O peridotite solidus (Fig. 5.12), liberating dissolved fluids and gases that interact with surrounding depleted wall rocks (mostly commonly harzburgite) of the lithosphere (Eggler 1987; Wyllie 1987; Haggerty 1989*a,b*). Metasomatism is considered to be laterally extensive, and metasome horizons at about 50–75 km and at 75–100 km have been proposed (Haggerty 1989*a,b*). As illustrated in Fig. 5.12, the upper metasomatic horizon is Na- and carbonate-enriched, whereas the lower metasome is K- and H_2O-enriched (Haggerty 1989*b*), consistent with the models of Wyllie (1987) and Eggler (1987). The lower metasome is constrained by oxide and silicate mineral stabilities (see summaries by Haggerty 1983, 1987), and both metasomes are more oxidized than the surrounding and depleted lithospheric harzburgite. Based on spinel compositions, redox states are estimated at ~NNO for the upper horizon and ~WM for the lower one.

The evidence for a relatively reduced (IW–WM) and depleted lithosphere at 150–200 km is supported by the ultramafic suite of diamond inclusions, as outlined above (see also Richardson, Chapter 3, this volume). However, trapped inclusions in *coated* eclogitic diamonds are enriched in H_2O, CO_3^{-2}, SiO_2, K_2O, CaO, and FeO, Ba, Sr, La, and Ce (Navon *et al.* 1988). These enriched concentrations resemble potassic melts with distinct affinities to metasomatic harzburgites typical of the lower metasome (Fig. 5.12). The coats on octahedral diamonds are diamond cubes, which implies that secondary solid-state crystallization and entrapment was at lower temperatures than initial octahedral diamond nucleation. Metastable growth of diamond in the upper mantle under fluid-rich conditions adds a new dimension to the redox equilibria of carbon species,

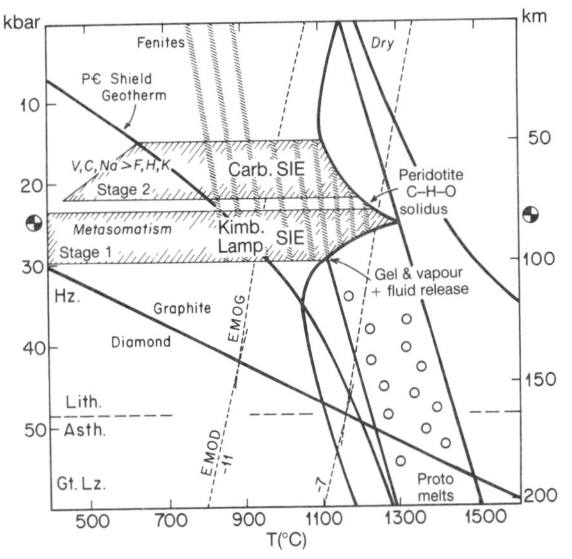

Fig. 5.12. *P–T* schematic diagram for the development of metasomes in the interval of 50 to 100 km in subcratonic upper mantle lithosphere. Proto-melts originating in the asthenosphere (Asth.) from garnet lherzolite (Gt. Lz.) may gel on intersection with the peridotite C–O–H solidus. Crystallization and fluid and vapour release takes place, metasomatizing previously depleted lithosphere (Lith.) which is composed dominantly of harzburgite (Hz). Two metasomes are proposed: a deeper horizon from which kimberlite and lamproite silicate-incompatible-element (SIE) signatures are attained; and a shallower horizon characterized by calcite. In the latter, vapour (V) + CO_2 (C) + Na_2O (Na) is greater than fluid (F) + H_2O (H) + K_2O (K). Stage 1 and Stage 2 refer to fractional distillation. The 'dry' peridotite solidus is labelled (on the right of the diagram), and a geotherm for a Precambrian shield is shown. Fenite development from melting of the carbonate metasome is illustrated at low pressures. The near vertical dashed lines labelled -7, -11, and EMOD–EMOG are oxygen fugacity estimates for the upper mantle (Eggler and Baker 1982). Refined fO_2 estimates are given in Fig. 5.13.

clearly deserving of further investigation. This is underscored by garnet-encapsulated diamond in metasomatized biotite gneisses (Sobolev and Shatsky 1987, 1990), and by the metastable growth of synthetic diamond on substrates at high temperatures and low ambient pressures (e.g. Angus and Hayman 1988).

The asthenosphere in subcratonic regions is dominated by fertile *garnet* lherzolite and, in adjacent thinner segments of lithosphere (rifts or accreted terrains), by fertile *spinel* lherzolite (see also Menzies, Chapter 4, this volume). In the models by Haggerty and Tompkins (1983) and Haggerty (1986), oxidized asthenosphere is defined as WM–FMQ. Megacryst ilmenite is fertile and is derived, along with other discrete mineral xenoliths (garnet, pyroxene, and olivine), from the asthenosphere. As emphasized from the T and fO_2 data sets in Figs 5.3, 5.4, and 5.6–5.8, ilmenite megacrysts and ilmenite–silicate xenoliths uniformly record WM–FMQ conditions of oxidation. The conclusion that the asthenosphere is more oxidized than the lithosphere is supported by low and high fO_2 values recorded respectively in Type 1 (depleted) and Type 2 (fertile) spinel lherzolites entrained in alkali basalts (Figs 5.3–5.5). In addition, garnets from low-temperature -generally depleted) lherzolites have lower $Fe^{3+}/(Fe^{3+}+Fe^{2+})$ ratios and are more reduced than garnets from fertile high-temperature lherzolites (Luth *et al.* 1988). Hence, samples of the asthenosphere record approximately similar redox states regardless of whether the source region is deep (>200 km) and the sampling media is kimberlite, or relatively shallow (<60 km) and sampled by alkali basalt.

The EMOG and EMOD fO_2 profiles illustrated in Fig. 5.12 are only schematically applicable. A more realistic distribution of fO_2 redox states for a typical subcratonic lithosphere is shown in Fig. 5.13 which is constrained by the numerous data sets presented in Figs 5.3–5.11. The highest state of oxidation is reached in the lithospheric metasome at ~50–100 km; the most reduced state is recorded in the region of ultramafic diamond formation at ~150–180 km. The abrupt and sharp contacts in fO_2 are more reasonably gradational, but are drawn to emphasize the lowest fO_2 condition recorded in samples thus far analysed. A second and deep metasome horizon (Wyllie 1989) is shown at the lithosphere–asthenosphere boundary (LAB); this zone is considered to be ($K+Ti+H_2O$)-enriched and oxidized (Haggerty 1989*a*,*b*). Diamond crystallization and growth in the lower lithosphere is illustrated by reduction of CO or CO_2 and/or oxidation of CH_4 from gases in melts derived from the underlying and fertile asthenosphere (Haggerty 1986). Melt-ponding and volatile release at the LAB are considered to be virtually continuous but particularly active at times of thermally-enhanced plume activity (indicated in Fig. 5.13 as COHNS plume), possibly related to plate tectonism (White and McKenzie 1989) and deep upper mantle eruptions (Crough *et al.* 1980; Le Roex 1986).

Cratonic lower crust is dominantly mafic garnet–granulite (Taylor and McLennan 1985; Dawson *et al.* 1986). Redox states recorded by lower crustal xenoliths from west Africa are WM to IW − 1, with metallic iron in

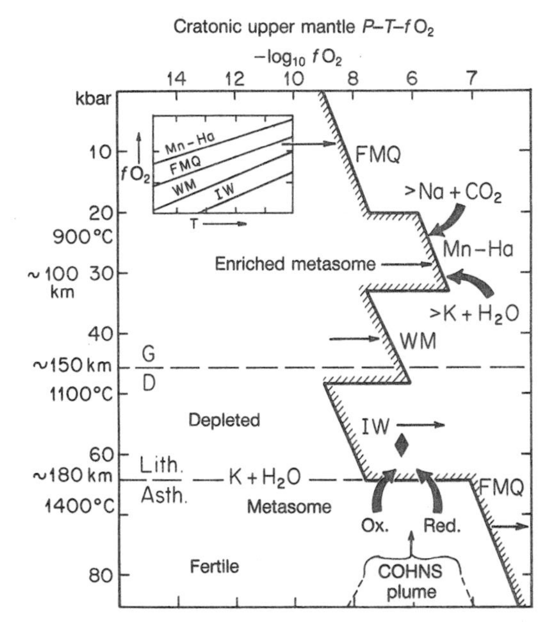

Fig. 5.13. Redox states in the lithosphere (Lith) and asthenosphere (Asth) as a function of T, P, and fO_2. The inset fO_2–T diagram serves to reference the fO_2 mantle profile by clockwise rotation through 90°. The arrows and inner or outer hatching represent minimum fO_2 estimates. Stratigraphic fO_2 values are more realistically gradational than abrupt. The lowest fO_2 values are recorded at the base of the lithosphere (IW); maximum fO_2 values are recorded in the enriched metasome (see Fig. 5.12). A deeper metasome at the lithosphere–asthenosphere boundary (see Fig. 5.14) is estimated at FMQ. Plume-activated metasomatism and eruption is invoked with an influx of carbon, oxygen, hydrogen, nitrogen, and sulphur (COHNS) gas species dissolved in asthenospherically-derived melts. Diamond formation is by either oxidation (OX.) of CH_4, for example, or reduction (RED.) of CO or CO_2, for example. G–D is the graphite–diamond stability curve; $K+H_2O$ and $Na+CO_2$ are characteristic of enrichment in the metasomes in association with Ba, Sr, Ti, Fe, LREE, NB, and Zr.

decomposed ilmenite and andradite (Haggerty and Toft 1985). Sapphirine-bearing granulites in kimberlites from South Africa are also highly reduced (Dawson and Smith 1987), whereas outcropping granulites are MW to FMQ−1 (Lamb and Valley 1984). The spectrum of oxidation (<IW to ~FMQ) is comparable to the lower portions of the lithosphere and contrasts with the upper crust which is at MH (magnetite–haematite) for granitic and related rocks (Haggerty 1976).

Summarizing all of the results above, Fig. 5.14 shows redox states in a model of a typical Archaean diamondiferous craton (stabilized by ~3 billion years ago) flanked by accreted mobile belts (Proterozoic) and palaeo-rift or active (post-Mesozoic) rift systems. Crustal segments thicken from rifts to the core of the craton, and the depleted lithospheric portion of the uppermost mantle thickens proportionately.

Metasomes merge to a maximum thickness at rift zones, in response to metasomatic fluid release from asthenospheric-derived melts underplated on to the crust and on to the lithosphere–asthenosphere boundary. A third barrier that induces volatile release from melts (see also Watson *et al.*, Chapter 6, this volume) is the impedance created by the protruding thermal maxima in the C–O–H peridotite solidus (Fig. 5.12). Rift zones give rise to alkali (Na+K) magmas that include carbonatites, melilitites, and basanites with fO_2 values of ~FMQ (Fig. 5.7). These magmas reflect the net accumulation of metasomites, from which the liquids are largely derived. Towards the craton interior the metasome is dominated by enriched, but previously depleted, harzburgite (Erlank *et al.* 1987) and MARID (mica, amphibole, rutile, ilmenite, diopside) melts (Dawson and Smith 1977; Waters and Erlank 1988).

The deepest portion of the lithosphere is at ~IW and is the source of diamonds of ultramafic affinity with octahedral diamonds forming at greater depths than cube-shaped diamonds (Fig. 5.14). Note that the diamond–graphite stability curve intersects the inferred lower limit of the lithospheric metasome, a region that is the likely source of the volatile and micro-inclusion-saturated cube diamond coatings described by Navon *et al.* (1988).

There are several additional important points in Fig. 5.14. The first is the inferred and continuous metasome at the LAB (Wyllie 1989; Haggerty 1989*a*,*b*); the second is the location and source region of conduits giving rise to alkalic, carbonatitic, kimberlitic, and lamproitic magmas, all with redox states at about FMQ, derived by melt mixing of oxidized asthenosphere with depleted and metasomatized lithosphere; and the third is that the redox state of lithospheric rifts is inferred to be more oxidized than lithospheres of accreted terrains (Proterozoic and Palaeozoic), all of which are substantially more oxidized than Archaean keels.

The complex redox figure unfolding for the subcontinental lithosphere should apply to some extent to oceanic upper mantle. Data in Fig. 5.1 for ocean floor rocks hint at contrasting redox states in the basalt source region; some are oxidized and others are potentially reduced. The evidence is circumstantial, but oceanic rift, plume-activated, and arc environments should show a substantially wider range in redox states

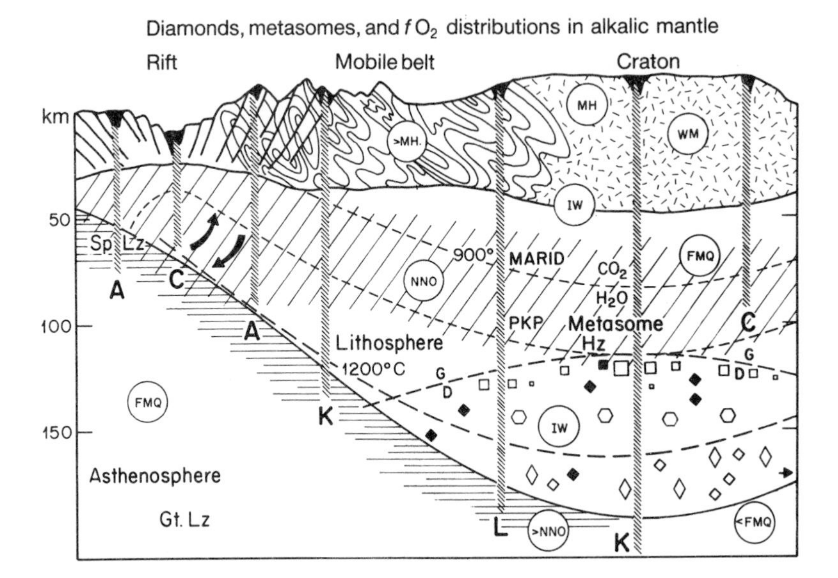

Fig. 5.14. Schematic cross-section of continental lithosphere from craton to mobile belt and rift zone. Possible diatreme paths for carbonatites (C), kimberlites (K), lamproites (L), and other alkali-rich (A) melts are shown originating in either the asthenosphere or the lithosphere. Metasomes (diagonal hatching) are divided into a shallower, CO_2-rich horizon, and a deeper, H_2O-rich horizon, but the interface is gradational. Curved arrows on lithospheric isotherms indicate possible temperature modifications during active and quiescent phases of activity in the asthenosphere. Note the thickening of metasomes in rift zones inferred from the cumulative effects of solidus intersection, underplating on the lower crust, and underplating at the lithosphere–asthenosphere boundary. Symbols below the graphite (G)–diamond (D) stability curve are indicative of diamond morphologies as a function of T (cubes are lower T than dodecahedra which are at lower T than octahedra). Sp Lz, spinel lherzolite; Gt Lz, garnet lherzolite; Hz, harzburgite; MARID, mica, amphibole, rutile, ilmenite, diopside; PKP, phlogopite, K-richterite peridotite. Circles enclose estimates of oxidation states. From Haggerty (1989*b*).

than is presently inferred. Redox recorders cannot be treated in isolation from geological setting, and upper mantle probes, such as basalts, require critical evaluation. Modified redox by subduction recyling is obvious. Less obvious are the redox effects at mid-ocean ridges. For example, if rifting is tectonically passive and magmas are generated by volatile-poor decompression melting of previously depleted asthenosphere, as might be the case for the Atlantic and Indian Oceans, then mantle source regions ought to be reduced. But is all ocean-rift melting dry and passive and equally depleted? The answer is probably no. Equally, what is the redox state of plume-activated volcanism and what are the factors governing fO_2?

Collectively and with these unanswered questions in mind, it is evident that neither a single nor an average redox state for the entire upper mantle can either be assigned or assumed, as illustrated schematically in Fig. 5.15. This is borne out by the convincing demonstration that mantle geochemical domains exist (e.g. Menzies 1989), by models that call upon excessively deep subduction (to the core–mantle boundary e.g. Silver *et al.* 1988), or geochemical data that imply significant plume-enrichment from sources in the lower mantle as summarized by Menzies and Hawkesworth (1987*a*).

5.6. CONCLUSIONS

Core formation in early Earth's history depleted the lower mantle in Fe and Ni. Crustal extraction was neither as complete nor as effective in fractionation—thus the presence of a lithosphere and an asthenosphere in the upper mantle. Geochemical depletion, volatile extraction, and diminishing iron activity created moderately reduced conditions in the continental lithosphere and the formation of lithospheric diamonds at >3 billion years ago. This inferred pristine condition, however, has evolved with time because the lithosphere has been extensively modified by melts from a reservoir of moderately fertile asthenosphere. These melts were

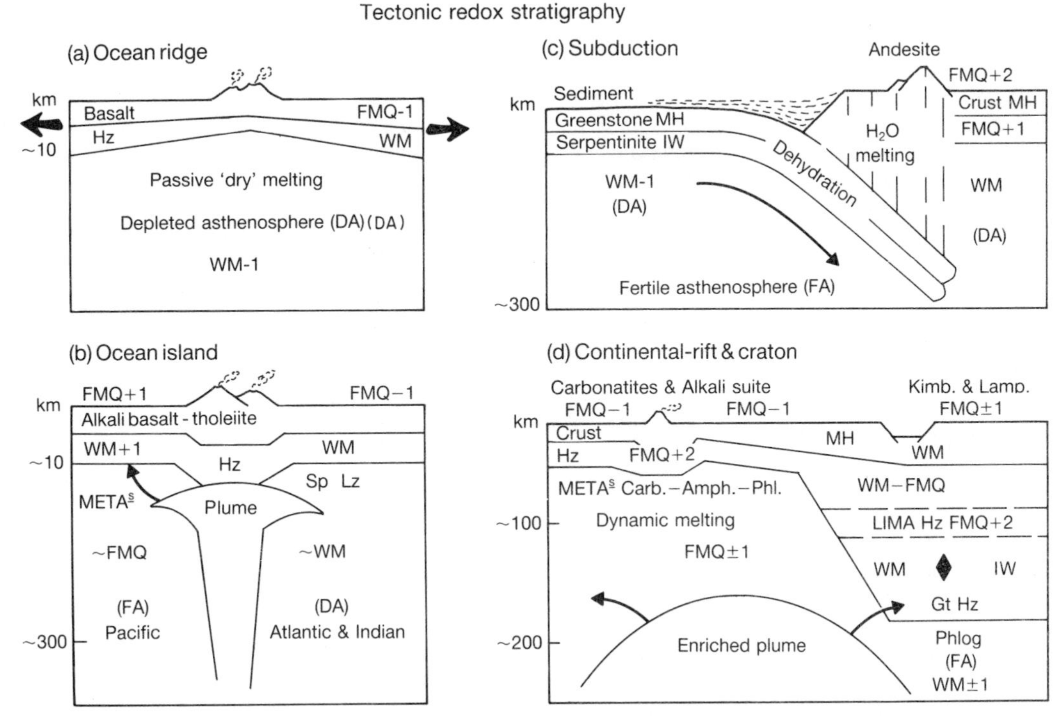

Fig. 5.15. Schematic cross-sections and contrasting fO_2 estimates for crusts, lithospheres, and asthenospheres in ocean ridge, ocean island, subduction, and continental rift and craton settings. Contrasting redox states are invoked for asthenospheres that have undergone previous depletion (Atlantic type), relative to moderately fertile asthenospheres (Indian and Pacific types). Passive rifting (a) and 'dry' melting produces an oceanic lithosphere of WM; plume-enrichment (b) in oceanic islands results in WM + 1; subducted oceanic lithosphere (c) with underlying depleted asthenosphere (DA) and deeper fertile asthenosphere (FA), coupled with dehydration and melting of the downgoing slab results in a complex but highly oxidized marginal continental lithosphere. Panel (d) is redrawn from Fig. 5.14 but illustrates the effect of plume activity, enrichment, and dynamic melting on the evolution of fO_2 in deep continental keels and subjacent rift zones. Metasomatism is pervasive throughout and elsewhere. Abbreviations in panel (d) are: phlog = phlogopite; LIMA = lindsleyite and mathiasite which are Ba- and K-specific chromian titanates characteristic of the lithospheric metasome; rift metasomatism is characteristically carbonate (carb.)-, amphibole (amph.)-, and phlogopite-enriched. Kimb. and Lamp. refer to kimberlites and lamproites.

impeded by thermal trauma, volatiles were released, and metasomatism and enrichment of previously depleted lithosphere took place. The resulting continental lithosphere thus contains only remnants of its ancient redox condition.

Overwhelming evidence for the latter-day redox state of the upper mantle favours fO_2 heterogeneity with a complexly stratified cratonic continental lithosphere and a moderately oxidized asthenosphere, the latter at WM–FMQ. The former ranges from IW to WM in the Archaean lithospheric keel (180–200 km), is modified to FMQ–NNO in metasome horizons (60–100 km), is at WM–FMQ at shallower stratigraphic depths, and has a possible reversion to WM–IW at the crust–mantle interface. Cratonic lower crust is at WM–IW, and appears, in general, to grade progressively through WM in the middle crust to MH in the upper crust.

There is a remarkably uniform redox state recorded in oceanic and continental basalts, namely WM–FMQ–NNO. Volatile degassing and wall–rock interaction of erupted melts is inevitable but redox conditions in the source regions of basalt genesis remains equivocal. Redox states should deviate; indeed this is inevitable given that any recorded fO_2 is the redox state of *melting*, not the intrinsic redox state of the mantle source. Passive decompression ridge melting of previously depleted asthenosphere should differ markedly from the thermally-enhanced plume-activated melting of fertile analogues. Furthermore, some departure in mantle redox states is expected, given the high solubilities of

H_2O and H_2, the moderate solubility of CO_2 in basaltic melts, and recognizing that $O \gg H \gg C$ because the upper mantle is dominated by oxygen-bearing minerals and volatiles; and basalt extraction is pervasive but not uniform. Notwithstanding the high buffering capacity of carbon, graphite is extremely rare, diamond is restricted, and both phases appear to derive by the disproportionation of C–O–H-rich fluids or melts.

Asthenospheric-derived xenoliths record redox conditions of WM–FMQ, metasomites are >NNO, and alkali-rich melts approach FMQ. This averaging is a reflection of mixing, in contrast to basalts for which the envelope around FMQ is the result of buffering imposed by the solubility limits of H_2O, H, and CO_2.

Moderation in the debate on redox states is in order given that the upper mantle has evolved with time and redox states are non-uniform. From physical and chemical constraints and from differences between and among continental and oceanic lithospheres and asthenospheres, homogeneous conditions of fO_2 should not, indeed, be expected. Although there is a convergence among data from different fO_2 determinative techniques, additional experimental research is required, specifically using the electrochemical method. In evaluating new data, greater emphasis must be given to the geological setting and source regions of xenoliths. In fact, much of the dispute surrounding redox states originates from the indiscriminant grouping of xenoliths of diverse origin; recognition of metasomatism and depletion are particularly important. Yet another substantial issue is the contribution of volatiles from lithospheric recycling into the asthenosphere; some contribution is expected, but the ultimate source of volatiles is considered to be the deeper upper mantle. Gas and fluid speciation, revised models for diamond genesis, and the redox condition of metasomites and plume-activated volcanism are fertile areas for future research that will enrich our understanding of the depleted continental lithosphere.

The tectonic approach adopted by Haggerty and Tompkins (1983) and followed in this chapter for lithospheric oxidation states is also attempted by Wood *et al.* (1990). Most of our conclusions are similar, but we differ on the matter of *ancient* and *depleted* subcratonic lithosphere.

ACKNOWLEDGEMENTS

Research was supported by the US National Science Foundation under grant EAR89-05046 which is gratefully acknowledged. Discussions over the past years with Moto Sato, Dick Arculus, John Delano, Dave Eggler, Gene Ulmer, Peter Wyllie, Ed Woermann, Hugh O'Neill, Bernie Wood, Glen Mattioli and many others on redox state have been most enlightening. I express my appreciation to these people and to Paul Toft, Marie Litterer, and Alice Bishko for assistance with manuscript preparation.

REFERENCES

Andersen, D. J. and Lindsley, D. H. (1988). Internally consistent solution models for Fe–Mg–Mn–Ti oxides: Fe–Ti oxides. *American Mineralogist*, **73**, 714–26.

Andersen, T., O'Reilly, S. Y., and Griffin, W. L. (1984). The trapped fluid phase in upper mantle xenoliths from Victoria, Australia: implications for mantle metasomatism. *Contributions to Mineralogy and Petrology*, **88**, 72–85.

Anderson, A. T. and Wright, T. L. (1972). Phenocrysts and glass inclusions and their bearing on oxidation and mixing of basaltic magmas, Kilauea volcano, Hawaii. *American Mineralogist*, **57**, 188–216.

Angus, J. C. and Hayman, C. C. (1988). Low-pressure, metastable growth of diamond and "diamondlike" phases. *Science*, **241**, 913–21.

Arculus, R. J. (1985). Oxidation status of the mantle: past and present. *Annual Reviews of Earth and Planetary Science*, **13**, 75–95.

Arculus, R. J. and Delano, J. W. (1980). Implications for the primitive atmosphere of the oxidation state of earth's upper mantle. *Nature*, **288**, 72–4.

Arculus, R. J. and Delano, J. W. (1981). Intrinsic oxygen fugacity measurements: techniques and results for spinels from upper mantle peridotites and megacryst assemblages. *Geochimica et Cosmochimica Acta*, **45**, 899–913.

Arculus, R. J. and Delano, J. W. (1987). Oxidation state of the upper mantle: present conditions, evolution and controls. In *Mantle xenoliths* (ed. P. H. Nixon), pp. 589–98. John Wiley and Sons, New York.

Arculus, R. J., Dawson, J. B., Mitchell, R. H., Gust, D. A., and Holmes, R. D. (1984). Oxidation states of the upper mantle recorded by megacryst ilmenite in kimberlite and type A and B spinel lherzolites. *Contributions to Mineralogy and Petrology*, **85**, 85–94.

Basaltic Volcanism Study Project (1981). *Basaltic volcanism on the terrestrial planets*. Pergamon, New York.

Bergmann, S. C. and Dubessy, J. (1984). CO_2–CO fluid inclusions in a composite peridotite xenolith: implications for upper mantle oxygen fugacity. *Contributions to Mineralogy and Petrology*, **85**, 1–13.

Boyd, F. R. and Finnerty, A. A. (1980). Conditions of origin of natural diamonds of peridotite affinity. *Journal of Geophysical Research*, **85**, 6911–18.

Brett, R., Huebner, J. S., and Sato, M. (1977). Measured oxygen fugacities of the Angra Dos Reis achondrite as a function of temperature. *Earth and Planetary Science Letters*, **35**, 363–8.

Brey, G. P. and Green, D. H. (1977). Systematic study of phase relations in olivine melilitite + H_2O + CO_2 at high pressures and petrogenesis of an olivine melilitite magma. *Contributions to Mineralogy and Petrology*, **61**, 141–62.

Buddington, A. F. and Lindsley, D. H. (1964). Iron–titanium oxide minerals and synthetic equivalents. *Journal of Petrology*, **5**, 310–57.

Carmichael, I. S. E. and Ghiorso, M. S. (1986). Oxidation–reduction relations in basic magma: a case for homogeneous equilibria. *Earth and Planetary Science Letters*, **78**, 200–10.

Carmichael, I. S. E. and Nicholls, J. (1967). Iron–titanium oxides and oxygen fugacities in volcanic rocks. *Journal of Geophysical Research*, **72**, 4665–87.

Chou, I. M. (1978). Calibration of oxygen buffers at elevated P and T using the hydrogen fugacity sensor. *American Mineralogist*, **63**, 690–703.

Christie, D. M., Carmichael, I. S. E., and Langmuir, C. H. (1986). Oxidation states of mid-ocean ridge basalt glasses. *Earth and Planetary Science Letters*, **79**, 397–411.

Crough, S. T., Morgan, W. J., and Hargraves, R. B. (1980). Kimberlites: their relationship to mantle hotspots. *Earth and Planetary Science Letters*, **50**, 260–74.

Daniels, L. R. M. and Gurney, J. J. (1990). Oxygen fugacity constraints on the southern African lithosphere. *Contributions to Mineralogy and Petrology* (in press).

Dawson, J. B. and Smith, J. V. (1977). The MARID (mica–amphibole–rutile–ilmenite–diopside) suite of xenoliths in kimberlite. *Geochimica et Cosmochimica Acta*, **41**, 309–23.

Dawson, J. B. and Smith, J. V. (1987). Reduced sapphire granulite xenoliths from the Lace Kimberlite, South Africa; implications for the deep structure of the Kaapvaal craton. *Contributions to Mineralogy and Petrology*, **95**, 376–83.

Dawson, J. B., Carswell, D. A., Hall, J., and Wedepohl, K. H. (eds) (1986). *The nature of the lower continental crust*. Geological Society Special Publication No. 24. Blackwell, London.

Deines, P. (1980). The carbon isotopic composition of diamond: relationship to diamond shape, color, occurrence and vapor composition. *Geochimica et Cosmochimica Acta*, **44**, 943–61.

Dingwell, D. B. and Virgo, D. (1987). The effect of oxidation state on the viscosity of melts in the system Na_2O–FeO–Fe_2O_3–SiO_2. *Geochimica et Cosmochimica Acta*, **51**, 195–205.

Duba, A. and Nicholls, I. A. (1973). The influence of oxidation state on the electrical conductivity of olivine. *Earth and Planetary Science Letters*, **18**, 59–64.

Duba, A., Heard, H. C., and Schock, R. N. (1974). Electrical conductivity of olivine at high pressure and under controlled oxygen fugacity. *Journal of Geophysical Research*, **79**, 1667–73.

Eggler, D. H. (1983). Upper mantle oxidation state: evidence from olivine–orthopyroxene–ilmenite assemblages. *Geophysical Research Letters*, **10**, 365–8.

Eggler, D. H. (1987). Solubility of major and trace elements in mantle metasomatic fluids: experimental constraints. In *Mantle metasomatism* (ed. M. A. Menzies and C. J. Hawkesworth), pp. 21–41. Academic Press, London.

Eggler, D. H. and Baker, D. R. (1982). Reduced volatiles in the system C–O–H: implications to mantle melting, fluid formation, and diamond genesis. In *High-pressure research in geophysics*. Advances in Earth and Planetary Sciences, Vol. 12 (ed. S. Akimoto and M. H. Manghnani), pp. 237–50. Center for Academic Publications, Tokyo.

Elliott, W. C., Grandstaff, D. E., Ulmer, G. C., Buntin, T., and Gold, D. P. (1982). An intrinsic oxygen fugacity study of platinum–carbon associations in layered intrusions. *Economic Geology*, **77**, 209–226.

Erlank, A. J., Waters, F. G., Hawkesworth, C. J., Haggerty, S. E., Allsopp, H. L., Rickard, R. S., and Menzies, M. (1987). Evidence for mantle metasomatism in peridotite nodules from the Kimberly pipes, South Africa. In *Mantle metasomatism* (ed. M. A. Menzies and C. J. Hawkesworth), pp. 221–311. Academic Press, London.

Eugster, H. P. and Wones, D. R. (1962). Stability relations of the ferruginous biotite, annite. *Journal of Petrology*, **3**, 81–125.

Finnerty, A. A. and Boyd, F. R. (1987). Thermobarometry for garnet peridotites: a basis for the determination of the

thermal and compositional structure of the upper mantle. In *Mantle xenoliths* (ed. P. H. Nixon), pp. 381–402. John Wiley and Sons, New York.
Foley, S. F. (1985). The oxidation state of lamproitic magmas. *Tschermaks Mineralogische und Petrographische Mitteilungen*, **34**, 217–38.
Foley, S. F. (1988). The genesis of continental basic alkaline magmas—an interpretation in terms of redox melting. *Journal of Petrology*. Special Volume on the Lithosphere, 139–61.
Freund, F., Kathrein, H., Wengler, H., Knobel, R., and Heinen, H. J. (1980). Carbon in solid solution in forsterite—a key to the untractable nature of reduced carbon in terrestrial and cosmogenic rocks. *Geochimica et Cosmochimica Acta*, **44**, 1319–33.
Friel, J. J. and Ulmer, G. C. (1974). Oxygen fugacity geothermometry of the Oka carbonatite. *American Mineralogist*, **59**, 314–18.
Gerlach, T. M. (1979). Evaluation and restoration of the 1970 volcanic gas analyses from Mt. Etna, Sicily. *Journal of Volcanology and Geothermal Research*, **6**, 165–78.
Gerlach, T. M. (1982). Interpretation of volcanic gas data from tholeiitic mafic lavas. *Bulletin Volcanologique*, **45**, 235–44.
Gerlach, T. M. and Nordlie, B. E. (1975*a*). The C–O–H–S gaseous system, part I: composition limits and trends in basaltic gases. *American Journal of Science*, **275**, 353–76.
Gerlach, T. M. and Nordlie, B. E. (1975*b*). The C–O–H–S gaseous system, part II: temperature, atomic composition, and molecular equilibria in volcanic gases. *American Journal of Science*, **275**, 377–94.
Giardini, A. A. and Melton, C. E. (1975). Gases released from natural and synthetic diamonds by crushing under high vacuum at 200°C, and their significance to diamond genesis. *Fortschritte der Mineralogie*, **52**, 455–64.
Green, D. H., Falloon, T. J., and Taylor, W. R. (1987). Mantle-derived magmas—roles of variable source peridotite and variable C–H–O fluid compositions. In *Magmatic processes: physicochemical principles*, Geochemical Society Special Publication No. 1 (ed. B. O. Mysen), pp. 139–54. Geochemical Society, Pennsylvania.
Green, H. W., II (1985). Coupled exsolution of fluid and spinel from olivine: evidence for O^- in the mantle? In *Point defects in minerals*, Geophysical Monograph No. 31 (ed. R. N. Schock), pp. 226–32. American Geophysical Union, Washington, DC.
Haggerty, S. E. (1976). Opaque mineral oxides in terrestrial igneous rocks. In *Oxide minerals*, Reviews of Mineralogy, Vol. 3 (ed. D. Rumble, III), pp. Hg1–Hg100. Mineralogical Society of America, Washington, DC.
Haggerty, S. E. (1978). The redox state of planetary basalts. *Geophysical Research Letters*, **5**, 443–6.
Haggerty, S. E. (1983). The mineral chemistry of new titanites from the Jagersfontein kimberlite, South Africa: implications for metasomatism in the upper mantle. *Geochimica et Cosmochimica Acta*, **47**, 1833–54.
Haggerty, S. E. (1986). Diamond genesis in a multiply-constrained model. *Nature*, **320**, 34–38.
Haggerty, S. E. (1987). Metasomatic mineral titanates in upper mantle xenoliths. In *Mantle xenoliths* (ed. P. H. Nixon), pp. 671–90. John Wiley and Sons, New York.
Haggerty, S. E. (1989*a*). Upper mantle opaque mineral stratigraphy and the genesis of metasomites and alkali-rich melts. *Journal of the Geological Society of Australia*, Special Publication No. 14, 687–99.
Haggerty, S. E. (1989*b*). Mantle metasomes and the kinship between carbonatites and kimberlites. In *Carbonatite—genesis and evolution* (ed. K. Bell), pp. 546–60. Allen and Unwin, London.
Haggerty, S. E. and Toft, P. B. (1985). Native iron in the continental lower crust: petrological and geophysical implications. *Science*, **229**, 647–9.
Haggerty, S. E. and Tompkins, L. A. (1983). Redox state of earth's upper mantle from kimberlitic ilmenites. *Nature*, **303**, 295–300.
Haggerty, S. E. and Tompkins, L. A. (1984). Subsolidus reactions in kimberlitic ilmenites: exsolution, reduction and the redox state of the mantle. In *Kimberlites I: kimberlites and related rocks* (ed. J. Kornprobst), pp. 335–57. Elsevier, Amsterdam.
Haggerty, S. E., Moore, A. E., and Erlank, A. J. (1985). Macrocryst Fe–Ti oxides in olivine melilitites from Namaqualand–Bushmanland, South Africa. *Contributions to Mineralogy and Petrology*, **91**, 163–70.
Harris, J. W. and Vance, E. R. (1974). Studies of the reaction between diamond and heated kimberlite. *Contributions to Mineralogy and Petrology*, **47**, 237–44.
Hervig, R. L., Smith, J. V., Steele, I. M., Gurney, J. J., Meyer, H. O. A., and Harris, J. W. (1980). Diamonds: minor elements in silicate inclusions: pressure–temperature implications. *Journal of Geophysical Research*, **85**, 6919–29.
Hoffman, P. F. (1988). United plates of America, the birth of a craton: early proterozoic assembly and growth of Laurentia. *Annual Reviews of Earth and Planetary Science*, **16**, 543–603.

Holloway, J. R. and Jakobsson, S. (1986). Volatile solubilities in magmas: transport of volatiles from mantles to planet surfaces. *Journal of Geophysical Research*, **91**, 505–8.

Huebner, J. S. and Sato, M. (1970). The oxygen-fugacity temperature relationships of manganese oxide and nickel oxide buffers. *American Mineralogist*, **55**, 934–52.

Jaques, A. L., O'Neill, H. St. C., Smith, C. B., Moon, J., and Chappell, B. W. (1990). Diamondiferous peridotite xenoliths from the Argyle (AK1) lamproite pipe, Western Australia. *Contributions to Mineralogy and Petrology*, **104**, 255–76.

Jordan, T. H. (1988). Structure and formation of the continental lithosphere. *Journal of Petrology*. Special Volume on the Lithosphere, 11–37.

Kilinc, A., Carmichael, I. S. E., Rivers, M. L., and Sack, R. O. (1983). Ferric–ferrous ratio of natural silicate liquids equilibrated in air. *Contributions to Mineralogy and Petrology*, **83**, 136–40.

Lamb, W. and Valley, J. W. (1984). Metamorphism of reduced granulites in low-CO_2 vapour-free environment. *Nature*, **312**, 56–8.

LeRoex, A. P. (1986). Geochemical correlation between southern African kimberlites and South Atlantic hotspots. *Nature*, **324**, 243–5.

Loureiro, D., Delano, J. W., Leblanc, M., Dautria, J.-M., Gurney, J. J., and Nixon, P. H. (1990). Electrochemical determinations of the oxygen fugacity in equilibrium with mantle-derived ilmenite megacrysts at 1 bar. *Contributions to Mineralogy and Petrology* (submitted).

Luth, R. W. and Boettcher, A. L. (1986). Hydrogen and the melting of silicates. *American Mineralogist*, **71**, 264–76.

Luth, R. W., Virgo, D., Boyd, F. R., and Wood, B. J. (1988). Iron in mantle-derived garnets: valence and structural state. *Annual Report of the Director of the Geophysical Laboratory, Carnegie Institution*, Washington, pp. 13–18. Carnegie Institution, Washington, DC.

Marx, P. C. (1972). Pyrrhotine and the origin of terrestrial diamonds. *Mineralogical Magazine*, **38**, 636–8.

Mathez, E. A. (1984). Influence of degassing on oxidation states of basaltic magmas. *Nature*, **310**, 371–5.

Mathez, E. A. and Delaney, J. R. (1981). The nature and distribution of carbon in submarine basalts and peridotite nodules. *Earth and Planetary Science Letters*, **56**, 217–32.

Matson, D. W., Muenow, D. W., and Garcia, M. O. (1984). Volatiles in amphiboles from xenoliths, Vulcan's Throne, Grand Canyon, USA. *Geochimica et Cosmochimica Acta*, **48**, 1629–36.

Mattioli, G. S. and Wood, B. J. (1986). Upper mantle oxygen fugacity recorded by spinel-lherzolites. *Nature*, **332**, 626–8.

Mattioli, G. S., Wood, B. J., and Carmichael, I. S. E. (1987). Ternary-spinel volumes in the system $MgAl_2O_4$–Fe_3O_4–$\gamma Fe_{8/3}O_4$: implications for the effect of P on intrinsic fO_2 measurements of mantle-xenolith spinels. *American Mineralogist*, **72**, 468–80.

Mattioli, G. S., Baker, M. B., Ritter, M. J., and Stolper, E. M. (1989). Upper mantle oxygen fugacity and its relationship to metasomatism. *Journal of Geology*, **97**, 521–36.

McMahon, B. M. (1984). Petrologic redox equilibria in the Benfontein sills and in the Allende meteorite and the T–fO_2 stability of kimberlitic ilmenite from the Monastery diatreme. D.Phil. dissertation, University of Massachusetts, Amherst.

McMahon, B. M. and Haggerty, S. E. (1984). The Benfontein kimberlite sills: magmatic reactions and high intrusion temperatures. *American Journal of Science*, **284**, 839–941.

Melton, C. E., Salotti, C. A., and Giardini, A. A. (1972). The observation of nitrogen, water, carbon dioxide, methane, and argon as impurities in natural diamonds. *American Mineralogist*, **57**, 1518–23.

Menzies, M. A. (1989). Cratonic, circumcratonic and oceanic mantle domains beneath the western United States. *Journal of Geophysical Research*, **94**, 7899–915.

Menzies, M. A. and Hawkesworth, C. J. (1987*a*). Upper mantle processes and composition. In *Mantle xenoliths* (ed. P. Nixon), pp. 725–38. John Wiley and Sons, New York.

Menzies, M. A. and Hawkesworth, C. J. (ed.) (1987*b*). *Mantle metasomatism*. Academic Press, London.

Meyer, H. O. A. (1985). Genesis of diamond: a mantle saga. *American Mineralogist*, **70**, 344–55.

Meyer, H. O. A. and McCallum, M. (1986). Mineral inclusions in diamonds from the Sloan kimberlites, Colorado. *Journal of Geology*, **94**, 600–12.

Murck, B. W., Burruss, R. C., and Hollister, L. S. (1978). Phase equilibria in fluid inclusions in ultramafic xenoliths. *American Mineralogist*, **63**, 40–6.

Mysen, B. O. and Boettcher, A. L. (1975). Melting of a hydrous mantle, II. Geochemistry of crystals and liquids formed by anatexis of mantle peridotite at high pressures and high temperatures as a function of controlled activities of water, hydrogen and carbon dioxide. *Journal of Petrology*, **16**, 549–90.

Mysen, B. O. and Virgo, D. (1985). Iron-bearing silicate melts: relations between pressure and redox equilibria. *Physics and Chemistry of Minerals*, **12**, 191–200.

Mysen, B. O., Virgo, D., Neumann, E., and Seifert, F. A. (1985). Redox equilibria and the structural states of ferric and ferrous iron in melts in the system $CaO–MgO–Al_2O_3–SiO_2–FeO$: relationship between redox equilibria, melt structure and liquidus phase equilibria. *American Mineralogist*, **70**, 317–31.

Navon, O., Hutcheon, I. D., Rossman, G. R., and Wasserburg, G. J. (1988). Mantle-derived fluids in diamond micro-inclusions. *Nature*, **335**, 784–9.

Nixon, P. H. (ed.) (1987). *Mantle xenoliths*. John Wiley and Sons, New York.

Nixon, P. H., Rogers, N. W., Gibson, I. L., and Grey, A. (1981). Depleted and fertile mantle xenoliths from southern Africa kimberlites. *Annual Reviews of Earth and Planetary Science*, **9**, 285–309.

O'Neill, H. C. and Wall, V. J. (1987). The olivine–orthopyroxene–spinel oxygen geobarometer, the nickle precipitation curve, and the oxygen fugacity of the earth's upper mantle. *Journal of Petrology*, **28**, 1169–91.

Osborne, E. F. (1962). Reaction series for subalkaline igneous rocks based on different oxygen pressure conditions. *American Mineralogist*, **47**, 211–26.

Pasteris, J. D. (1987). Fluid inclusions in mantle xenoliths. In *Mantle xenoliths* (ed. P. H. Nixon), pp. 691–707. John Wiley and Sons, New York.

Pasteris, J. D. and Wanamaker, B. J. (1988). Laser Raman microprobe analysis of experimentally re-equilibrated fluid inclusions in olivine: some implications for mantle fluids. *American Mineralogist*, **73**, 1074–88.

Phaal, C. (1965). Surface studies of diamond. *Industrial Diamond Review*, **25**, 486–9, 591–5.

Pollack, H. N. (1986). Cratonization and thermal evolution of the mantle. *Earth and Planetary Science Letters*, **80**, 175–82.

Richardson, S. H., Gurney, J. J., Erlank, A. J., and Harris, J. W. (1984). Origin of diamonds in old enriched mantle. *Nature*, **310**, 198–202.

Robinson, D. N. (1978). The characteristics of natural diamond and their interpretation. *Mineral Science Engineering*, **10**, 55–72.

Roedder, E. (1965). Liquid CO_2 inclusions in olivine-bearing nodules and phenocrysts from basalts. *American Mineralogist*, **50**, 1746–82.

Roedder, E. (ed.) (1984). *Fluid inclusions*, Reviews in Mineralogy, Vol. 12. Mineralogical Society of America, Washington, DC.

Ryabchikov, I. D. (1983). Oxidation–reduction equilibria in the upper mantle [translation]. *Doklady Akademii Nauk SSR*, **268** (3), 703–6.

Ryabchikov, I. D., Green, D. H., Wall, V. J., and Brey, G. P. (1981). The oxidation state of carbon in the environment of the low velocity zone. *Geokhimiya*, **2**, 221–32.

Ryerson, F. J., Durham, W. B., Cherniak, D. J., and Lanford, W. A. (1989). Oxygen diffusion in olivine: effect of oxygen fugacity and implications for creep. *Journal of Geophysical Research*, **94**, 4105–18.

Sack, R. O., Carmichael, I. S. E., Rivers, M., and Ghiorso, M. S. (1980). The ferric–ferrous equilibria in natural silicate liquids at 1 bar. *Contributions to Mineralogy and Petrology*, **75**, 369–76.

Sato, M. (1970). An electrochemical method of oxygen fugacity control of furnace atmosphere for mineral synthesis. *American Mineralogist*, **55**, 1424–37.

Sato, M. (1972). Intrinsic oxygen fugacities of iron-bearing oxide and silicate minerals under low total pressure. *Geological Society of American Memoirs*, **135**, 289–307.

Sato, M. (1976). Oxygen fugacity and other thermochemical parameters of Apollo 17 high-Ti basalts and their implications on the reduction mechanism. *Proceedings of the Seventh Lunar Science Conference* Supplement 2, Pergamon Press, New York, pp. 1323–44.

Sato, M. (1978). Oxygen fugacity of basaltic magmas and the role of gas-forming elements. *Geophysical Research Letters*, **5**, 447–9.

Sato, M. and Valenza, M. (1980). Oxygen fugacities of the layered series of the Skaergaard intrusion, east Greenland. *American Journal of Sciences*, **280A**, 134–58.

Sato, M. and Wright, T. C. (1966). Oxygen fugacity directly measured in magmatic gases. *Science*, **153**, 1103–5.

Saxena, S. K. (1989). Oxidation state of the mantle. *Geochimica et Cosmochimica Acta*. **53**, 89–95.

Sen, G. and Jones, R. E. (1988). Exsolved silicate and oxide phases from clinopyroxenes in a single Hawaiian xenolith: implications for oxidation state of the Hawaiian upper mantle. *Geology*, **16**, 69–72.

Silver, P. G., Carlson, R. W., and Olson, P. (1988). Deep slabs, geochemical heterogeneity, and the large scale structure of mantle convection: Investigation of an enduring paradox. *Annual Reviews of Earth and Planetary Science*, **16**, 477–541.

Smith, D. (1987). Genesis of carbonate in pyrope from ultramafic diatremes on the Colorado Plateau, Southwestern United States. *Contributions to Mineralogy and Petrology*, **97**, 389–96.

Sobolev, N. V. and Shatsky, V. S. (1987). Inclusions of carbon minerals in garnets of metasomatic rocks. *Geologiya i Geofizika*, **28**, 77–80.

Sobolev, N. V. and Shatsky, V. S. (1990). Diamond inclusions in garnets from metamorphic rocks. *Nature*, **343**, 742–6.

Sobolev, N. V., Efimova, E. S., and Pospelova, L. N. (1981). Native iron in Yakutia diamonds, its paragenesis. *Journal of Geology and Geophysics* (*SSR*), **12**, 25–59.

Spencer, K. J. and Lindsley, D. H. (1981). A solution model for coexisting iron titanium oxides. *American Mineralogist*, **66**, 1189–201.

Taylor, S. R. and McLennan, S. M. (1985). *The continental crust: its composition and evolution*. Blackwell, London.

Taylor, W. R. and Green, D. H. (1987). The petrogenetic role of methane: effect on liquidus phase relations and the solubility mechanism of reduced C-H volatiles. In *Magmatic processes: physicochemical principles*, Geochemical Society Special Publication No. 1 (ed. B. O. Mysen), pp. 121–38. Geochemical Society, Pennsylvania.

Tingle, T. N. and Green, H. W., II (1987). Carbon solubility in olivine: implications for upper mantle evolution. *Geology*, **15**, 324–6.

Treiman, A. H. and Essene, E. J. (1984). A periclase–dolomite–calcite carbonatite from the Oka complex, Quebec, and its calculated volatile composition. *Contributions to Mineralogy and Petrology*, **85**, 149–57.

Ulmer, G. C., Rosenhauer, M., Woermann, E., Ginder, J., Drory-Wolff, A., and Wasilewski, P. (1976). Applicability of electrochemical oxygen fugacity measurements to geothermometry. *American Mineralogist*, **61**, 653–60.

Ulmer, G. C., Grandstaff, D. E., Weiss, D., Moats, M. A., Buntin, T. A., Gold, D. P., Hatton, C. J., Kadik, A., Koseluk, R. A., and Rosenhauer, M. (1987). The mantle redox state; an unfinished story. In *Mantle metasomatism and alkaline magmatism*, Geological Society of America Special Paper No. 215 (ed. E. M. Morris and J. D. Pasteris), pp. 5–23. Geological Society of America, Boulder, Colorado.

Venturelli, G., Mariani, E. S., Foley, S. F., Capedri, S., and Crawford, A. J. (1988). Genesis and conditions of crystallization of Spanish lamproitic rocks. *Canadian Mineralogist*, **26**, 67–79.

Virgo, D., Luth, R. W., Moats, M. A., and Ulmer, G. C. (1988). Constraints on the oxidation state of the mantle: an electrochemical and ^{57}Fe Mossbauer study of mantle-derived ilmenites. *Geochimica et Cosmochimica Acta*, **52**, 1781–94.

Waters, F. G. and Erlank, A. J. (1988). Assessment of the vertical extent and distribution of mantle metasomatism below Kimberley, South Africa. *Journal of Petrology*, Special Lithosphere Issue, 185–204.

Watkins, N. D. and Haggerty, S. E. (1968). Oxidation and magnetic polarity in single Icelandic lavas and dikes. *Geophysical Journal of the Royal Astronomical Society*, **15**, 305–15.

White, R. and McKenzie, D. (1989). Magmatism at rift zones: The generation of volcanic continental margins and flood basalts. *Journal of Geophysical Research*, **94**, 7685–7730.

Wood, B. J., Bryndzia, L. T., and Johnson, K. E. (1990). Mantle oxidation state and its relationship to tectonic environment and fluid speciation. *Science*, **248**, 337–45.

Woermann, E. and Rosenhauer, M. (1985). Fluid phases and the redox state of the earth's mantle. *Fortschritte der Mineralogie*, **63** (2), 263–349.

Woermann, E., Hirschberg, A., and Lamprecht, A. (1970). Das System Hamatit–Ilmenit–Geikielith unter hohen Temperaturen und hohen Drucken. *Fortscchritte der Mineralogie*, **47** (1), I–II, 79–80.

Wones, D. R. and Gilbert, M. C. (1969). The fayalite–magnetite–quartz assemblage between 600° and 800°C. *American Journal of Science*, **267A**, 480–8.

Wyllie, P. J. (1980). The origin of kimberlite. *Journal of Geophysical Research*, **85**, 6902–10.

Wyllie, P. J. (1987). Metasomatism and fluid generation in mantle xenoliths. In *Mantle xenoliths* (ed. P. H. Nixon), pp. 609–21. John Wiley and Sons, New York.

Wyllie, P. J. (1989). *The genesis of kimberlites and some low-SiO_2, high-alkali magmas*. Geological Society of Australia Special Publication No. 14, pp. 603–15. Geological Society of Australia. Blackwell Scientific Publishers, Carlton, Victoria, Australia.

Yamaoka, S., Kanda, H., and Setaka, N. (1980). Etching of diamond octahedrons at high temperatures and pressure with controlled oxygen partial pressure. *Journal of Materials Science*, **15**, 332–6.

6

Distribution of fluids in the continental mantle

E. Bruce Watson, James M. Brenan, and Don R. Baker

6.1. INTRODUCTION

The existence of low-viscosity, non-silicate fluids in the subcontinental mantle is made apparent not only by the chemical (metasomatic) imprint they may leave on upper mantle samples, but also by more direct expressions such as volcanic emissions and inclusions in mantle minerals (see also Menzies, Chapter 4, this volume; Richardson, Chapter 3, this volume). In most cases, the identity, source, and transport mechanism of the fluid responsible for any particular metasomatic signature remains obscure. There does exist, however, a sufficient experimental and theoretical base from which to make some generalizations regarding plausible composition, distribution, and transport mechanism of fluid in a given region of the upper mantle. The purpose of this chapter is first to review briefly the constraints that can be placed on the nature of fluids at various depths (i.e. the large-scale distribution), which is determined by the components available and the governing phase equilibria. We will then move on to consider in more detail the behaviour and spatial distribution of upper mantle fluids on the scale of individual mineral grains in the host rock. This 'small-scale' distribution is affected primarily by what we loosely call the 'wetting behaviour', which ultimately is determined by the relative magnitudes of the mineral–mineral and fluid–mineral interfacial energies in the rock of interest. The last major section of the chapter summarizes the implications of data on wetting behaviour for fluid transport and upper mantle metasomatism.

The only type of fluid deliberately omitted from this discussion is silicate melt. We devote no space to this fluid not because we consider it insignificant as a metasomatic agent, but because its behaviour has been discussed at length elsewhere in the context of *magmatic* processes. The reader will immediately detect a focus upon C–O–H fluids, particularly CO_2–H_2O mixtures, for which the best data base on small-scale distribution exists. Molten carbonate is discussed in some detail because new, heretofore unpublished, wetting data are available and because this fluid has the potential to be a particularly effective metasomatic agent.

6.2. LARGE-SCALE DISTRIBUTION: NATURE OF FLUIDS AT VARIOUS DEPTHS

6.2.1. General nature of upper mantle fluids

Fluids in the upper mantle are probably dominated by the elements hydrogen, carbon, and oxygen with subordinate amounts of sulphur, nitrogen, fluorine, and chlorine (see also Kyser, Chapter 7, this volume). The elemental composition and molecular speciation of subsolidus mantle fluids is controlled both by oxygen fugacity and by crystal–fluid partitioning.

Oxygen fugacity and its control on fluid speciation

The ambient oxygen fugacity in most of the upper mantle is probably buffered at a relatively high value by the coexistence of enstatite, magnesite, olivine, and graphite (EMOG; or EMOD if diamond is the stable polymorph of carbon; see Eggler and Baker 1982; Haggerty, Chapter 5, this volume) although there is evidence that at least some portions of the upper mantle have oxygen fugacities near the iron–wüstite buffer (Arculus 1985; Haggerty, Chapter 5, this volume). Using the modified Redlich–Kwong equation of state (Holloway 1977, with corrections by D. H. Eggler, personal communication), fluid speciation can be calculated along the solidus of a slightly hydrated and carbonated peridotite (Fig. 6.1). Fluids in the C–O–H–N–S system buffered by EMOG are predominantly H_2O and CO_2; for any given total nitrogen content, N_2 is approximately one to two orders of magnitude more abundant than NH_3 and SO_2 is about six orders of magnitude more abundant than H_2S. In these mantle fluids at EMOG-buffered oxygen fugacities there is little H_2 and CH_4 near the peridotite solidus; however, at oxygen fugacities near the iron–wüstite

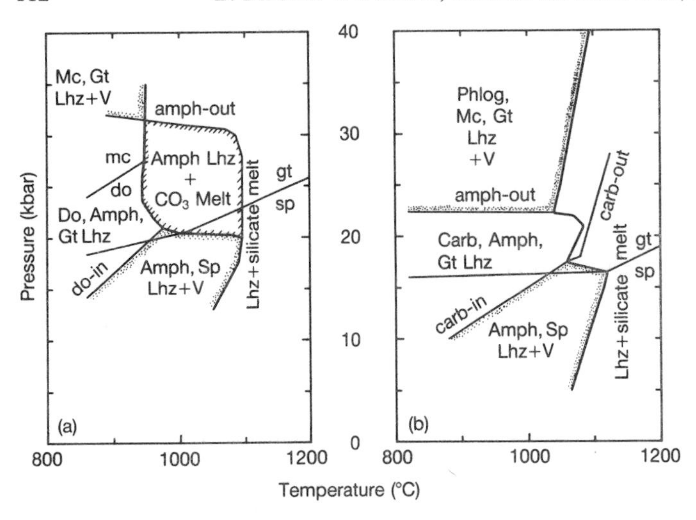

Fig. 6.1. Near-solidus phase relations of two peridotites at pressures 10–40 kbar: (a) after Wallace and Green (1988); and (b) after Olafsson and Eggler (1983). Although these two phase diagrams are broadly similar, attention is drawn to the differences in amphibole and carbonate stability. Of particular interest to mantle metasomatism is the extensive stability field of carbonate melt in (a).

buffer the fluids become CH_4 and H_2O mixtures, dominated by CH_4.

6.2.2. Peridotite phase equilibria and volatile reservoirs in the upper mantle

Water and carbon dioxide

Whether a free fluid phase is present in the upper mantle is determined by the temperature and pressure stability of mineral phases that are suitable hosts for various fluid components. Although olivine, clinopyroxene, and garnet can contain water, the amounts are small, hundreds of p.p.m. at most (Wilkins and Sabine 1973); in the upper mantle the primary mineral reservoirs for H_2O and CO_2 are amphibole and phlogopite for water, and dolomite and magnesite for CO_2. Many phase equilibrium studies on peridotite starting materials have been performed over the past three decades, but most of these addressed the location of the anhydrous and water-saturated solidi. Given as examples for near-solidus peridotite phase relations (Fig. 6.1) are the recent work by Olafsson and Eggler (1983) and by Wallace and Green (1988). In both studies the authors added only small amounts of volatiles, attempting to reproduce natural conditions as closely as possible. Both the pyrolite starting material of Wallace and Green (1988) and the natural lherzolite of Olafsson and Eggler (1983) contained only 0.3 Wt % water and only a few Wt % CO_2. Although the results of the two studies are broadly similar, important differences do exist.

At pressures below the stability of dolomite (see Fig. 6.1 and note the differences in minimum pressure required for dolomite to be stable in (a) and (b)), CO_2 is strongly partitioned into the fluid phase. Water, on the other hand, is strongly partitioned into amphibole. Thus, for fluids in the shallowest upper mantle—10 to 20 kbar or 33 to 66 km—the fluid is CO_2-rich, H_2O-poor (Eggler and Holloway 1977; Wyllie 1978). At 16 to 23 kbar in Fig. 6.1(b) (or 21 to 32 kbar in Fig. 6.1(a)) amphibole *and* a carbonate—dolomite at lower pressures and magnesite at higher pressures—coexist. The combination of amphibole and carbonate can absorb all the H_2O and CO_2 in the system, so it is unlikely that a free vapour phase exists in the upper mantle over this depth range. At pressures above the stability of amphibole (>23 kbar according to Olafsson and Eggler (1983) or >32 kbar according to Wallace and Green (1988)), the stable hydrous phase is phlogopite. Because of the low concentration of K_2O in the upper mantle (hundreds of p.p.m.) little phlogopite can be formed and much of the H_2O is partitioned into the fluid phase which, consequently, becomes water-rich.

The major difference between the studies of Olafsson and Eggler (1983) and Wallace and Green (1988) is the presence of the phase field for lherzolite + carbonate melt between 20 and 32 kbar and 975 and 1100°C in the latter study. This carbonate melt is hydrous and may or may not coexist with a separate water-rich fluid, depending upon the total content of H_2O and CO_2 in the system.

For more reducing conditions (e.g. near iron–wüstite), much less is known about the phase relations of peridotite. As discussed above, the fluids will be almost exclusively CH_4–H_2O mixtures (see also Haggerty,

Chapter 5, this volume). The CH_4/H_2O ratios are difficult to predict even qualitatively, but, because a carbonate is unlikely to form at these low oxygen fugacities (Eggler and Baker 1982), the fluid will probably be CH_4-rich at pressures in the amphibole stability field and poorer in CH_4 in the phlogopite stability field.

Nitrogen, sulphur, fluorine, and chlorine

Compared with the dominant components H_2O and CO_2, little is known about the amounts of these elements in mantle fluids. Nevertheless, a few comments concerning their distribution are in order. Nitrogen is found in upper mantle micas and in diamonds. The micas contain 100 to 200 p.p.m. N (Sutton *et al.* 1990) and diamonds an average of 800 p.p.m. N (Sellschop *et al.* 1980). Sulphide minerals are found in peridotite nodules and diamonds from kimberlites (Sharp 1966; Bishop *et al.* 1975; Haggerty and Tompkins 1982) indicating a finite sulphur fugacity in the upper mantle. Fluorine and chlorine are found in upper-mantle amphiboles, phlogopites, and apatites (Dawson 1980; Delaney *et al.* 1980). Mean fluorine and chlorine contents of the micas studied by Delaney *et al.* (1980) are 0.43 and 0.08 Wt %, respectively, certainly not inconsequential concentrations. The existence of all these elements in minerals requires their presence in any coexisting fluid; unfortunately, however, the concentrations in the fluid cannot be estimated because mineral–fluid partition coefficients for these elements are unknown.

In summary, for EMOG-controlled oxygen fugacities and at temperatures below the formation of carbonatitic melts, fluids in the upper mantle are CO_2-rich at pressures in the amphibole stability field; fluids are probably absent at P, T conditions of both amphibole and carbonate stability (but see Wyllie (1987) for an alternative viewpoint); at pressures above amphibole stability where phlogopite is the stable hydrous phase, the fluids become H_2O-rich due to the limitation on phlogopite abundance imposed by bulk K_2O content. Above 950°C and in the pressure range 20–32 kbar the dominant fluid is a carbonate melt.

Although mantle fluids in general are dominated by H_2O and CO_2, it is expected that they also contain N, NH_3, SO_2, H_2S, F, and Cl in amounts that may be geologically significant. The abundances of these 'impurities' in a given region of the upper mantle are controlled by equilibrium with the solid assemblage (and hence by the large-scale distribution of minerals). Interestingly, these minor components may in turn affect the small-scale distribution of the fluids that contain them. As we shall see later, dissolved halides have a small but possibly important effect on the grain-scale geometry of aqueous fluids in peridotite.

6.3. WETTING CHARACTERISTICS: GRAIN-SCALE DISTRIBUTION AS DICTATED BY INTERFACIAL ENERGY CONSIDERATIONS

6.3.1. Overview of theory and experimental results

General principles

Until recently (Watson and Brenan 1987) little factual information was available concerning the small-scale distribution of fluids in upper mantle rocks, although a common misconception was that any fluid not actually included within mineral grains would form a thin film along all grain boundaries. Because of the long-standing interest among ceramists in porous and/or partially-molten materials, however, the theoretical framework for understanding fluid distribution in rocks has existed for many years. This theory is well represented by the work of Beere (1975), who discused in general terms the principles governing the nature of 'porosity' (either liquid- or gas-filled) in polycrystalline materials approaching textural equilibrium. H. S. Waff and co-workers (Bulau *et al.* 1979; Waff and Bulau 1979, 1982) recognized the significance of the early studies by ceramists and pioneered the application of the same concepts to peridotite in which the fluid was basaltic melt.

The equilibrium distribution of fluid in a rock develops as the rock–fluid system evolves toward a state of minimum interfacial energy, or 'textural equilibrium'. For the present purposes, the most important aspect of the approach toward textural equilibrium is that well-defined angular relationships are established wherever three interfaces (grain boundaries or interphase boundaries) intersect along a line. If the three interfaces all separate the same phase, then the angle formed between any two of them is 120°—a value arrived at by considering each interface to be a membrane under tension ('surface' tension) and balancing the forces about the line of intersection. Introduction of a second phase into the system (in our case, a fluid) complicates matters by creating a new kind of interface, i.e. one separating unlike phases, fluid and mineral. This fluid/mineral interface will have an interfacial 'tension' (energy) different from that of the mineral/mineral boundaries. Consequently, at locations where two crystals and a 'parcel' of fluid come together (i.e. at a pore), the force balance about the line of intersection will not yield equal angles between all three interfaces. The angle between the two intersecting fluid/mineral interfaces, usually called the dihedral or wetting angle and designated θ, is given by

$$\theta = 2 \arccos[\gamma_{mm}/2\gamma_{fm}] \qquad (1)$$

where γ refers to the interfacial energy per unit area of the subscripted interface, either mineral/mineral (mm) or fluid/mineral (fm) (see Fig. 6.2(a)). Two specific values of θ are of special significance in considering the distribution of fluids in rocks. One of these is 0°, which is a critical value for the following reason: If θ drops to zero—a condition that arises if $\gamma_{fm} \leq 0.5\ y_{mm}$ – the fluid will penetrate along the mineral/mineral interface, producing what might be described as a grain-boundary film (see Fig. 6.2(b)). As long as $\gamma_{fm} > 0.5\gamma_{mm}$, θ will assume a positive value in rocks approaching textural equilibrium, and a grain-boundary film will not exist. It will become clear shortly that, for all fluids and minerals investigated thus far, $\gamma_{fm} \gg 0.5\gamma_{mm}$.

The absence of a grain-boundary film does not necessarily preclude the existence of a fluid phase that is continuous in three dimensions (an important condition in considering metasomatic transport), because interconnectivity can be achieved along grains edges even if the surfaces between grains remain dry. As in the case of grain-boundary wetting, it is the value of θ, as a manifestation of the relative magnitudes of γ_{fm} and γ_{mm}, that determines whether grain edges will be penetrated (wet) by any fluid present in the rock.

Figure 6.3 is a schematic representation of the contrasting nature of a rock in which the grain edges are dry and one in which they are penetrated by fluid. Which configuration is thermodynamically stable can be assessed by examining a small segment of grain edge, shown in the middle frame of Fig. 6.3, and determining whether the total interfacial energy embodied in that segment is lower for the dry or the wet case. This assessment can be made for various values of θ without specific knowledge of the interfacial energies, because if values for θ and one γ are assumed, eqn (1) gives the other γ (see Watson 1982). The bottom frame of Fig. 6.3 reveals that the penetration of fluid along grain edges lowers the overall interfacial energy of the system *if* $\theta < 60°$. This angle, then, is the other key θ value (besides 0°) because it constitutes the cut-off criterion for interconnectivity of the fluid phase (for this reason, 60° is commonly referred to as the 'pinch-off' angle). Strictly speaking, the pinch-off angle is 60° only for low fluid fractions in the rock ($< \sim 1$ per cent); see Von Bargen and Waff 1986), but we will proceed with the reasonable assumption that the upper mantle contains only small amounts of non-silicate fluid.

The significance of the 60° pinch-off angle to the behaviour of fluids in the mantle cannot be overemphasized: *the value of* θ *(< or > 60°) is the first-order determinant of whether a fluid phase is continuous in three*

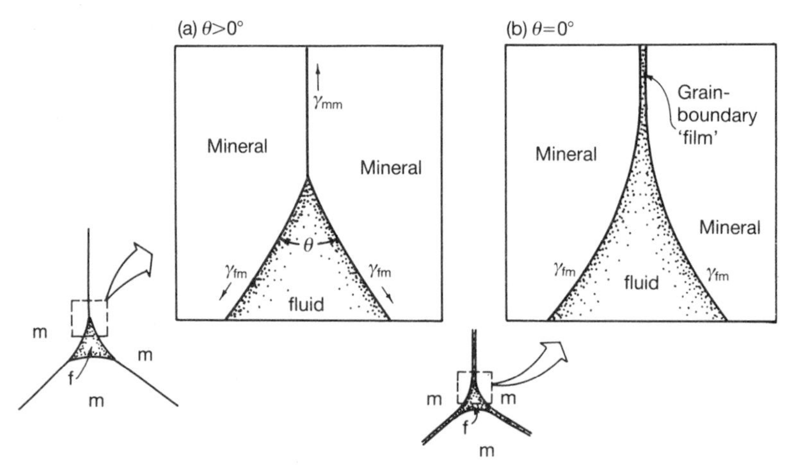

Fig. 6.2. (a) Schematic illustration of the angular relationship between two fluid–mineral (fm) interfaces along their line of intersection with a mineral–mineral (mm) boundary. The wetting angle θ is given by eqn (1), which is a balance of the three interfacial energies (or 'tensions'), designated by γs, about their line of intersection. In (b) is shown the special (and unrealistic) case of $\theta = 0°$, in which the fluid moves in along the mineral–mineral interface to form a grain-boundary 'film'. See text and Fig. 6.3.

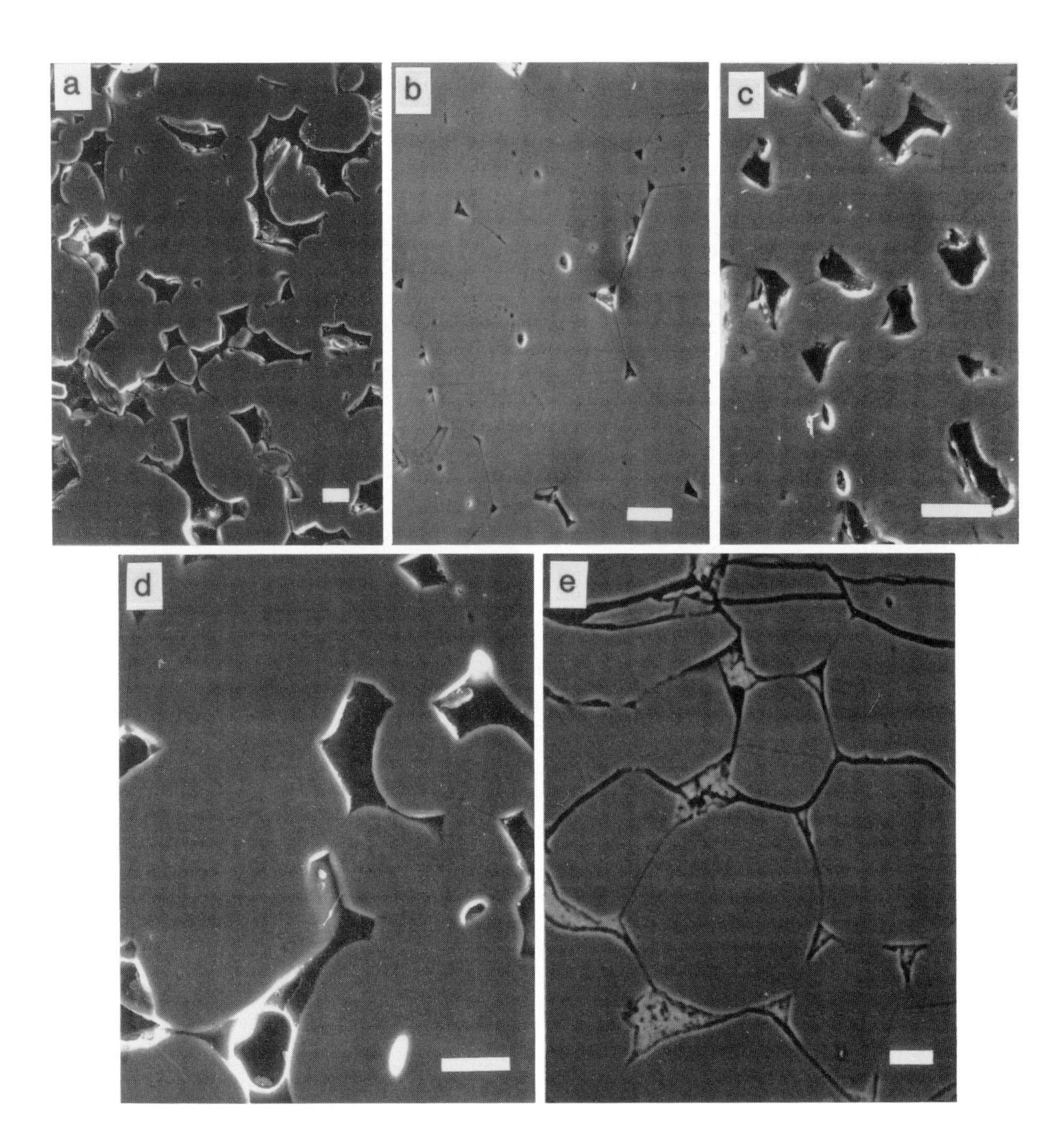

Plate 1. Secondary-electron photomicrographs of porosity features (i.e. fluid distribution) in synthetic dunites: (a) polished surface of sample from run 34, which contained 5.1 Wt % aqueous CaF_2 solution at run conditions, now preserved as void space (black); (b) dunite showing porosity features produced by a more geologically realistic amount of water (0.9 per cent) at 1200°C and 10 kbar; (c) porosity resulting from 1.5 Wt % CO_2-rich fluid (run 24; $X_{CO_2} \sim 0.86$); (d) porosity features (now partially filled with epoxy) produced by 4.9 wt % H_2O (run 14); (e) polished surface of dunite containing triangular patches of quench carbonate melt resulting from the presence of 7.8 Wt % $CaCO_3$ at 30 kbar and 1400°C (run C6). Note that in (a) and (c) most olivine grain boundaries are not discernible; in (d) some boundaries are faintly visible; and in (b) and (e) they are generally apparent. All photos except (b) (and possibly (e)) represent fluid abundances deliberately higher than one would expect in the upper mantle, so the overall textures are somewhat misleading. The significance of the photos lies mainly in the θ values obtained from them (see text for discussion). White scale bars are 10 μm in (a)–(e).

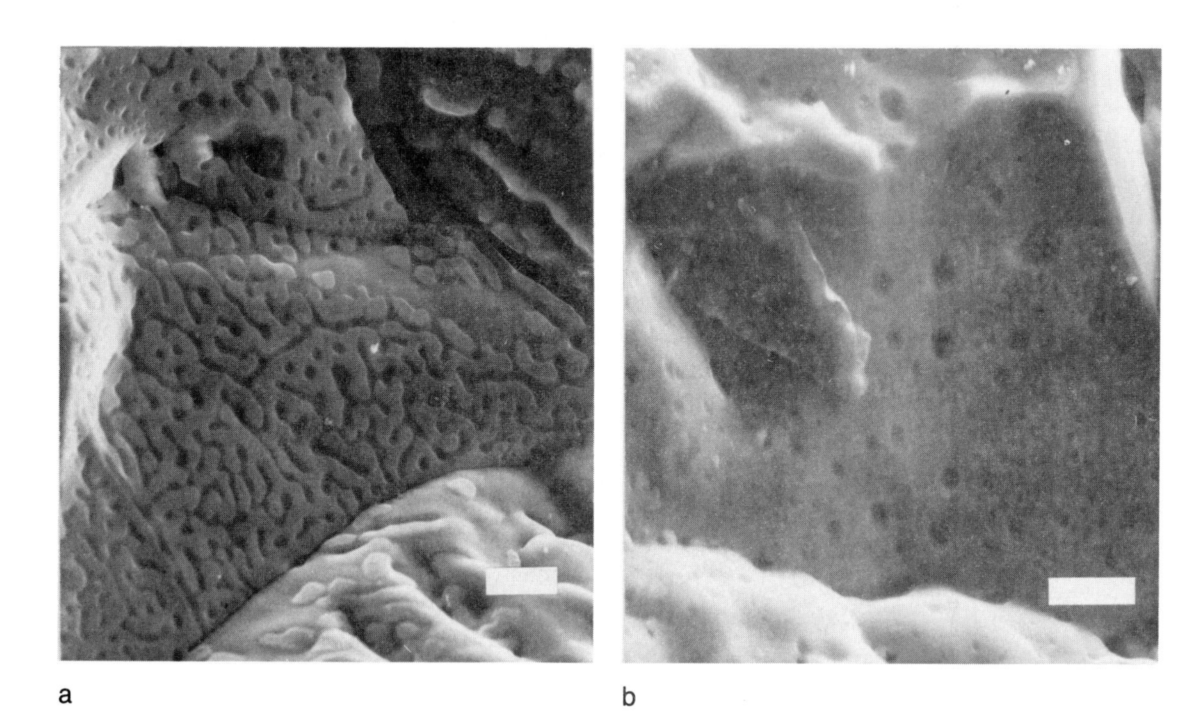

a b

Plate 2. Secondary-electron images of olivine grain boundaries in dunite samples that were subjected to (a) CO_2 and (b) CO_2–H_2O penetration at 1200°C and 10 kbar. Cracks resulting from fluid penetration along grain boundaries were allowed to heal for 24 hours at run conditions before quenching; the partially-healed cracks were then broken apart and the pieces mounted directly in epoxy for examination of the exposed surfaces with the SEM. Depressions on the grain boundary surfaces correspond to fluid-filled porosity at run conditions. Note 'brain texture' in (a) and isolated, spheroidal pores in (b). The difference in extent of healing is probably due to the difference in fluid composition ($X_{CO_2} = 1$ in (a) and 0.4 in (b)). The scale bars are 10 μm. See text for discussion.

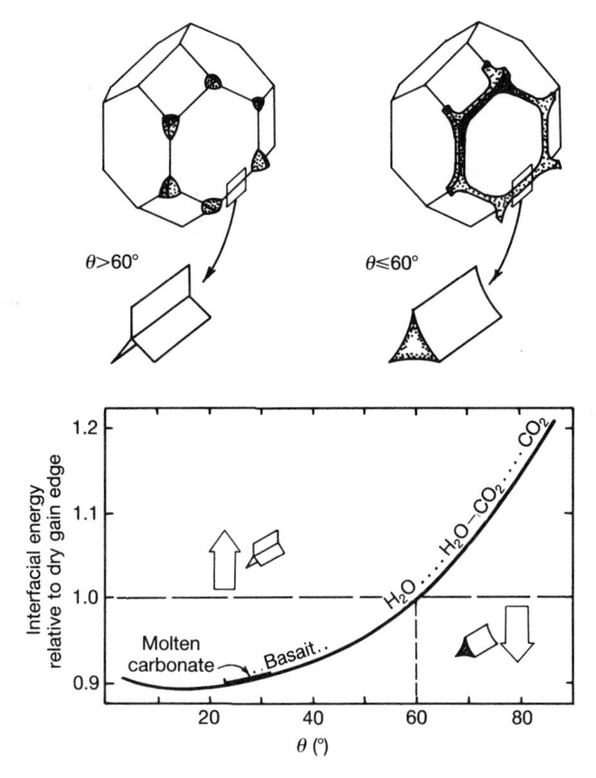

Fig. 6.3. At the top is shown the contrasting distribution of fluid about a single grain in a rock for the case of $\theta > 60°$ (left) and $\theta \leq 60°$ (right). For $\theta > 60°$, the fluid is present as isolated pores at grain corners; if $\theta \leq 60°$, the fluid forms a three-dimensionally continuous network of triangular grain-edge tubules, a short segment of which is enlarged at right centre. The graph at the bottom, modified from Watson (1982), shows how the interfacial energy in a grain edge region (relative to the fluid-absent case) changes as a function of θ. A value greater than 1 on the vertical axis means that the dry grain edge region is energetically favoured—i.e. fluid will not penetrate. Values of θ for various fluids in equilibrium with olivine are indicated along the curve. The range for H_2O–CO_2 fluids is from Watson and Brenan (1987; see also Fig. 6.5); the range for 'basalt' (which actually encompasses komatiite melts) includes data of Waff and Bulau (1982), Toramaru and Fujii (1986), and Walker *et al.* (1988); information on carbonate melts is from the present work (see Fig. 6.5) and that of Hunter and McKenzie (1989).

dimensions, and therefore of whether a fluid-bearing mantle has significant permeability. The importance of knowing θ for all plausible mantle fluids and *P–T* conditions thus seems obvious.

Ideally, one would like to know γ values for all conceivable interfaces between mantle minerals and fluids, and, from these, be able to calculate θs using eqn (1). Unfortunately, however, first-principles computation or experimental measurement of γ for even a single mineral–fluid pair is a formidable task: γ depends not only upon the identity of the phases but also upon their major- and trace-element composition, temperature, pressure *and* crystal orientation. Moreover, very accurate knowledge of interfacial energies is needed for even a rough estimate of θ. Consequently, the traditional approach in characterizing wetting behaviour has been to measure θ directly, a procedure we have adopted for mantle fluid–mineral assemblages.

Experimental approach

The experimental techniques used to characterize θ are described in detail by Watson and Brenan (1987). Because the focus of this chapter is on the upper mantle, we will restrict the description here to experiments involving Mg-rich olivine specifically.

The starting material was prepared from single crystals of natural forsteritic olivine (San Carlos, Arizona) which were pre-selected for clarity and lack of inclusions. These were crushed and sieved to pass 230 mesh, and aliquots of ~15 mg were loaded into platinum or gold capsules with a desired amount (usually 2–5 Wt %) of fluid (H_2O) or fluid-producing compound (silver oxalate for CO_2, oxalic acid dihydrate for $0.5CO_2$–$0.5H_2O$, or alkali or calcium carbonate). The capsules were then sealed in a carbon arc and subjected to high *P–T* conditions (900–1400°C, 5–30 kbar) in a piston-cylinder device using pressure cells consisting of NaCl, Pyrex, crushable alumina, and graphite (intrinsic fO_2 slightly below FMQ). After maintaining conditions for 1–6 days (depending on temperature), the experiments were rapidly quenched (~50°C s^{-1}) and the capsules were recovered for analysis. The run products generally consisted of fully 'welded' olivine aggregates of somewhat larger grain size than the starting material (to a degree depending mainly upon run temperature and fluid composition). What existed as fluid-filled porosity at run conditions was preserved upon quench either as intergranular void space (in the case of C–O–H fluids) or as fine-grained quench carbonate containing minor amounts of amorphous quench silicate material. Some typical run-product textures are shown in Plate 1.

We used a modernized version of a traditional procedure (e.g. Riegger and Van Vlack 1960) to estimate θ values. The critical step was to prepare a well-polished surface on which we could obtain good secondary-electron photomicrographs at 1000–3000 ×. To characterize a single apparent θ value, polynomials were fitted to the digitized traces of two intersecting mineral/fluid interfaces, and the angle between them at the point of intersection was computed.

Because the θ measurements obtained from any two-dimensional section are only apparent values, a good statistical basis must be established in order to estimate a true θ for a given sample. Our procedure was to record 100 ± 15 apparent angles for each sample and use the median as the best estimate of the true angle (see Riegger and Van Vlack 1960). It is important to note, however, that the observed distribution of apparent angles is always broader than is expected for a single-valued true angle. We believe this broadening reflects the variation of the olivine/fluid interfacial energy with crystallographic orientation.

6.3.2. Fluids in the C–O–H system

General systematics

Figure 6.4(a) summarizes the results of θ determinations on San Carlos olivine run at 1000–1150°C and 10 kbar in the presence of various CO_2–H_2O mixtures (see also Table 6.1). The most important message of this figure is that for fluids ranging between H_2O and CO_2, θ is universally higher than 60°, varying smoothly from 65 to 70° for pure H_2O to ~90° for pure CO_2. A number of experiments run at similar conditions (1000°C and 10 kbar) but involving aqueous solutions of NaCl, KCl, CaF_2, and Na_2CO_3 yielded θ values indistinguishable from one another (all 60–63°) and only slightly lower than the value for pure H_2O at 1000°C (see Fig. 6.5(b)). This was a somewhat surprising result in view of the pronounced effect on θ of adding alkali halide salts to water coexisting with quartz (see Watson and Brenan 1987). It should be noted, though, that the small 'halide effect' on θ in dunite could be just sufficient to establish fluid interconnectivity.

We encountered difficulty in obtaining convincing equilibrium textures in the presence of more reduced C–O–H fluids. It seems clear, however, that θ for CO is well above 60° and probably close to 90°. A tentative value of ~70° was obtained for CH_4, but in this case we have been unable to verify the fluid composition.

Effects of temperature and pressure

Because our initial series of experiments on C–O–H fluid wetting was done over a relatively narrow range in temperature and at only one pressure, nothing could be concluded regarding the dependence of θ on these intensive variables (although some negative tempera-

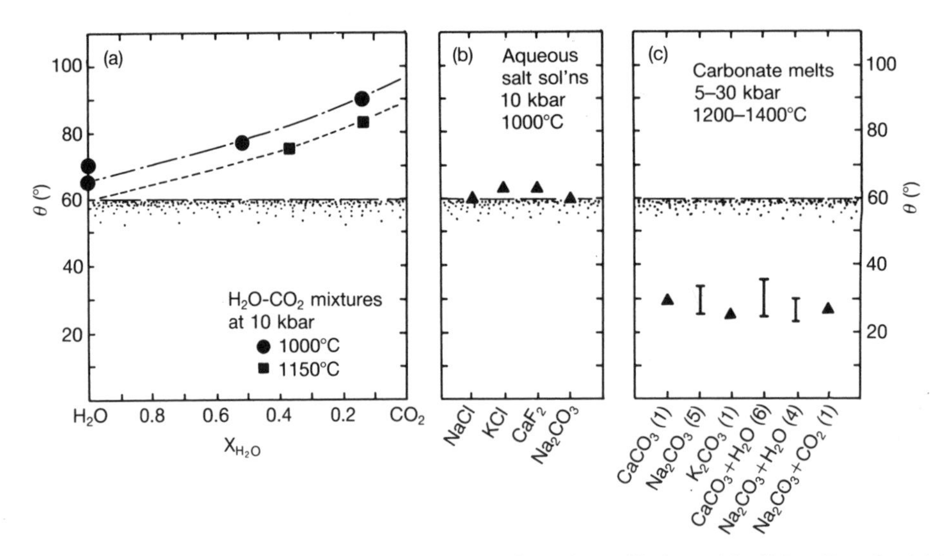

Fig. 6.4. Graphical summary of θ values for various non-silicate fluids in equilibrium with olivine. Data for (a) CO_2–H_2O mixtures and (b) aqueous salt solutions are from Watson and Brenan (1987); (c) shows new data for carbonate melts. The values in parentheses following the carbonate melt identifiers are the number of experiments to which the indicated range applies. The stippled band at and below 60° emphasizes that θ values $\leq 60°$ lead to fluid-phase interconnectivity even at fluid volume fractions less than 0.01. See Table 6.1 for a complete description of run conditions and initial fluid compositions. Note that all fluids shown contain variable (and generally uncharacterized) amounts of dissolved olivine.

Table 6.1. Summary of run information.

Run no.	*T* (°C)	*P* (GPa)	Time (days)	Fluid[a]/melt comp.	Wt % Fluid/melt	θ
6	1000	1.0	3.0	0.14[b]	1.7	90
7	1000	1.0	3.0	0.52	4.6	77
8	1000	1.0	3.0	1.0	3.0	65
14	1000	1.0	6.0	1.0	4.9	70
24	1150	1.0	5.0	0.14	1.5	83
28	1150	1.0	5.0	0.37	5.4	75
33	1000	1.0	5.0	NaCl (0.37)[c]	3.1	60
34	1000	1.0	5.0	CaF_2 (0.2)	5.1	63
35	1000	1.0	5.0	Na_2CO_3 (0.27)	5.9	60
36	1000	1.0	5.0	KCl (0.27)	5.8	63
CO1	1000	1.0	4.0	0.4[d]	6.1	nd (60)[i]
T1	1250	1.0	2.0	0.14	10.0	70
T2	1200	1.0	1.0	1.0	0.5	50
T3	1200	2.0	1.5	1.0	7.4	45
T4	1200	0.5	1.9	1.0	5.0	65
T5	900	1.0	7.2	1.0	5.0	70
T6	1200	1.0	1.0	CH_4–H_2O	nd	68
C1a	1250	1.0	2.6	Na[e]	5.2	25
C1b	1250	1.0	2.6	K	4.2	25
C6	1400	3.0	2.3	Ca	7.8	30
S4a	1250	0.9	2.0	Na	14.7	25
S4b	1250	0.9	2.0	Na	13.4 (9.6)[h]	23
S4c	1250	0.9	2.0	Na	15.4 (1% CO_2)	27
S6	1250	1.0	2.0	Na	13.2 (5.0)	26
S7	1350	1.0	1.8	Na	15.4 (4.8)	27
S15a	1250	2.0	1.3	Ca	15.8 (4.9)	36
S15b	1250	2.0	1.3	Na	15.8 (5.3)	30
S18	1200	0.5	1.9	Ca (Fo)[f]	15.0 (5.0)	35
S19	1200	1.0	2.0	Ca (Fo)[f]	15.0 (5.0)	34
S20	1300	1.0	1.5	Na (Fo)	19.6	30
S22a	1200	2.0	2.1	Ca[g]	6.4 (5.0)	27
S22b	1200	2.0	2.0	Na	15.0	33
S23	1300	0.5	1.5	Na	14.4	33
S24a	1300	2.0	1.2	Ca[g]	6.4 (5.0)	25
S24b	1300	2.0	1.2	Ca[g]	6.4 (5.0)	27

[a] All runs performed on San Carlos olivine unless otherwise specified.
[b] X_{H_2O} FMQ fluid composition estimated from Holloway and Reese (1974).
[c] Salt concentration (mg. salt/mg. solution) in H_2O solution.
[d] X_{CO_2} for CO–CO_2 mixture.
[e] Na, K, Ca refers to Na–, K–, and Ca–carbonate melt.
[f] Synthetic forsterite used in place of San Carlos olivine.
[g] H_2O added as $Mg(OH)_2$ (+SiO_2 to make forsterite).
[h] Wt % H_2O added to carbonate.
[i] nd = not determined; no. in parentheses is visual estimate.

ture dependence was suggested by the data; see Fig. 6.5(a)). In the interest of more thoroughly characterizing fluid behaviour in the upper mantle, we have now acquired additional data at higher temperature and at pressures above and below 10 kbar. The new measurements are summarized in Fig. 6.5, which reveals that, at 10 kbar, θ decreases with increasing temperature for both H_2O and CO_2, such that the critical 60° value is intersected at about 1100°C for H_2O and (with some extrapolation) 1400°C for CO_2.

As shown in Fig. 6.5(b), there is also substantial negative dependence of θ upon pressure in the olivine–H_2O system: at 1200°C, θ drops from ~65° to ~50° as pressure increases from 5 to 10 kbar, and an additional 5° in the 10–20 kbar interval. We note that the observed diminution in θ as pressure increases may actually

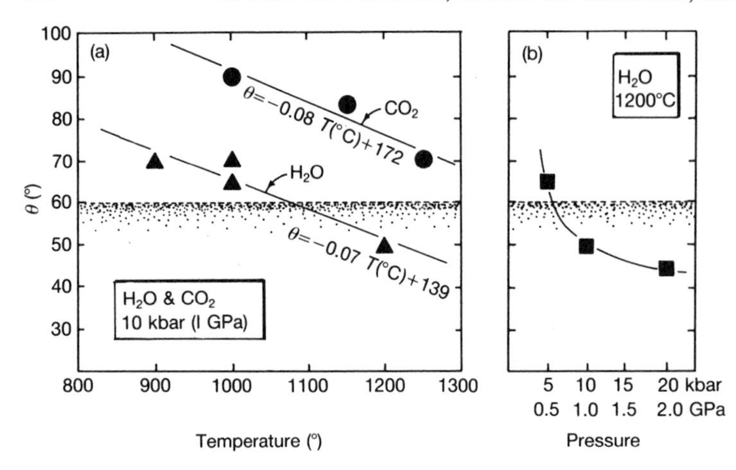

Fig. 6.5. (a) Summary of the temperature dependence of θ for H_2O and CO_2 in dunite at 10 kbar. (b) Measured pressure effect on θ in dunite at 1200°C.

derive from a change in fluid composition (as opposed to an intrinsic pressure effect), inasmuch as the solubility of olivine in H_2O is likely to be pressure-dependent. The actual cause of the change in θ is immaterial in considering the behaviour of fluids in the mantle, though the change itself is potentially very important.

The key conclusion concerning C–O–H fluids in the mantle is that, for the range of P–T conditions over which a free fluid phase is likely to exist (see earlier section on large-scale fluid distribution), θ is generally $>60°$. Consequently, low-viscosity/density fluids will rarely form an interconnected network. Close inspection of Fig. 6.5(a) and comparison with Fig. 6.1 does, however, leave open the possibility that water-rich C–O–H fluids could achieve interconnectivity at temperatures just below the peridotite solidus. In fact, bulk diffusion experiments now underway in our laboratory show that the range in θ resulting from interfacial-energy anisotropy causes partial interconnectivity of H_2O at temperatures as low as 900°–1000°C (cf. Fig. 6.4. and the last paragraph of the section on 'Experimental approach').

6.3.4. Carbonate melts

Using methods similar to those described previously for C–O–H fluids, we have measured θ values for molten calcium-, sodium-, and potassium carbonate coexisting with San Carlos olivine or synthetic forsterite (with and without added H_2O or CO_2) over the temperature range 1200°–1400°C at pressures of 5 to 30 kbar (see Table 6.1 and Fig. 6.4(c)). Despite the variations in both fluid composition and P–T conditions, θ is restricted to the narrow range 23–36° and shows no systematic dependence upon intensive variables. These results confirm the low θ value reported for molten carbonatite by Hunter and McKenzie (1989), and they indicate that any conceivable carbonate melt in an olivine-rich mantle, even when present in very small amounts, will be interconnected in three dimensions. As discussed at a later point, this is an important result when considering carbonate melts as potential metasomatic agents.

6.4. IMPLICATIONS OF WETTING BEHAVIOUR FOR UPPER MANTLE METASOMATISM

6.4.1. Characteristics of effective metasomatic agents

An upper mantle fluid must meet two requirements in order to be an effective metasomatic agent. First, it must be capable of dissolving the components of interest to a sufficient extent that realistic fluid/rock ratios can accomplish the desired or observed chemical or isotopic changes (see also Menzies, Chapter 4, this volume). Considerable attention has been given to solubilities of various components in mantle fluids (e.g. Holloway 1971; Schneider and Eggler 1986; see summary by Eggler 1987); thus, we will not dwell on this topic, noting only that a general ranking can be made of the ability of non-silicate fluids to dissolve major oxides and trace elements: molten carbonate > H_2O-rich fluid > CO_2-rich fluid. This ranking must be qualified in the respect that solubilities are pressure- and species-dependent and may also be strongly affected by the presence of anionic volatile components such as Cl and F. Unfortunately, the experimental data base on solubilities of geochemically interesting elements is sparse due to inherent difficulties in making measurements at mantle conditions.

The second requirement of an effective metasomatic fluid is that it be mobile. The question of fluid mobility in the upper mantle has been addressed previously using theoretical approaches that incorporate the concepts of both fracture propagation (e.g. Shaw 1980; Emerman *et al.* 1986; Spera 1987) and Darcy flow (i.e. flow through an already porous medium under the influence of a pressure gradient—e.g. Sleep, 1974; Walker *et al.* 1978; McKenzie 1985). Although the resulting models are valuable as guides to the plausible scales and rates of fluid transport in the upper mantle, they leave some important questions unanswered (e.g. under what circumstances is pervasive, possibly cryptic, metasomatism favoured over modal metasomatism confined largely to veins?). They also leave the uninitiated reader with the impression that only the physical properties of fluids (e.g. viscosity, compressibility) affect their ability to migrate in the mantle. In fact, very low-viscosity fluids such as CO_2–H_2O mixtures may under some circumstances lack the ability to pervasively penetrate solid rock, while a more viscous melt might readily do so under the same conditions. This contrast in behaviour is counterintuitive, but probably real.

In the following paragraphs, we will review, from the perspective of experimentalists concerned mainly with fluid and rock chemistry, the implications of studies in our laboratory for fluid migration in the mantle. The discussion will include both surface energy considerations and results of simple crack propagation experiments on mantle materials.

6.4.2. Infiltration driven by interfacial energy reduction

Watson (1982) hypothesized that under some circumstances magma bodies might tend to disperse by infiltrating their host rocks, even when the two are in chemical equilibrium. The unorthodox aspects of this idea were: (1) that no fluid overpressure was called upon to drive the process; and (2) the penetration of melt into the host rock was envisioned to occur on an intergranular scale, not as relatively large, discrete veins. The driving force for infiltration invoked by Watson was a reduction in interfacial energy of the system that occurs as a fluid having a wetting angle $<60°$ penetrates the grain edges of a previously fluid-free rock. From the earlier work of Waff and Bulau (1979), basaltic melts were known to form wetting angles against olivine of considerably less than 60°, so the infiltration process seemed reasonable (Fig. 6.3). Experiments in a synthetic dunite–haplobasalt system appeared to confirm the proposed infiltration tendency, and yielded a penetration rate of 1–2 mm day^{-1} at 1290°C and 10 kbar.

Despite the existence of a clear driving force for infiltration (see the description of 'general principles' in previous section), the *mechanism* of haplobasalt penetra-

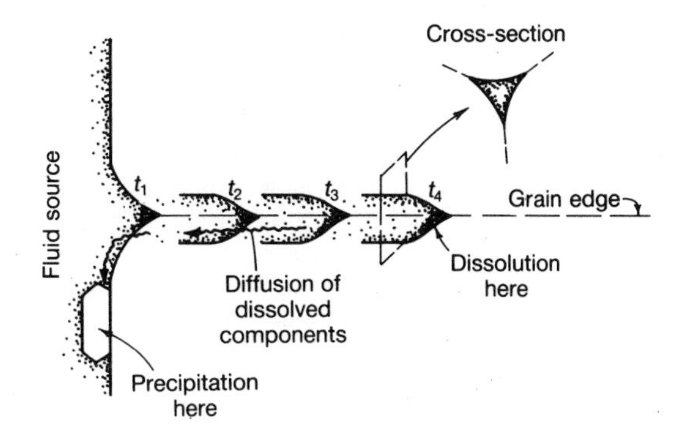

Fig. 6.6. Schematic illustration of the penetration of fluid along a previously dry grain edge, showing 'snapshots' at progressive times t_1 to t_4. The driving force for penetration (infiltration) is reduction in interfacial energy, and the mechanism is dissolution of solid material at the tip of the penetrating tubule with concurrent precipitation in the fluid source region. The process is probably rate-limited by diffusion of dissolved fluid components in the direction opposite that of fluid penetration. See text for discussion.

tion into dunite is less obvious. However, because no mechanical work or volumetric free energy changes are involved, the possibilities can quickly be narrowed down to a dissolution/precipitation process in which olivine is dissolved along grain edges (triple junctions) and precipitated elsewhere, most likely within the fluid source reservoir (see Fig. 6.6). The infiltration rate might, therefore, be limited by either dissolution/precipitation kinetics or diffusion in the fluid. At the high temperatures under consideration, diffusion is the more likely rate-limiting step.

The infiltration process envisioned by Watson (1982) to operate in the margins of magma bodies in hot surroundings is perhaps even more plausible as a mechanism for moving non-silicate fluid in the mantle. The one prerequisite, of course, is that θ be less than 60°. If the infiltration process is indeed rate-limited by diffusion in the fluid, transport rates much higher than the 1–2 mm day^{-1} measured by Watson (1982) might well apply to the low-viscosity fluids (C–O–H or molten carbonate) presently under consideration. Assuming a characteristic diffusivity in silicate melt of 10^{-7} cm^2 s^{-1} and one four orders of magnitude higher in aqueous fluid and applying the diffusional transport approximation $X^2 \sim D \cdot t$, one arrives at a penetration rate two orders of magnitude higher for aqueous fluid than for basaltic melt, i.e. ~ 150 mm day^{-1} or ~ 55 m $year^{-1}$. We believe this is a plausible rate for aqueous fluid infiltration in the upper mantle, but it certainly should not be taken as a hard number, because its veracity depends on the validity of some key assumptions.

(1) The wetting angle for aqueous fluid is indeed less than 60°. As discussed in the previous section, this will not be the general case in the upper mantle, although conditions just below the peridotite solidus are permissive.

(2) The infiltration process is diffusion-controlled, not reaction-controlled. Data from our laboratory on diffusion in fluid-bearing dunites (Watson 1988) give no indication of reaction control on transport, but the possibility of its existence cannot be ruled out.

(3) Scaling of infiltration rates for two fluids can be done simply by multiplying by the square root of the diffusivity ratio. The problem here is that the amount of dissolved material transported by diffusion in a fluid depends not only on the diffusivity but on the solubility as well. Olivine is far less soluble in aqueous fluid than in basaltic melt, so the fluid may actually have less capacity to transport dissolved olivine despite having a much higher diffusivity.

The above assumptions notwithstanding, there does exist experimental confirmation of our calculated infiltration rate of ~ 55 m year^{-1}: At somewhat lower temperature and higher pressure (850°C and 20 kbar), Mysen *et al.* (1978) measured an aqueous fluid infiltration rate in peridotite of ~ 23 m year^{-1}. It is also worth noting that Stevenson (1986), in a theoretical model of interfacial energy-driven fluid transport, predicted similar maximum rates for low-viscosity fluids (though his model dealt specifically with the redistribution of fluid in a region characterized by spatially variable fluid abundance, not with the 'end-member' case of fluid moving into a previously dry region).

In our view, the one non-silicate fluid most susceptible to interfacial energy-driven transport in the upper mantle is carbonate melt. Our reasons are basically twofold: (1) the wetting angle of 25–30° typical of carbonate melts results in a near-maximum driving force for infiltration (see graph in Fig. 6.3); and (2) carbonate melts are characterized not only by very high component diffusivities (e.g. Janz and Lorenz 1961; Janz and Saegusa 1963) but also by an ability to dissolve considerable amounts of silicate material (e.g., Wallace and Green 1988). (Ongoing experiments in our lab on bulk diffusion in dunite containing various fluids show that molten carbonate, relative to CO_2–H_2O and basaltic melt, causes by far the greatest enhancement in bulk diffusion.) Given no other driving force for transport than interfacial energy reduction, *carbonate melts are potentially the most effective metasomatic agents in the upper mantle.*

6.4.3. Crack propagation and related topics

Overview

As mentioned previously, except at P–T conditions at or near the peridotite–CO_2–H_2O solidus, θ values exceeding 60° restrict the equilibrium distribution of small amounts of CO_2–H_2O fluid in olivine-rich assemblages to be one of isolated pockets at grain corners and grain edge intersections. In the absence of any pre-existing fractures, fluid dispersal under these circumstances cannot be facilitated by interfacial energy-driven infiltration, and is likely to be limited to transient events where P_{fluid} exceeds the confining pressure by some amount and results in propagation of cracks.

In an elastic material, the process of crack growth represents a competition between the increase in surface energy that accompanies the formation of new crack surface and the decrease in elastic strain energy arising from the rupture of strained bonds (Tetelman and McElevy 1967). Stable crack growth in such materials involves a continuous increase in the stress tending to displace the crack surfaces until the overall decrease in elastic energy overwhelms the increase in surface energy and further crack growth becomes energetically favourable. The unstable crack growth that ensues at this point may reach velocities that, in theory, are limited only by the viscosity of the stress-transmitting medium (Anderson 1979). As might be expected, however, crack growth in natural materials is likely to be influenced by a host of additional factors that serve to complicate this simple theory: (1) at elevated P and T, rocks can experience plastic deformation which, in turn, contributes an element of non-recoverable work to the crack extension process (Tetelman and McElevy 1967); (2) the fluid phase in the extending crack may lower the crack surface energy and thus lower the stress required for unstable crack extension (i.e. stress corrosion—cf. Anderson and Grew 1977; Atkinson 1984); and (3) material flaws (e.g. grain or subgrain boundaries, large grains, fluid inclusions, etc.), which serve as sites for crack nucleation and growth, will have a profound effect on both the crack extension pathways and the overall fracture resistance of a material (Rice 1977). Because of these complicating factors, it is difficult to predict not only under what mantle conditions (if any) CO_2–H_2O fluids may be transported by crack extension, but also by what pathways (and to what extent) fluid dispersal by this process might be achieved. Results of recent experiments by Brenan and Watson (1988) do suggest, however, that fluid penetration by crack extension at high P–T conditions is plausible and that, at least for the conditions investigated, fluid penetration may be pervasive on the grain scale.

Experiments and their implications

Brenan and Watson (1988) monitored the penetration of CO_2 in synthetic dunite with experiments conducted mostly at 10 kbar and 1200°C. Their results indicated that fluid penetration occurs principally via a system of grain boundary cracks and that penetration velocities probably exceed 0.01 mm s^{-1}. The spatial extent of fluid transport, however, was found to be dependent on the dunite fabrication method, suggesting a microstructural and/or crystal-chemical control on the fracture resistance of their samples. In considering the application of such results to real rocks, it should be noted that the overall fine-grained nature of the synthetic rocks used in the Brenan–Watson study suggests that even the largest strength-controlling flaws in their samples are likely to represent a lower bound for flaws in natural rocks. Because the fracture resistance of a material may be largely controlled by the size and orientation of flaws (Rice 1977), the results of Brenan and Watson imply that fluid-driven crack extension is possible in some mantle environments.

Although the fluid that drives the crack extension process may itself be an important chemical transport agent, it should also be noted that the residual porosity remaining after a crack propagation event may provide a pathway for later diffusive or advective transport. Recent experiments in our laboratory involving fluid penetration into cores of natural dunite reveal partial healing textures on crack surfaces that preserve pore interconnectivity for significant lengths of time.

The experimental procedure was first to pack into a nickel capsule a previously devolatilized natural dunite core surrounded by a mixture of fine-grained olivine and silver oxalate or hydromagnesite (CO_2 and H_2O sources, respectively). The capsule was then subjected to 1200°C and 10 kbar for a predetermined length of time. At run conditions, a free fluid phase was generated around the dunite core and penetrated it primarily along pre-existing grain boundaries. The cracks produced by this process were observed to undergo various amounts of healing in experiments of prolonged duration. In Plate 2 secondary electron images of grain boundary surfaces from samples treated in the above manner are shown. Plate 2 (a) shows characteristic island-like features (regions of grain–grain contact) separated by numerous 'bays' and 'inlets' of porosity ('brain texture'). Such features are expected during the crack healing process because the continuous fluid film present along the crack immediately after fluid penetration is energetically unstable ($\theta \gg 0°$; see Fig. 6.2 and 6.4(a)) and is eventually displaced by olivine to produce an isolated pore structure like that shown in Plate 2 (b). (See Wanamaker and Evans (1985) for a more detailed description of this process.) The development of 'brain texture' could be important in metasomatism because the characteristic channel-ways may constitute transient pathways of rapid mass transport (until the stable, isolated pore geometry develops). The longevity of pore interconnectivity on a brain-textured surface is unknown; however, recent experiments at 1 atmosphere on healing of CO_2-filled microcracks in olivine suggest that only minor healing (as gauged by the extent to which unconnected spheroidal pores develop) occurs after annealing times as long as 64 hours at 1400°C (Wanamaker and Evans 1985). Because mass transport through the fluid and surface diffusion are both likely to be important in the crack healing process, increases in both P and X_{H_2O} of the fluid will probably accelerate healing. This conclusion is supported by Plate 2, which contrasts the extent of healing in the presence of (a) pure CO_2 and (b) an H_2O–CO_2 mixture.

On the basis of observations made on natural peridotites, Rovetta (1981) has also put forth the idea that permeability can be enhanced by microcracking. He and co-workers (Rovetta 1981; Rovetta *et al.* 1986) interpreted CO_2-filled microcracks in their samples as having arisen from stress inhomogeneities around fluid inclusions present during deformation. Although the circumstances of crack formation may be different from those in our experiments, the implications for enhanced mass transport are the same.

Regarding *when* and *where* in the mantle fluid transport by crack propogation might occur, there exist certain conditions that may lead to the production of a pressurized volatile phase. The scenarios include heating of fluid inclusion-rich rock (e.g. Knapp and Knight 1977), mineral devolatilization reactions, and volatile exsolution from silicate or carbonate magma. Of particular geochemical interest is the situation involving magma devolatilization. Not only are deep-seated magmas likely to contain substantial amounts of dissolved volatiles, but upon exsolution the volatile phase is also likely to take up significant concentrations of trace and major components from the magma (see summary of solubilities in Eggler 1987). The nature of alteration zones in regions invaded by exsolved fluids will be complex and dependent upon such factors as the nature of the mantle wall-rock, the composition of the fluid, the ambient P–T conditions, and the bulk composition of the magma prior to fluid exsolution.

Cryptic versus vein metasomatism

The factors summarized in this and the preceding section lead to the following conclusions. Metasomatic fluids can penetrate non-porous rock by both crack propagation and interfacial energy-driven infiltration. The former migration mechanism is the only one available to CO_2-rich fluids ($\theta \gg 60°$), but carbonate melts and (under the right circumstances) H_2O-rich fluids may take advantage of the latter mechanism. In view of the tendency for carbonate melts and (occasionally) H_2O to wet grain edges, these fluids seem the more likely agents of cryptic metasomatism that is pervasive on the scale of individual grains in a rock. Vein metasomatism, on the other hand, is by its very nature suggestive of a crack propagation origin. No mantle fluids are specifically precluded from migrating by this mechanism—even those having $\theta < 60°$—but a plausible means of either sustaining or repeatedly generating a fluid overpressure is needed.

6.5. CONCLUDING REMARKS

The nature and large-scale distribution of fluids in the continental upper mantle is broadly constrained by existing experimental and observational data in combination with molecular speciation theory. In a mantle consisting of slightly hydrated and carbonated peridotite, the critical dependence of fluid composition upon the stability and identity of volatile-bearing minerals leads to a general depth stratification with respect to fluid composition and abundance. At subsolidus conditions ($T < \sim 1000$–$1050°C$) the following generalizations apply: At pressures below ~15–20 kbar where amphibole is stable but carbonate is not, any fluid present will be CO_2-rich. If amphibole and carbonate coexist, which they can do at intermediate pressures of ~15–25 kbar, all potential C–O–H fluid components are locked up in the solid assemblage, thus precluding the existence of a free C–O–H fluid phase. At high pressures (> ~30 kbar) where carbonate is stable but amphibole is not, the fluid will be water-rich to an extent determined by the bulk-rock abundance of K_2O, which stabilizes the alternative and high-pressure hydrous mineral, phlogopite.

The recent experimental results of Wallace and Green (1988) on slightly hydrated (0.3 Wt % H_2O) and carbonated pyrolite show a field in which carbonate melt coexists with amphibole lherzolite at ~20–30 kbar and 950–1100°C; thus, even though a free C–O–H fluid is not likely to exist over this pressure range, a sufficiently high temperature can still produce a non-silicate metasomatic fluid (namely, carbonatite).

As a complement to the above guidelines regarding large-scale fluid distribution, the wetting characteristics of C–O–H and carbonate fluids can be used not only to characterize the grain-scale distribution of upper mantle fluids, but also to assess their effect on bulk-rock physical properties such as electrical conductivity and shock-wave velocity. Under no circumstances of chemical and mechanical equilibrium will an upper mantle fluid form a grain-boundary film. Moreover, only under relatively restricted conditions will a C–O–H fluid form a continuously interconnected phase (in particular, the fluid would have to be water-rich and the pressure above about 30 kbar, with temperature just below the solidus). Carbonate melts, in contrast, will form a phase that is continuously interconnected along grain edges over any P–T conditions at which they can coexist with peridotite (see also Richardson, Chapter 3, this volume). In combination with their ability to dissolve trace and major rock components, this property of carbonate melts makes them potentially very effective metasomatic agents.

The same characteristic that enables a fluid already present in a rock to continuously wet grain edges (namely, relatively low interfacial energy) also affords a means of fluid penetration into initially non-porous rock, even in the absence of a pressure gradient. This interfacial energy-driven infiltration may occur at velocities up to 20–50 m year^{-1}—considerably slower than velocities associated with crack propagation driven by fluid overpressure, but nevertheless significant. The grain-edge infiltration process may be especially important in facilitating metasomatism that is pervasive on the scale of individual grains in a peridotite.

Fluids that do not wet grain edges (e.g. C–O–H compositions under most circumstances) will tend to form isolated pores at grain corners. Unless $P_{fluid} > P_{total}$, such fluids generally will be immobile and contribute little or nothing to bulk-rock transport. Transient events in which fluid overpressures trigger crack propagation can contribute not only directly to metasomatic transport but also indirectly by leaving a residual porosity that temporarily enhances diffusion or fluid flow.

The principles governing large- and small-scale fluid distribution in the upper mantle derive from phase-equilibrium and interfacial-energy considerations, respectively. These have been presented here in review/synthesis form not as an alternative to physical models of fluid transport, but as factors to be considered in conjunction with fluid dynamics to understand better the behaviour of fluids in the mantle.

ACKNOWLEDGEMENTS

The experimental studies done at Rensselaer Polytechnic Institute were supported by the National Science Foundation, Division of Earth Sciences, under grant nos EAR84-06200 and EAR87-17341.

REFERENCES

Anderson, O. L. (1979). The role of fracture dynamics in kimberlite pipe formation. In *Kimberlites, diatremes, and diamonds; their petrology, mineralogy and geochemistry* (ed. F. R. Boyd and H. O. A. Meyer), pp. 344–53. American Geophysical Union, Washington, DC.

Anderson, O. L. and Grew, P. C. (1977). Stress corrosion theory of crack propagation with application to geophysics. *Reviews of Geophysics and Space Physics*, **15,** 77–104.

Arculus, R. J. (1985). Oxidation status of the mantle: past and present. *Annual Reviews of Earth and Planetary Science*, **13,** 75–95.

Atkinson, B. K. (1984). Subcritical crack growth in geological materials. *Journal of Geophysical Research*, **89,** 4077–114.

Beere, W. (1975). A unifying theory of the stability of penetrating phases and sintering pores. *Acta Metallurgica*, **23,** 131–8.

Bishop, F. C., Smith, J. V., and Dawson, J. B. (1975). Pentlandite–magnetite intergrowths in De Beers spinel lherzolite: review of sulphides in nodules. *Physics and Chemistry of the Earth*, **9,** 323–37.

Brenan, J. M. and Watson, E. B. (1988). Fluids in the lithosphere. 2: Experimental constraints on CO_2 transport in dunite and quartzite at elevated *P–T* conditions with implications for mantle and crustal decarbonation processes. *Earth and Planetary Science Letters*, **91,** 141–58.

Bulau, J. R., Waff, H. S., and Tyburczy, J. A. (1979). Mechanical and thermodynamic constraints on fluid distribution in partial melts. *Journal of Geophysical Research*, **84,** 6102–8.

Dawson, J. B. (1980). *Kimberlites and their xenoliths.* Springer-Verlag, Berlin.

Delaney, J. S., Smith, J. V., Carswell, D. A., and Dawson, J. B. (1980). Chemistry of micas from kimberlites and xenoliths. II. Primary- and secondary-textured micas from peridotite xenoliths, **44,** 857–72.

Eggler, D. H. (1987). Solubility of major and trace elements in mantle metasomatic fluids: experimental constraints. In *Mantle metasomatism* (ed. M. A. Menzies and C. J. Hawkesworth), pp. 21–42. Academic Press, New York.

Eggler, D. H. and Baker, D. R. (1982). Reduced volatiles in the system C–O–H: implications to mantle melting, fluid formation, and diamond genesis. In High-pressure research in geophysics (ed. S. Akimoto and M. Manghani). *Advances in Earth and Planetary Sciences*, **12,** 237–50.

Eggler, D. H. and Holloway, J. R. (1977). Partial melting of peridotite in the presence of H_2O and CO_2: principles and review. In *Magma genesis* (ed. H. J. B. Dick), pp.15–36. State of Oregon Department of Geology and Mineral Resources Bulletin no. 96. State of Oregon Department of Geology and Mineral Resources, Portland, Oregon.

Emerman, S. H., Turcotte, D. L., and Spence, D. A. (1986). Transport of magma and hydrothermal solutions by laminar and turbulent fluid fracture. *Physics of the Earth and Planetary Interiors*, **41,** 249–59.

Haggerty, S. E. and Tompkins, L. A. (1982). Sulfur solubilities in mantle derived nodules from kimberlites. *EOS, Transactions of the American Geophysical Union*, **63,** 463.

Holloway, J. R. (1971). Composition of fluid-phase solutes in a basalt–H_2O–CO_2 system. *Geological Society of America, Bulletin*, **82,** 233–8.

Holloway, J. R. (1977). Fugacity and activity of molecular species in supercritical fluids. In *Thermodynamics in geology* (ed. D. G. Fraser), pp. 161–82, Reidel, Boston.

Holloway, J. R. and Reese, R. L. (1974). The generation of N_2–CO_2–H_2O fluids for use in hydrothermal experimentation. I. Experimental method and equilibrium calculations in the C–O–H–N system. *American Mineralogist*, **59,** 587–97.

Hunter, R. H. and McKenzie, D. (1989). The equilibrium geometry of carbonate melts in rocks of mantle composition. *Earth and Planetary Science Letters*, **92,** 347–56.

Janz, G. J. and Lorenz, M. R. (1961). Molten carbonate electrolytes: physical properties, structure and mechanisms of electrical conductance. *Journal of the Electrochemical Society*, **108,** 1052–8.

Janz, G. J. and Saegusa, F. (1963). Molten carbonates as electrolytes: Viscosity and transport properties. *Journal of the Electrochemical Society*, **110,** 452–6.

Knapp, R. B. and Knight, J. E. (1977). Differential thermal expansion of pore fluids: fracture propagation and microearthquake production in hot pluton environments. *Journal of Geophysical Research*, **82,** 2515–22.

McKenzie, D. (1985). Extraction of magma from the crust and mantle. *Earth and Planetary Science Letters*, **74,** 81–91.

Mysen, B. O., Kushiro, I., and Fujii, T. (1978). Preliminary experimental data bearing on the mobility of H_2O in crystalline upper mantle. *Carnegie Institution of Washington, Yearbook*, **77,** 793–7.

Olafsson, M. and Eggler, D. H. (1983). Phase relations of amphibole, amphibole–carbonate, and phlogopite–carbonate peridotite: petrologic constraints on the asthenosphere. *Earth and Planetary Science Letters*, **64,** 305–15.

Rice, R. W. (1977). Microstructural dependence of mechanical behavior. In *Treatise in materials science.* 11: *Properties and microstructure* (ed. R. K. MacCrone), pp. 200–369. Academic Press, New York.

Riegger, O. K. and Van Vlack, L. H. (1960). Dihedral angle measurement. *AIME Transactions*, **218,** 933–5.

Rovetta, M. R. (1981). Melt and vapor migration in the upper mantle: Ultramafic nodule permeability estimated from fluid/melt inclusions. *EOS, Transactions of the American Geophysical Union*, **62,** 1062.

Rovetta, M. R., Delaney, J. R., and Blacic, J. D. (1986). A record of high-temperature embrittlement of peridotite in CO_2-permeated xenoliths from basalt. *Journal of Geophysical Research*, **91,** 3841–8.

Schneider, M. E. and Eggler, D. H. (1986). Fluids in equilibrium with peridotite minerals: Implications for mantle metasomatism. *Geochimica et Cosmochimica Acta*, **50,** 711–24.

Sellschop, J. P. F., Madiba, C. C. P., and Annegarn, H. J. (1980). Light volatiles in diamond: physical interpretation and genetic significance. *Nuclear Instruments and Methods*, **168,** 529–34.

Sharp, W. E. (1966). Pyrrhotite: a common inclusion in South African diamonds. *Nature*, **211,** 402–3.

Shaw, H. R. (1980). The fracture mechanisms of magma transport from the mantle to the surface. In *Physics of magmatic processes* (ed. R. B. Hargraves), pp. 201–64. Princeton University Press, Princeton, New Jersey.

Sleep, N. H. (1974). Segregation of magma from a mostly crystalline mush. *Geological Society of America, Bulletin*, **85,** 1225–32.

Spera, F. J. (1987). Dynamics of translithospheric migration of metasomatic fluid and alkaline magma. In *Mantle metasomatism* (ed. M. A. Menzies and C. J. Hawkesworth), pp. 1–20. Academic Press, New York.

Stevenson, D. J. (1986). On the role of surface tension in the migration of melts and fluids. *Geophysical Research Letters*, **13,** 1149–52.

Sutton, S. R., Smith, J. V., Jones, K. W., and Dawson, J. B. (1990). Nitrogen microanalysis of silicates using the $^{14}N(d, a)^{12}C$ nuclear reaction and initial results on upper mantle and peralkaline magmatic micas. *Nuclear Instruments and Methods* (submitted).

Tetelman, A. S. and McElevy, A. J. (1967). Fracture of structural materials. *Progress in Material Science*, **21,** 171–442.

Toramaru, A. and Fujii, N. (1986). Connectivity of melt phase in a partially-molten peridotite. *Journal of Geophysical Research*, **91,** 9239–52.

Von Bargen, N. and Waff, H. S. (1986). Permeabilities, interfacial areas and curvatures of partially-molten systems: results of numerical computations of equilibrium microstructures. *Journal of Geophysical Research*, **91,** 9261–76.

Waff, H. S. and Bulau, J. R. (1979). Equilibrium fluid distribution in an ultramafic partial melt under hydrostatic stress conditions. *Journal of Geophysical Research*, **84,** 6109–14.

Waff, H. S. and Bulau, J. R. (1982). Experimental studies of near-equilibrium textures in partially-molten silicates at high pressure. *Advances in Earth and Planetary Sciences*, **12,** 229–36.

Walker, D., Jurewicz, S. R., and Watson, E. B. (1988). Adcumulus dunite growth in a laboratory thermal gradient. *Contributions to Mineralogy and Petrology*, **99,** 306–19.

Walker, D., Stolper, E., and Hays, J. F. (1978). A numerical treatment of melt/solid segregation: size of the Eucrite parent body and stability of the terrestrial low-velocity zone. *Journal of Geophysical Research*, **83,** 6005–13.

Wallace, M. E. and Green, D. H. (1988). An experimental determination of primary carbonatite composition. *Nature*, **335,** 343–5.

Wanamaker, B. J. and Evans, B. (1985). Experimental diffusional crack healing in olivine. In *Point defects in minerals*, American Geophysical Union Monograph No. 31 (ed. R. N. Schock), pp. 194–210. American Geophysical Union, Washington, DC.

Watson, E. B. (1982). Melt infiltration and magma evolution. *Geology*, **10,** 236–40.

Watson, E. B. (1988). Diffusional transport in fluid-bearing rocks: Experiments involving Fe in synthetic dunite. *EOS, Transactions of the American Geophysical Union*, **69,** 1512.

Watson, E. B. and Brenan, J. M. (1987). Fluids in the lithosphere. 1: Experimentally-determined wetting characteristics of CO_2–H_2O fluids and their implications for fluid transport, host-rock physical properties, and fluid inclusion formation. *Earth and Planetary Science Letters*, **85,** 497–515.

Wilkins, R. W. T. and Sabine, W. (1973). Water content of some nominally anhydrous silicates. *American Mineralogist*, **58**, 508–16.

Wyllie, P. J. (1978). Mantle fluid compositions buffered in peridotite–CO_2–H_2O by carbonate, amphibole, and phlogopite. *Journal of Geology*, **86**, 687–713.

Wyllie, P. J. (1987). Discussion of recent papers on carbonated peridotite, bearing on mantle metasomatism and magmatism. *Earth and Planetary Science Letters*, **82**, 391–7.

7

Stable isotopes in the continental lithospheric mantle

T. Kurtis Kyser

7.1. INTRODUCTION

Much of the effort of geochemists studying the evolution of the lithospheric mantle has been directed at understanding the radiogenic isotope systematics in mantle xenoliths and mafic lavas. These data have been partially successful because the disparity in the chemical behaviour of some radioactive parent isotopes with their daughter nuclides is so substantial as to result in variable isotopic compositions of the daughter elements that are time-dependent and, thus, reveal age relations among possible reservoirs in the asthenosphere, lithosphere, and crust. In many instances, however, the results are not unique or permit multiple interpretations. For example, there is no consensus as to whether flood basalts represent partial melts of the continental mantle lithosphere, or another reservoir (Carlson 1983, 1984; Nelson 1983*a,b*; see also Menzies and Kyle, Chapter 8, this volume). Moreover, with most geochemical tracers, the utility of radiogenic isotopes as tracers is limited to either silicate liquids or to only a few phases in mantle xenoliths because most radiogenic isotopes are incompatible trace elements whose distributions are controlled primarily by those phases which concentrate incompatible elements. These properties of radiogenic isotopes have been exploited successfully to trace the origin of trace elements in metasomatic fluids in the lithospheric mantle and to determine the relative ages and relations between mafic magmas and phlogopite, pyroxene, and garnet (e.g. Kramers *et al.* 1983; Menzies and Wass 1983; Cohen *et al.* 1984; Menzies 1987; see also Menzies, Chapter 4, this volume).

The incompatible element characteristics of most radiogenic isotopes renders them of limited use in the study of the petrogenesis of olivine and orthopyroxene, the major phases in the continental lithosphere, or the origin of H_2O- or CO_2-rich fluids in the lithospheric mantle, or the origin of sulphides which are important for determining the relation between metallogenesis and mafic magmas. These aspects of the lithosphere usually are investigated using experimental phase relations which normally provide information about equilibrium conditions but not about the dynamic processes which may operate in the lithosphere. Stable isotopes of hydrogen, carbon, oxygen, and sulphur are directly relevant to the study of the lithospheric mantle because hydrogen and carbon are major components in H_2O- and CO_2-rich metasomatic fluids (see also Watson *et al.*, Chapter 6, this volume), sulphur is the major anion to which many economically important metals are bound in the lithosphere, and oxygen is the major component in all phases other than sulphides or metallic alloys. Although stable isotopes cannot provide time-dependent information about the evolution of the continental lithospheric mantle, when used with radiogenic isotopes, both the origin of fluids and phases as well as their history eventually may be discerned. For example, trace elements and oxygen isotopes may be decoupled when elements such as Sr, Nd, Pb, sulphur, and hydrogen are concentrated in a volumetrically minor melt or fluid that mixes with another reservoir which is the source for oxygen and other major elements.

This chapter is a review of what is known about the isotopic composition of oxygen, hydrogen, and sulphur in the continental lithospheric mantle. The continental lithospheric mantle is non-convecting, consisting primarily of peridotites to a depth of about 150 km under the continental crust. It is distinct from the oceanic lithospheric mantle in that the continental lithospheric mantle is cooler and extends to greater depths and distinct from the asthenosphere which is deeper, higher in temperature, and convecting (Anderson 1987; Boyd and Mertzman 1987; see also Anderson, Chapter 1, this volume). Some of the data and interpretations presented have only recently been obtained because stable isotope studies of materials from the continental lithospheric mantle are not numerous. It is clear that our understanding of the isotope systematics of these elements in the mantle as a whole is somewhat rudimentary but it is also clear that they should be an integral part of studies on the petrology of the continental lithospheric mantle.

7.2. OXYGEN ISOTOPES

Oxygen has three stable isotopes: ^{16}O, which is the most abundant isotope comprising about 99.76 per cent of all oxygen; ^{17}O, which is 0.04 per cent; and ^{18}O, which makes up about 0.2 per cent of all oxygen (Table 7.1). Oxygen isotopic compositions, and in fact all light stable isotope compositions, are reported as the ratio of the heavy minor isotope to the more abundant light isotope relative to this same ratio in a standard. For oxygen, the compositions are reported in the delta notation (δ) in units of per mil (parts per thousand) relative to Standard Mean Ocean Water (SMOW) such that

$$\delta^{18}O(\text{sample}) = [(^{18}O/^{16}O)\text{sample}/(^{18}O/^{16}O)\text{SMOW} - 1] \times 1000.$$

Differences in the $\delta^{18}O$ values between two phases, A and B, are quoted as $\Delta^{18}O(A-B) = \delta^{18}O_A - \delta^{18}O_B$, which is approximately analogous to the fractionation of oxygen isotopes between A and B. The total variation in the $\delta^{18}O$ values of terrestrial materials is about 100 per mil although the continental lithospheric mantle has a rather restricted range of $\delta^{18}O$ values from 4.5 to 7.5 per mil (Table 7.1). In contrast, the $\delta^{18}O$ value of lunar samples are all near 5.5 per mil (Taylor and Epstein 1970; Clayton *et al.* 1971) which coincides with the best estimate for the average composition of the Earth's upper mantle.

Oxygen is by far the most abundant element in the continental lithospheric mantle. Approximately 50 per cent by weight of most silicate minerals and melts, 92 weight per cent of water, and 75 weight per cent of CO_2 is oxygen. The reservoir of oxygen in the continental lithosphere is so vast that only very significant processes, such as the addition of substantial amounts of material with $\delta^{18}O$ values different from lithosphere, would be expected to affect the isotopic composition of the lithospheric mantle. Measured variations in the $\delta^{18}O$ values of the lithospheric mantle are 4.5 to 7.5 per mil but the causes of these variations are unclear. The following sections critically examine some of the possible causes and include a discussion of equilibrium silicate liquid–mineral fractionations, and the effect of open-system exchange and Soret diffusion on the isotope systematics of the continental mantle.

7.2.1. Oxygen isotope effects at high temperatures

Equilibrium processes

The fractionation of oxygen isotopes between any two phases varies primarily because of differences in the vibrational energies of molecules containing different isotopes. The vibrational energies of isotopic compounds vary as a function of temperature such that, at infinite temperature, the differences become vanishingly small as does the fractionation of oxygen isotopes between any two phases (Bigeleisen and Mayer 1947; Urey 1947; Bottinga and Javoy 1973, 1975). Two phases with similar physical properties, such as olivine and pyroxene, both of which are silicate minerals, have

Table 7.1. Abundance of stable isotopes of selected elements (Hoefs 1980 and references therein) and their typical range of isotopic compositions and average value in the continental lithospheric mantle

Element	Isotope	Relative abundance (%)	Range of values in continental lithospheric mantle	Average value
Hydrogen	^{1}H	99.9844	$\delta D = -120$ to -50	$\delta D = -80$
	D	0.0156		
Carbon	^{12}C	98.89	$\delta^{13}C = -25$ to -5	$\delta^{13}C = -10$
	^{13}C	1.11		
Nitrogen	^{14}N	99.64	$\delta^{15}N = -12$ to 0	$\delta^{15}N = 0$
	^{15}N	0.36		
Oxygen	^{16}O	99.7630	$\delta^{18}O = 4.5$ to 7.5	$\delta^{18}O = 6.2$
	^{17}O	0.0375		
	^{18}O	0.1995		
Sulphur	^{32}S	95.02	$\delta^{34}S = -2$ to $+10$	$\delta^{34}S = +3$
	^{33}S	0.75		
	^{34}S	4.21		
	^{36}S	0.02		

similar vibrational energy states and, consequently, have similar $^{18}O/^{16}O$ ratios even at temperatures of 900–1300°C in the continental lithospheric mantle.

In marked contrast, two phases with very distinct material properties, like olivine and a tholeiitic or CO_2-rich liquid may have vibrational energy states that differ enough such that the fractionation of oxygen isotopes between them may be several per mil even at magmatic temperatures. Thus, given equilibrium, the $\delta^{18}O$ values of coexisting olivine, pyroxene, spinel, and garnet in the continental mantle should be similar, whereas fluids may have distinctly different values from the dominant mineral phases.

Fractionation of oxygen isotopes between phases in the continental lithospheric mantle usually have been inferred from theoretical extrapolation of data from hydrothermal exchange experiments at temperatures less than 800°C or from empirical data on natural samples. Few isotope exchange experiments have been performed at high temperatures because of the enormous complexities involved with hydrothermal runs on large charges at temperatures in excess of 1000°C. Muehlenbachs and Kushiro (1974) attempted to circumvent these difficulties and examine the rate of diffusion of oxygen by exchanging oxygen isotopes in CO_2 and O_2 gas with basaltic liquid, plagioclase, and enstatite at 1100–1500°C. By combining the various CO_2–silicate fractionations, they found enstatite to be less ^{18}O-rich than basaltic liquid or plagioclase at temperatures less than about 1300°C, but ^{18}O-rich relative to basaltic liquid or plagioclase above 1300°C. At 1400°C, the difference between the $\delta^{18}O$ value of basaltic liquid and enstatite is about -0.5 per mil. Although these experiments were not reversed, they are significant in that they suggest that fractionations of oxygen isotopes between phases having very different properties in the continental mantle may be small, but are finite, measurable, and temperature-dependent.

The existence of small but significant temperature-dependent fractionations of oxygen isotopes between two distinctly different phases such as silicate liquid and enstatite at high temperatures is not surprising from theoretical considerations because only at infinite temperature should fractionations approach zero. Kyser and Walker (in preparation) have examined the possible equilibrium fractionation of oxygen isotopes between komatiitic and tholeiitic liquid and olivine and pyroxene by melting natural samples doped with Mg-silicates to vary the liquidus temperatures in isotopically distinct CO_2/CO atmospheres at quartz–fayalite–magnetite (QFM). Although the data are scattered because of the differential exchange rates of oxygen isotopes among CO_2, silicate liquid, and minerals, a temperature-dependent fractionation between silicate liquid and olivine or pyroxene is implied by the data (Fig. 7.1(a),(b)). As expected the structure of the silicate liquid also affects the fractionation as evidenced by the different temperatures at which the $\delta^{18}O$ values of different silicate liquids and minerals become identical. These data, and those of Muehlenbachs and Kushiro (1974), demonstrate that small but significant fractionations may exist at high temperatures between phases with very different properties. If the melt–mineral fractionations are valid, extraction of a mafic melt from the lithosphere may result in a residue enriched in ^{18}O by an amount that depends on the degree of partial

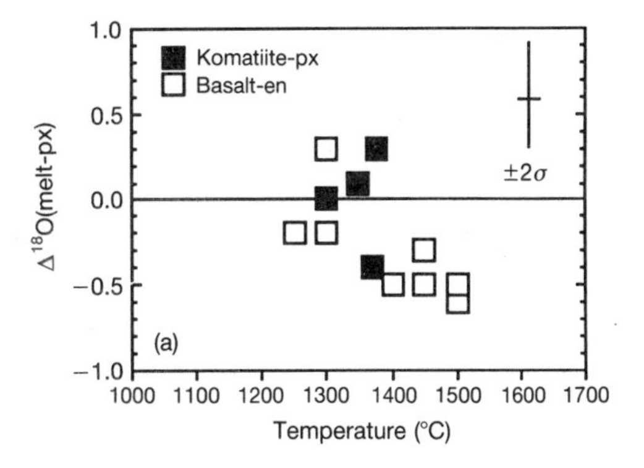

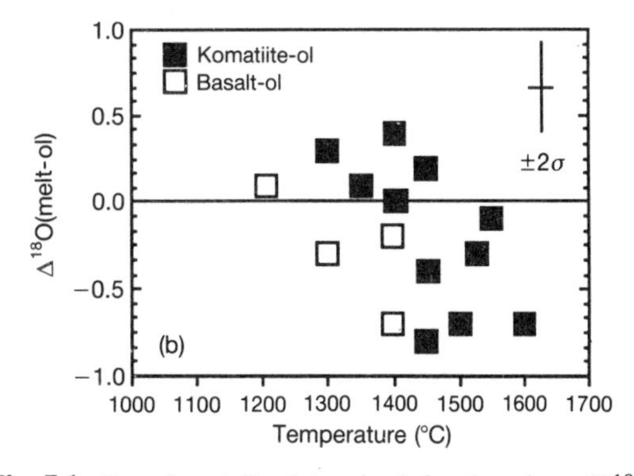

Fig. 7.1. Experimentally determined fractionations ($\Delta^{18}O$ melt-px, ol) between (a) komatiitic liquid-orthopyroxene (closed squares) and basaltic liquid-enstatite (open squares) and (b) komatiitic liquid-olivine (closed squares) and tholeiitic liquid-olivine (open squares) between temperatures of 1200 and 1600°C. Data suggest that melt-mineral fractionations change sign at temperatures that depend on the structure of the silicate liquid. Basalt-enstatite data from Muehlenbachs and Kushiro (1974) and all other results from Kyser and Walker (in preparation).

melting. Partial melting may then induce oxygen isotope heterogeneities in residual and refractory peridotite source regions.

Temperatures within the continental lithospheric mantle range from 900 to 1300°C and are high enough so that differences between the $\delta^{18}O$ values of coexisting silicate minerals should be small. Kyser *et al.* (1981) measured the chemical compositions and $\delta^{18}O$ values of coexisting olivine, clinopyroxene, orthopyroxene, and spinel in ultramafic xenoliths from both the continental and oceanic lithosphere and of glass and phenocrysts in mafic lavas. They argued that any differences in the $\delta^{18}O$ values of these phases should result solely from temperature because the assemblages form and reside in high-temperature environments where exchange rates are rapid. Using a variety of cation geothermometers and assuming that the assemblages represented equilibrium, they found that the per mil fractionations between coexisting clinopyroxene and orthopyroxene were all near zero (Fig. 7.2(a)), whereas apparent fractionations between pyroxene and olivine varied from $+1.2$ per mil at 900°C to -1.4 per mil at 1300°C (Fig. 7.2(b)). The 'cross-over' in the $\Delta^{18}O$(pyroxene–olivine) values, is determined primarily by data from xenoliths, which Kyser *et al.* (1981) incorrectly assumed were equilibrium assemblages. Gregory and Taylor (1986*a*,*b*) noted from the data of Kyser *et al.* (1981) that the $\delta^{18}O$ values of olivine are more variable than pyroxenes in the same way that the $\delta^{18}O$ values of feldspar are more variable than quartz in hydrothermally altered plutonic rocks or plagioclase values are more variable than pyroxene in ophiolites. Inasmuch as hydrothermal alteration results in non-equilibrium $\Delta^{18}O$(quartz–feldspar) and $\Delta^{18}O$(pyroxene–feldspar) values because feldspar exchanges isotopes more readily with water than does quartz or pyroxene, Gregory and Taylor (1986*a*,*b*) argued that many mantle xenoliths represent non-equilibrium assemblages because olivine exchanges preferentially with a metasomatic fluid. Further, they suggest that equilibrium $\Delta^{18}O$(pyroxene–olivine) values in mantle peridotites should be near zero.

The conjecture by Gregory and Taylor (1986*a*,*b*) that many mantle xenoliths represent non-equilibrium assemblages and that there is no 'cross-over' in pyroxene–olivine fractionations has only recently been verified experimentally. Clayton *et al.* (1989) have developed a technique whereby mineral–calcite fractionations are determined at high temperatures and 9–12 kbar. Chiba *et al.* (1988) reported no apparent cross-over in diopside–forsterite fractionations up to 1300°C using preliminary data from this technique. Rosenbaum, Kyser, and Walker (in preparation) employed an analogous technique using oxygen isotope exchange

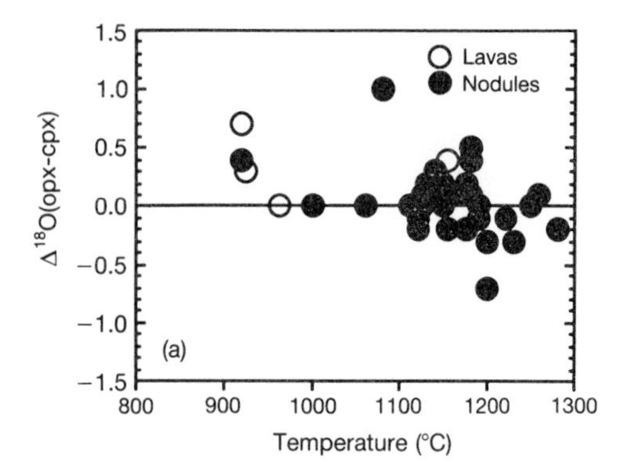

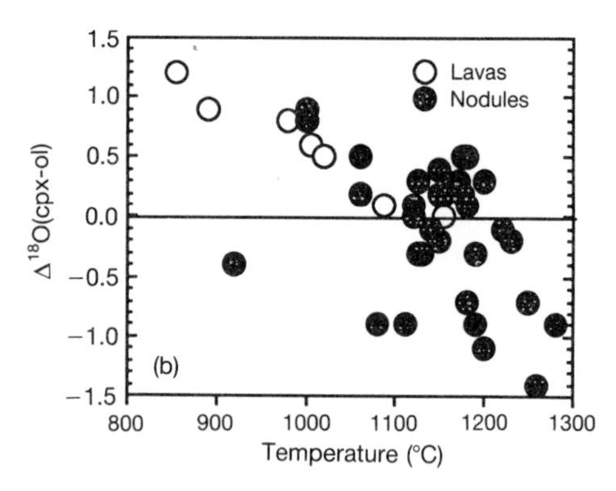

Fig. 7.2. Differences between $\delta^{18}O$ values of coexisting (a) orthopyroxene (opx) and clinopyroxene (cpx) and (b) clinopyroxene and olivine (ol) in mafic lavas (open symbols) and mantle peridotites (closed symbols) as a function of cation equilibration temperatures. These data suggest that coexisting orthopyroxene and clinopyroxene have the same $\delta^{18}O$ values when in equilibrium whereas clinopyroxene and olivine exhibit differences that vary with apparent cation equilibration temperatures. Gregory and Taylor (1986*a*) interpret the cpx–ol apparent reversal to reflect disequilibrium as a result of open-system exchange of olivine with a metasomatic fluid (data from Kyser *et al.* (1981) and Kyser and Stern (1988)).

between olivine–$BaCO_3$ and enstatite–$BaCO_3$ at temperatures of 1000 to 1400°C (Fig. 7.3(a)), from which pyroxene–olivine fractionations can be obtained (Fig. 7.3(b)).

Although the data have a substantial error, these experiments indicate two important phenomena: silicate–carbonate fractionations at temperatures in the lithosphere are about -1 per mil (Fig. 7.3(a)), and pyroxene–olivine and clinopyroxene–orthopyroxene

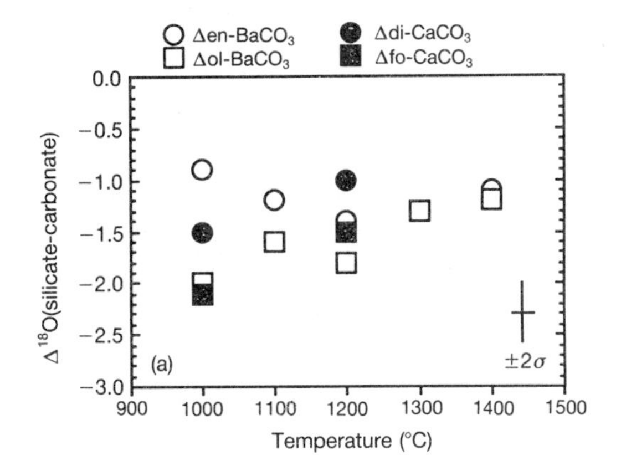

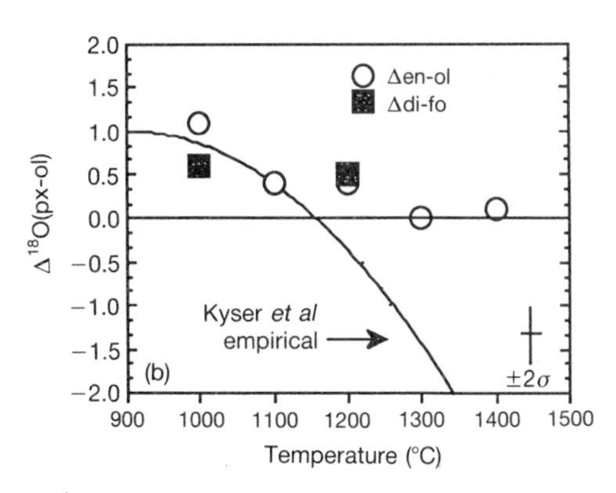

Fig. 7.3. (a) Experimentally determined fractionations between silicates (enstatite (en), diopside (di), and olivine (ol)) and carbonates over the temperature range 1000–1400°C using synthetic and natural pyroxene and olivine starting materials run at pressures of 15 kbar in a matrix of $BaCO_3$ or $CaCO_3$. Results for $BaCO_3$ systems from Rosenbaum, Kyser, and Walker (in preparation) and for $CaCO_3$ systems from Chiba *et al.* (1988). (b) Pyroxene–olivine (px–ol) fractionations determined by combining the appropriate silicate–carbonate fractionations of part (a). Shown for reference are the uncertainties and the empirical $\Delta^{18}O$(px–ol) curve derived from mantle peridotites by Kyser *et al.* (1981).

fractionations at these temperatures are near zero per mil.

Although the empirical fractionations derived for clinopyroxene–orthopyroxene by Kyser *et al.* (1981) agree with these experimental results, the empirical fractionations proposed by Kyser *et al.* (1981) for pyroxene–olivine obviously are in error and must have been derived from mantle peridotites having non-equilibrium mineral assemblages. Thus, silicate minerals in equilibrium in the continental lithospheric mantle should have similar $\delta^{18}O$ values, whereas silicate liquids, carbonates, and, possibly, oxides may have slightly different $\delta^{18}O$ values than coexisting silicate minerals. Consequently, equilibrium fractionations will result in only small disparities in the $\delta^{18}O$ values of the continental lithosphere and the melts generated therein.

Non-equilibrium processes—open-system exchange

Gregory and Criss (1986) and Gregory and Taylor (1986*a,b*) have detailed the consequences for the $\delta^{18}O$ values of minerals exchanging isotopes in open and closed systems. At temperatures near 1000°C, pyroxene and olivine have equilibrium fractionations of about 1 per mil (Fig. 7.3(b)), whereas at temperatures >1200°C the $\delta^{18}O$ values of clinopyroxene, orthopyroxene, and olivine approach the same values. Gregory and Taylor (1986*b*) showed that pyroxene and olivine having $\delta^{18}O$ values of 6.0 and 5.5 per mil, respectively, will experience only a small change in their $\delta^{18}O$ values during closed system exchange as shown in Fig. 7.4. Generally, the $\delta^{18}O$ values of pyroxene and olivine are more variable than 6.0 and 5.5 per mil, respectively in the continental mantle, so that a plot of $\delta^{18}O$(olivine) versus $\delta^{18}O$(pyroxene) for closed systems should fall within $\Delta^{18}O$(olivine–pyroxene)$=0$ to -1.0 per mil (i.e. >1200 to 1000°C; Fig. 7.4).

Under open-system conditions, during which an external fluid interacts at different rates with olivine, pyroxene, and spinel in a peridotite, the $\delta^{18}O$ value of one phase, such as olivine, will change more rapidly with time than other phases until equilibrium between the minerals and fluid is obtained at high fluid/rock ratios. Differential rates of exchange between any fluid and olivine and pyroxene will result in steeply sloping curves on a $\delta^{18}O$(olivine) versus $\delta^{18}O$(pyroxene) plot as shown in Fig. 7.4. This diagram represents open-system exchange of fluids having $\delta^{18}O$ values of $+5$ and $+8.5$ and temperatures of 1100°C with olivine and pyroxene having initial $\delta^{18}O$ values of 5.5 and 6.0 per mil, respectively. It is assumed that the fractionation between fluid and pyroxene at 1100°C is 1 per mil and the rate of exchange between the olivine and fluid is an order of magnitude faster than between pyroxene and fluid so that the olivine has fully equilibrated with the fluid when only about 20 per cent of the pyroxene has exchanged. Open-system exchange of fluids having $\delta^{18}O$ values not in equilibrium with olivine and pyroxene in the continental lithospheric mantle can result in $\delta^{18}O$ values of olivine and pyroxene in peridotites that do not fall within $\Delta^{18}O$(olivine–

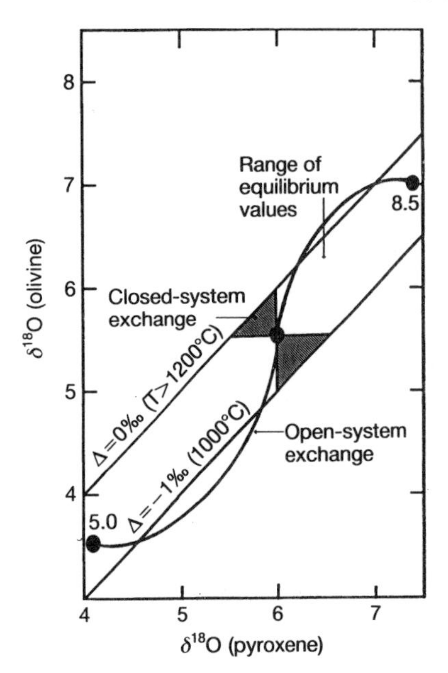

Fig. 7.4. Changes in the $\delta^{18}O$ values of olivine and pyroxene in a peridotite during closed-system exchange (cross-hatched area) and during open-system exchange with fluids having $\delta^{18}O$ values of 8.5 and 5.0 per mil (solid curves). The curves represent the changes with relative time in the $\delta^{18}O$ values of olivine and pyroxene having initial values of 5.5 and 6.0, respectively, assuming that the rate of exchange of oxygen isotopes between olivine and the fluid is 10 times faster than between pyroxene and fluid and that the equilibrium fractionation between the fluid and olivine and pyroxene is 1.5 and 1.0 per mil, respectively. Olivine and pyroxene in isotopic equilibrium at mantle temperatures and pressures should have $\delta^{18}O$ values that plot between $\Delta^{18}O$ (ol–px)$=0$ to -1 per mil (i.e. 1200 and 1000°C) (modified from Gregory and Taylor 1986*a*).

pyroxene) values of 0 to -1.0 per mil if one of the minerals exchanges isotopes at a faster rate with the fluid. If the exchange rates are similar or the fluid/rock ratio high enough to approach equilibrium, the $\delta^{18}O$ values of the olivine and pyroxene will move nearly diagonally on a $\delta^{18}O$(olivine) versus $\delta^{18}O$(pyroxene) diagram and all $\Delta^{18}O$(olivine–pyroxene) values should fall within 0 and -1 per mil.

The amount of external fluid needed to produce disequilibrium in the $\delta^{18}O$ values of minerals in the continental lithosphere during open-system exchange is an important consideration for the rheology and physical and chemical characteristics of the lithosphere. Kyser *et al.* (1986*b*) have modelled the amount of a CO_2-rich metasomatic fluid needed to change the $\delta^{18}O$ value of an olivine from 5.5 to 6.5 per mil assuming a fractionation of 1 per mil between CO_2 and olivine. The highest $\delta^{18}O$ value measured in phlogopite of probable metasomatic origin is 7.5 per mil (Sheppard and Dawson 1975), suggesting 8.5 per mil as an upper limit for the $\delta^{18}O$ value of metasomatic fluids in the mantle. If the $\delta^{18}O$ value of the CO_2-rich fluid is 8.5 per mil, a minimum of between 15 and 25 Wt % CO_2 must affect the olivine to increase the $\delta^{18}O$ value of the olivine by 1.0 per mil (Fig. 7.5). Similar results would be obtained using an H_2O-rich fluid. An interesting corollary of the open-system exchange model is that the $\delta^{18}O$ value of an isotopically anomalous fluid passing through a homogeneous lithosphere will achieve isotopic equilibrium with the lithosphere after encountering a minimum of about three volume equivalents of the lithosphere. Thus, unless the volume of the metasomatic fluid is substantial, the oxygen isotopic composition of a fluid will be buffered by that of the lithosphere through which it is passing during open-system exchange.

Soret diffusion

Another disequilibrium process that may affect the oxygen isotopic composition of the continental lithospheric mantle is Soret diffusion. The Soret effect involves the internal redistribution of species within an initially homogeneous phase in response to a temperature gradient. The temperature gradient imposes a

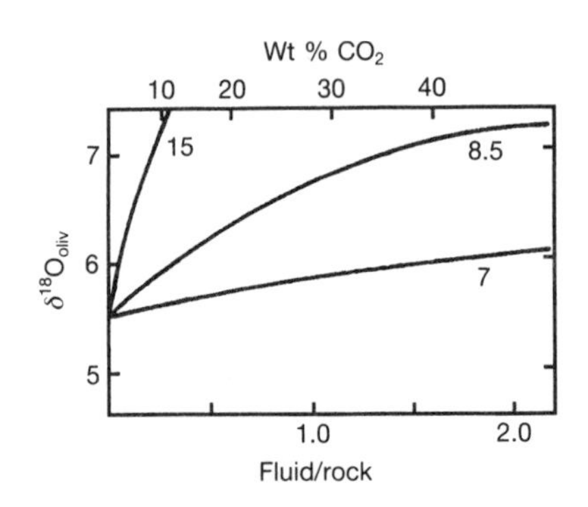

Fig. 7.5. Calculated changes in the $\delta^{18}O$ value of olivine having an initial value of 5.5 during open-system exchange with CO_2-rich metasomatic fluids having $\delta^{18}O$ values of 7, 8.5, and 15 per mil. Wt % CO_2 calculated from atomic fluid/rock ratio assuming that olivine comprises 80 per cent of the rock, that the fluid is pure CO_2, and that the fractionation between CO_2 and olivine is 1 per mil. The fluid/rock ratios represent minimum values. $\delta^{18}O$ values of metasomatic fluids as reflected by phlogopites in the mantle are less than 8.5 per mil (after Kyser *et al.* 1986*b*).

heterogeneous internal energy distribution which results in diffusion of species to equalize the internal energy. Substantial chemical gradients as a result of Soret separation have been experimentally observed in silicate liquids with temperature gradients of 250°C (Walker *et al.* 1981; Walker and DeLong 1982) and the results extrapolated to explain the chemical diversity of mid-ocean ridge basalts and lunar samples. During Soret separation, light species tend to accumulate at the hot end and more massive species at the cold end of a temperature gradient.

As with any chemical component, the effect of Soret diffusion on $^{18}O/^{16}O$ ratios in a temperature gradient in a silicate liquid depends on the structure of oxygen-bearing species in the melt. Although the exact nature of oxygen-containing species in complex silicate melts is presently unknown, one would expect from what is known about relative isotope fractionations that those containing network-former structural components such as SiO_2 and Al_2O_3 would be enriched in ^{18}O relative to network-modifier components. Walker and DeLong (1982) reported that both Si and Al concentrations were enhanced at the hot end of their experimental charges so that, if there is a Soret effect for oxygen isotopes dependent only on cation–oxygen compounds, $^{18}O/^{16}O$ ratios should be higher in the hottest portions of silicate liquids where Si and Al are concentrated.

The effect of Soret diffusion on the distribution of ^{18}O in silicate melts under conditions of an imposed temperature gradient has been investigated experimentally by Lesher, Kyser, and Walker (in preparation). Their results, shown in Fig. 7.6, indicate that the Soret effect on oxygen isotopes is unexpectedly substantial, with $\delta^{18}O$ values varying by almost 9 per mil along an imposed temperature gradient of 250°C over 4 mm. The magnitude of the Soret distribution of oxygen isotopes apparently is not greatly affected by the composition of the silicate liquid: ^{18}O is enriched in cooler portions of the melt in contrast to what is expected if oxygen isotopes are affected primarily by the distribution of network formers. However, the results are consistent with the general observation that larger species (i.e. ^{18}O relative to ^{16}O) migrate toward cooler regions as a result of Soret diffusion.

The effect of Soret diffusion on the chemical composition of geological systems usually is minimal because the rates of Soret separation are limited by chemical diffusion and the magnitude of temperature gradients under most geological conditions is small. Walker *et al.* (1981) and Walker and DeLong (1982) have discussed the geological implications of Soret separation and concluded that thermal boundary layers, such as between a silicate liquid and cooler country rock, provide the best natural environment to enhance the effects of Soret diffusion. A silicate liquid passing through the continental lithospheric mantle may enhance Soret effects because the system can involve temperature contrasts between a limited quantity of liquid and the lithosphere. Changes in the distribution of oxygen isotopes in the ascending liquid and formation of cumulate phases, with $\delta^{18}O$ values different from either the liquid or the surrounding mantle, may result because of the Soret effect. Open-system exchange between selected minerals in wall-rock peridotites and ^{18}O-rich liquids at the cooler boundary layer also may occur resulting in oxygen isotope disequilibrium of small portions of the lithosphere. The Soret effect would not only result in wall-rock minerals and cumulates which may have variable $\delta^{18}O$ values, but also provide a mechanism for progressively changing the $\delta^{18}O$ values of fluids as they ascend and exchange with the lithosphere. Although the separation of oxygen isotopes in silicate liquids as a result of the Soret effect is real, the magnitude of the effect is presently unknown in natural systems and must await further research.

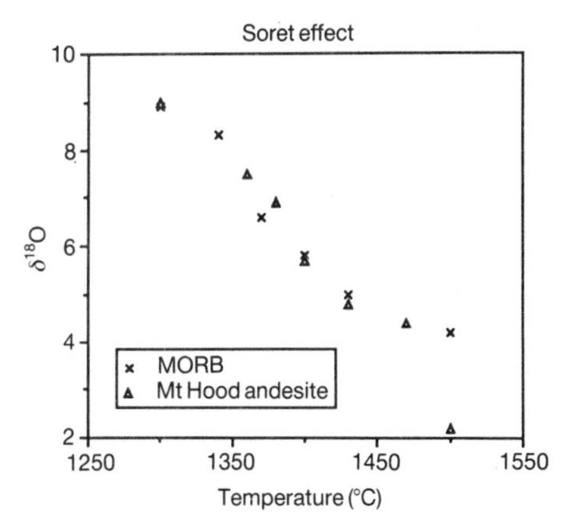

Fig. 7.6. Variations in the $\delta^{18}O$ values of two silicate liquids (MORB, crosses; Mt Hood andesite, triangles) in charges 4 mm in length over a temperature gradient of 200°C. The silicate liquids were held in the temperature gradient for longer than 200 hours and all of the chemical components were distributed in response to the Soret effect, with light components like SiO_2 diffusing preferentially to the hotter end of the charge. The effect of Soret diffusion at high temperatures on oxygen isotopes is substantial and may be applicable to fluid flow in the mantle where temperature gradients may exist between the fluid and the wall-rock peridotite (data from Lesher, Kyser, and Walker, in preparation).

7.2.2. Oxygen isotopes in eclogites

Eclogite xenoliths from diamondiferous kimberlites constitute an important suite of xenoliths because they are among the deepest samples of the continental lithospheric mantle. Although eclogites represent less than 10 per cent of the mantle xenolith suite in diamondiferous kimberlites, they range dramatically in major, trace, and isotopic composition, are extremely fertile in the components needed to generate basalts (MacGregor and Carter 1970; Jagoutz *et al.* 1984; MacGregor and Manton 1986), and include examples with Archaean model ages (Manton and Tatsumoto 1971; Kramers 1979; Jagoutz *et al.* 1984; McCulloch 1986 (see also Richardson, Chapter 3, this volume)). As such, eclogites represent samples of the continental lithospheric mantle with extreme compositions and model ages that must be accounted for in any model of the evolution of the Earth's mantle.

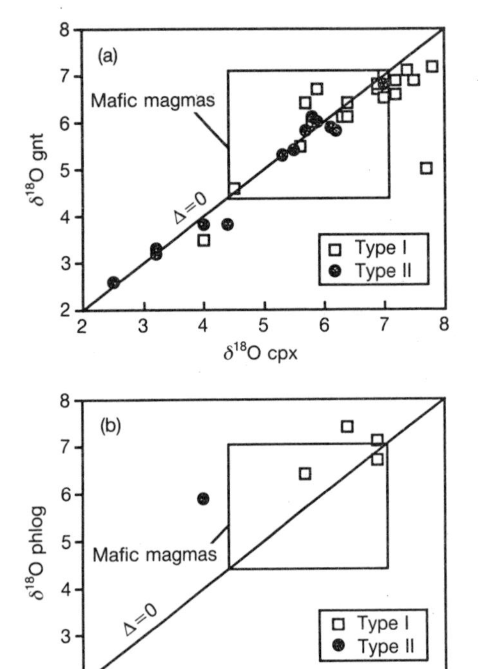

Fig. 7.7. Relations between the $\delta^{18}O$ values of (a) garnet (gnt) and clinopyroxene and (b) phlogopite (phlog) and clinopyroxene (cpx) in type I (squares) and type II (circles) eclogitic nodules from kimberlites in southern Africa. Type of eclogite is determined using the petrographic and mineralogic criteria of MacGregor and Carter (1970). Equilibrium $\delta^{18}O$ values of clinopyroxene, garnet, and phlogopite should be similar at higher temperatures. Most eclogites have garnet and clinopyroxene in isotopic equilibrium whereas phlogopite, which occurs along veins and was introduced by modal metasomatism, is not usually in isotopic equilibrium with the host minerals. Many eclogites have mineral $\delta^{18}O$ values similar to uncontaminated mafic magmas and some samples have isotopic compositions outside the range of normal silicate melts but similar to the range measured in hydrothermally altered oceanic crust (Fig. 7.8). Radiogenic isotope studies on selected samples have Archaean model ages but variable and younger internal ages (data from Vogel and Garlick 1970; Jagoutz *et al.* 1984; Ongley *et al.* 1987; Caporuscio *et al.* 1987).

As a group, eclogite xenoliths have the most diverse range in isotopic compositions of mantle materials, with limiting whole-rock $\delta^{18}O$ values of 2.2 to 7.9 per mil as inferred from the values of clinopyroxene and garnet (Fig. 7.7(a)). This range in $\delta^{18}O$ values for eclogites is about twice that of all other types of mantle xenoliths and clearly indicates that the oxygen isotopic composition of the continental lithosphere varies substantially in any region of the mantle where eclogite survives. In addition, the clinopyroxene and garnet in most eclogites have similar $\delta^{18}O$ values regardless of whether whole-rock values are high or low and, thus, are in isotopic equilibrium as might be expected for high-temperature systems in which the minerals are also in textural and chemical equilibrium (cf. Figs 7.2(a) and 7.3(b)). However, the $\delta^{18}O$ values of phlogopite, whose textural relations are indicative of formation by secondary infiltration metasomatism of the eclogites (Ongley *et al.* 1987), are greater than 5.9 mil and differ substantially from some of the $\delta^{18}O$ values of associated clinopyroxene (Fig. 7.7(b)) and garnet.

The chemical and oxygen isotope variations of mantle eclogites have been ascribed to either: (1) magmatic fractionation of clinopyroxene and garnet from silicate liquids at high pressures (e.g. MacGregor and Carter 1970; Kyser *et al.* 1982); or (2) metamorphism of subducted Archaean oceanic crustal cumulates and basalts that were altered by surficial processes (e.g. Helmstaedt and Doig 1975; Jagoutz *et al.* 1984; MacGregor and Manton 1986; Gregory and Taylor 1986*b*; Ongley *et al.* 1987). Evidence that many eclogites form from high-pressure magmatic processes includes: (1) experimental observations that mafic magmas have clinopyroxene and garnet as near solidus phases (Ito and Kennedy 1970); (2) many (but not all) eclogites have $\delta^{18}O$ values of 5 to 7 per mil, similar to the range in pristine mafic magmas (Kyser *et al.* 1982); and (3) the chemical compositions of composite and single types of eclogite can be related to trends expected from high-pressure fractionation of melts produced from metasomatized peridotites (Hatton and Gurney 1987).

However, the wide range in $\delta^{18}O$ values of eclogites is the most compelling evidence that at least some eclogite nodules represent metamorphic equivalents of hydrothermally altered oceanic crust. The $\delta^{18}O$ values of eclogites are within the range of intact sections of hydrothermally altered oceanic crust (Fig. 7.8), which varies from about 2 to 15 per mil (Spooner *et al.* 1974; Muehlenbachs and Clayton 1976; Heaton and Sheppard 1977; Gregory and Taylor 1981). Furthermore, Gregory and Taylor (1986*b*) have demonstrated that, if the source region of the eclogites was in the mantle and had $\delta^{18}O$ values of 5.7 per mil as reflected by the isotopic composition of the moon and mid-ocean ridge basalts, it would be impossible to generate the extreme $\delta^{18}O$ values of the eclogites from closed-system exchange.

MacGregor and Carter (1970), Jagoutz *et al.* (1984), and MacGregor and Manton (1986) divided eclogite xenoliths into two major groups on the basis of their $^{18}O/^{16}O$ ratios and mineralogical, chemical, and petrographical compositions. Type I (or A) eclogites have $\delta^{18}O$ values greater than the supposed mantle value of 5.7 per mil, are granular and layered, and contain pyroxenes with exsolution features, high K_2O contents, light rare earth element (LREE) enrichments, and occasional positive Eu anomalies. These properties were interpreted by MacGregor and Manton (1986) as reflecting metamorphism of crustal cumulates. Type II (or B) eclogites have $\delta^{18}O$ values less than 5.7 per mil, gneissic textures, pyroxenes with low K_2O contents and an absence of exsolution features, LREE depletions, modal phlogopite, and chemical compositions which vary somewhat systematically with $\delta^{18}O$ values of the eclogites. Systematic variations among the chemical compositions, $\delta^{18}O$ values, and modal phlogopite in type II eclogites are attributed to the effects of hydrothermal alteration on oceanic volcanic rocks and subsequent metamorphism to eclogites (MacGregor and Manton 1986). However, there is no reason to expect distinct $\delta^{18}O$ values for different types of eclogites if eclogites represent different portions of the oceanic crust because hydrothermal alteration does not affect all rock types systematically but varies with water/rock ratio and temperature. In fact, $\delta^{18}O$ values of hydrothermally altered oceanic crust trend from higher (+15) to lower (+2) values with depth (Fig. 7.8), but in an extremely irregular manner. Thus, $\delta^{18}O$ variations in eclogites may relate more to the processes during subduction, not the submarine environment.

The specific characteristics of eclogites which have been cited as evidence for hydrothermal alteration of the precursor are an inverse relation between $\delta^{18}O$ values and $^{87}Sr/^{86}Sr$ ratios, which is predicted for exchange with sea water, and a direct correlation between $\delta^{18}O$ values, K_2O contents, and Mg/Fe ratios (MacGregor and Manton 1986). Jagoutz *et al.* (1984) have used rare earth element (REE) patterns in eclogites to argue that the precursor was oceanic crust. However, Caporuscio *et al.* (1987) did not find the same relationships among

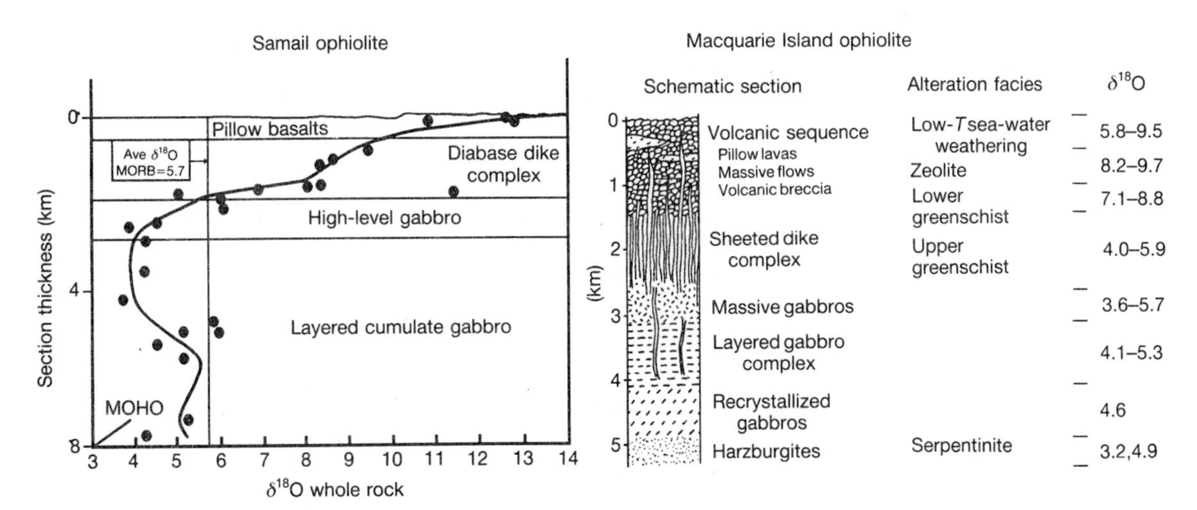

Fig. 7.8. Variation in the $\delta^{18}O$ value of whole-rock samples from the Samail ophiolite (after Gregory and Taylor 1981) and the Macquarie Island ophiolite (after Cocker *et al.* 1982). The range is similar to, but larger than, the range observed in eclogites from kimberlites. The trend in $\delta^{18}O$ values shown by the Macquarie Island ophiolite, wherein $\delta^{18}O$ values generally decrease irregularly with depth into the crust, is typical of most ophiolites (e.g. Stern *et al.* 1976; Sheppard 1980; Schiffman *et al.* 1984).

$\delta^{18}O$ values, petrography, K_2O contents of the pyroxenes, and REE as a function of eclogite type but did note correlations between $\delta^{18}O$ values and Al_2O_3, MnO, and FeO contents. On the basis of chemical data alone, Caproruscio *et al.* (1987) argued that most eclogites could be interpreted as originating by liquid–crystal fractionation at high pressures and only the range in $\delta^{18}O$ values indicates the involvement of lower-temperature processes.

The coexisting clinopyroxene and garnet in most eclogite nodules are in oxygen isotope equilibrium regardless of the whole-rock $\delta^{18}O$ values. Secondary effects are ubiquitous to eclogite, occurring as modal metasomatism in the form of veins (Hatton and Gurney 1987; Ongley *et al.* 1987), late-stage deuteric alteration, and infiltration of kimberlitic fluids (Caporuscio *et al.* 1987). In some samples, the $\delta^{18}O$ values of the phlogopites differ significantly from clinopyroxene (Fig. 7.7(b)) implying that the source of the fluids that produced the phlogopites may be unrelated to the eclogite. There is no evidence from the $\delta^{18}O$ values of the clinopyroxene and garnet that the eclogites have undergone open-system exchange with the metasomatic fluids associated with the phlogopites. Instead, the lowest $\delta^{18}O$ value of the phlogopites is in an eclogite with a low $\delta^{18}O$ value and the most ^{18}O-rich phlogopites are in ^{18}O-rich eclogites. The isotopic composition of the phlogopites appears to be affected by the isotopic composition of the primary garnet and pyroxene rather than the metasomatic fluids affecting the composition of the garnet and pyroxene, signifying a rock- rather than a fluid-buffered system (Ongley *et al.* 1987).

Eclogite nodules represent a real enigma to mantle petrologists. The major and trace element compositions can be modelled effectively as either near-surface processes or high-pressure differentiation effects; the wide range in the $\delta^{18}O$ values of eclogites is convincing evidence that near-surface effects are integral in the formation of at least some eclogites and eclogite-facies pillow basalts rule out high-pressure differentiation as the sole mechanism for generating eclogites. The radiogenic isotope systematics imply that the eclogites originally formed during the Archaean from a depleted reservoir and were affected by later fluids prior to incorporation in the kimberlite. The initial Archaean age and variable $\delta^{18}O$ values of eclogites have great implications for the structure, composition, and origin of the continental lithosphere: hydrothermally altered oceanic crust can be subducted into the mantle below the continental crust, be converted to eclogite, and survive incorporation, metasomatism, and melting for billions of years. This process operates in North America as well as Africa, is not restricted to the Archaean (Helmstaedt and Doig 1975; Ater *et al.* 1984), and produces eclogites from depths in excess of 120 km that contain diamonds and graphite (e.g. Hatton and Gurney 1979). That such rock types survive in the continental mantle attests to the extreme heterogeneity of the continental lithosphere and its non-convecting nature (see also Richardson, Chapter 3, this volume).

Eclogites have been cited by Gregory and Taylor (1986*b*) as reservoirs for metasomatic fluids which may affect peridotites during open-system exchange. Although large variations in the $\delta^{18}O$ values of eclogites and the substantial quantity of oceanic crust subducted beneath the continents throughout time make eclogites attractive reservoirs for metasomatic fluids, some problems do exist. These include: (1) the transition from gabbro or basalt to eclogite most likely occurs after or concurrent with the breakdown of most hydrous minerals in subduction zones so most eclogites are essentially anhydrous; (2) hydrous fluids in a downgoing slab of oceanic crust may be effectively removed at shallow levels via metamorphic reactions and the melts that are produced at convergent margins; (3) many eclogites themselves have been metasomatized by an extraneous and deep large ion lithophile element (LILE)-enriched fluid; and (4) no evidence exists for metasomatic fluids or melts having low $\delta^{18}O$ values as would be expected from the eclogites. Nevertheless, eclogite nodules are not only proof that oceanic crust can be subducted, but also that the oxygen isotopic and chemical composition of the continental lithospheric mantle is variable and that this variability can survive intact on a time-scale of billions of years (see also Richardson, Chapter 3, this volume).

7.2.3. Oxygen isotopes in peridotites

Alpine peridotites and peridotite xenoliths in alkalic lavas and kimberlites, in addition to eclogite nodules in kimberlites, represent intact portions of the continental lithospheric mantle. The oxygen isotopic composition of the continental lithosphere as inferred from studies of peridotite xenoliths and alpine peridotites has been the focus of several studies. Javoy (1980) measured the stable isotope systematics in spinel lherzolites from the alpine peridotite at Beni Bousera and was the first to report variations in the $\delta^{18}O$ values of minerals in peridotites. Subsequently, Kyser *et al.* (1981, 1982) examined the significant variations in $\delta^{18}O$ values of minerals in spinel lherzolite xenoliths in alkalic lavas from several localities in the western USA (Fig. 7.9(a)) and the Massif Central in France (Fig. 7.9(b)) and garnet peridotites from kimberlites in southern Africa (Fig. 7.9(c)). Kempton *et al.* (1988) reported constant

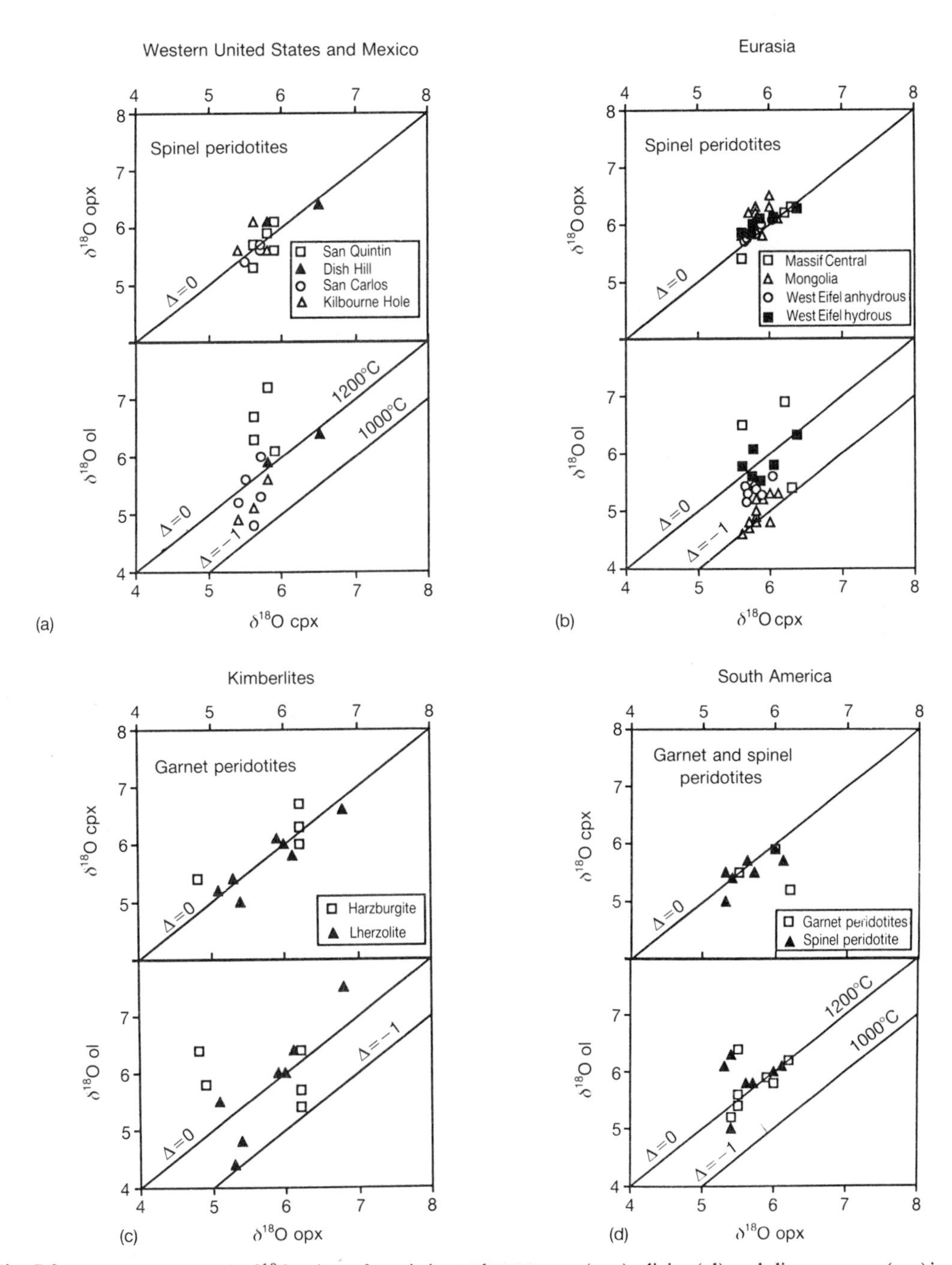

Fig. 7.9. Relation between the δ^{18}O values of coexisting orthopyroxene (opx), olivine (ol), and clinopyroxene (cpx) in: (a) spinel peridotite xenoliths from alkali basalts in the western USA (Dish Hill, CA; San Carlos, AZ; Kilbourne Hole, NM) and Mexico (San Quintin, Baja CA); (b) spinel peridotite xenoliths from alkali basalts in Europe (Massif Central and West Eifel) and Asia (Mongolia); (c) garnet peridotite xenoliths from South African kimberlites; and (d) garnet and spinel peridotite xenoliths from alkali basalts of the Patagonia Plateau of Chile and Argentina. Equilibrium fractionations for opx and cpx should be zero (i.e. $\Delta=0$) at temperatures above 1000°C, whereas differences between the δ^{18}O values of ol and cpx and opx at equilibrium varies with temperature from $\Delta=0$ at 1200°C to $\Delta=-1$ at 1000°C. Values of Δ^{18}O(ol–px) outside these limits indicate disequilibrium. Data for Mongolia from Harmon *et al.* (1986/87), West Eifel from Kempton *et al.* (1988), South America from Kyser and Stern (1988), and all other data from Kyser *et al.* (1981, 1982).

orthopyroxene–clinopyroxene fractionations but variable and reversed $^{18}O/^{16}O$ fractionations between pyroxene and olivine in anhydrous and hydrous spinel lherzolites from West Eifel, Germany (Fig. 7.9(b)). Kyser and Stern (1988) found oxygen isotope disequilibrium between pyroxene and olivine in a few spinel and garnet peridotites from the Patagonian Plateau in South America (Fig. 7.9(d)) but no relation between the degree of disequilibrium and the presence of hydrous phases. Harmon *et al.* (1986/87) reported consistent and sizeable differences of 1 per mil between the $\delta^{18}O$ values of clinopyroxene and olivine in relatively undepleted spinel lherzolites from a basanitic cinder cone in the Tariat Depression in Mongolia (Fig. 7.9(b)). They attributed these large fractionations and overall low $\delta^{18}O$ values to closed-system equilibrium exchange in a diapir that originated in a 'primitive' region of the asthenosphere and was trapped in the continental lithosphere.

Compilation of the oxygen isotope data from these studies reveals that the continental lithosphere, as reflected by spinel and garnet peridotites, is heterogeneous with mineral $\delta^{18}O$ values varying between 4.5 and 7.5 per mil (Fig. 7.9a–d). Further, the $\delta^{18}O$ values of coexisting clinopyroxene and orthopyroxene in most xenoliths usually are similar, suggesting isotopic equilibrium, whereas the $\delta^{18}O$ values of olivine in some samples differ from those of coexisting pyroxene, indicating that the olivine is not in oxygen isotope equilibrium with pyroxene (cf. Fig. 7.2).

Oxygen isotope disequilibrium between olivine and pyroxene is most pronounced in spinel lherzolites from the western United States and Mexico. When taken together the oxygen isotope systematics of these samples define a steeply dipping trend on a $\delta^{18}O$(olivine) versus $\delta^{18}O$(clinopyroxene) plot which transects the equilibrium fractionation lines corresponding to 1200 and 1000°C (Fig. 7.9(a)). Gregory and Taylor (1986*b*) modelled this trend as open-system exchange of olivine having an initial $\delta^{18}O$ value of 5.0 with a metasomatic fluid having a high $\delta^{18}O$ value of *c.* 8.5 per mil to explain positive $\Delta^{18}O$(olivine–pyroxene) values, and a fluid with a low $\delta^{18}O$ value of *c.* 5 per mil to explain the very negative $\Delta^{18}O$(olivine–pyroxene) values (Fig. 7.4). It is not clear, however, how fluids with end-member compositions of 5 to 8 per mil would remain physically separated or unbuffered by lithosphere (cf. Fig. 7.5). In fact, if $\Delta^{18}O$(olivine–pyroxene) values of 0 to -1 reflect the temperature of last equilibration, rather than disequilibrium in the presence of a low ^{18}O fluid (5 per mil), then the only disequilibrium $\Delta^{18}O$(olivine–pyroxene) values are those greater than 0 and there is no need to invoke low ^{18}O fluids. Two lherzolites from Dish Hill, California, have whole-rock $\delta^{18}O$ values that vary by 1 per mil but olivine and pyroxene in apparent isotopic equilibrium. In effect, only a few samples from the USA and Mexico have olivines with aberrant $\delta^{18}O$ values, and most of these are very depleted lherzolites from San Quintin, Baja Mexico, an area of recent subduction.

In marked contrast to most other suites of xenoliths, the clinopyroxene-rich spinel lherzolites from Mongolia have consistent pyroxene–olivine fractionations of about 1 per mil, corresponding to an oxygen isotope equilibration temperature of about 1000°C (Fig. 7.3). Harmon *et al.* (1986/87) reported pyroxene cation equilibration temperatures for these samples of 890–1094°C, in good agreement with the oxygen isotope temperatures. They also reported slight ^{18}O-enrichments of orthopyroxene relative to clinopyroxene in some samples which is predicted from theoretical calculations (Kieffer 1982) but not observed in lavas, peridotite xenoliths (Kyser *et al.* 1981), or peridotite massifs (Javoy 1980).

Because Harmon *et al.* (1986/87) did not find a correlation between $\delta^{18}O$ values and major and trace element compositions, they suggested that the lherzolitic xenoliths from Mongolia represent closed-system exchange in a diapir originally from a primitive portion of the asthenosphere that was emplaced into the lithosphere. Low degrees of partial melting occurred at *c.* 2 billion years ago with subsequent trace element metasomatism based on models using Sm–Nd isotope systematics. Remarkably, this metasomatism did not produce any hydrous phases nor were the oxygen isotope systematics of the olivines affected according to Harmon *et al.* (1986/87).

Kempton *et al.* (1988) adopted the open-system exchange model of Gregory and Taylor (1986*a*), wherein olivine preferentially exchanges with an ^{18}O-rich metasomatic fluid, to explain the oxygen isotope systematics in spinel peridotites from West Eifel, Germany (Fig. 7.9(b)). Xenoliths from West Eifel containing amphibole or phlogopite (hydrous peridotites), have clinopyroxenes with LREE enrichments, low ε_{Nd} values, and positive $\Delta^{18}O$(olivine–clinopyroxene) values, the latter indicative of open-system exchange. Kempton *et al.* (1988) attribute the ^{18}O enrichment in the olivines, and the LREE enrichment and low ε_{Nd} values of the clinopyroxenes to interaction of anhydrous lithosphere with an ^{18}O-rich aqueous fluid from subducted oceanic crust and sediments. However, there is no geological evidence for the presence of subducted oceanic crust in this area.

In support of the existence of an ^{18}O-rich metasomatic fluid in the Eifel area, they cite $\delta^{18}O$ values of 7.0 and

8.1 for two amphiboles, 6.0 for phlogopite in a wehrlite coexisting with the amphibole having the value of 7.0 and δD values of -50 to -39, which are between the mantle δD value of -80 and the sea-water δD value of 0 per mil. The phlogopite and amphibole from the same wehrlite have different $\delta^{18}O$ values, which is interpreted by Kempton *et al.* (1988) to signify the existence of more than one metasomatic fluid both of which presumably came from the same subducted material, at nearly the same depth, but released at different times. These metasomatic fluids must affect preferentially the $\delta^{18}O$ value of the olivines and, simultaneously, the trace element content but not the oxygen isotopic composition of the pyroxenes. In most systems, oxygen isotope exchange usually is concurrent with chemical exchange so that affecting the chemical but not the isotopic composition of the clinopyroxene by an ^{18}O- and LREE-rich fluid is unlikely. Furthermore, there is no *a priori* reason to expect the ^{18}O-rich fluid associated with amphibole to exchange more efficiently with the olivine, than the fluid with a $\delta^{18}O$ value near 6.0 associated with the phlogopite.

Garnet harzburgite and lherzolite xenoliths in kimberlites have the most variable $\delta^{18}O$ values for peridotites, with mineral $\delta^{18}O$ values ranging from 4.4 to 7.5 per mil (Fig. 7.9(c)). This range in $\delta^{18}O$ values encompasses the values of most eclogites from kimberlites (Fig. 7.7), although some eclogites have much lower $^{18}O/^{16}O$ ratios. There does not appear to be a difference in the range of $\delta^{18}O$ values of garnet harzburgites or garnet lherzolites (Fig. 7.9(c)). As will be shown, the range in oxygen isotopic composition of peridotite xenoliths from kimberlites is similar to that of continental basalt.

Garnet-bearing and garnet-free lherzolites and harzburgites from the Pleistocene Pali-Aike alkali basalts of the Patagonian Plateau lavas of southern South America afford an opportunity to study the effect of metasomatism on the oxygen isotope systematics of spatially-associated garnet and spinel peridotites and mafic lavas at convergent plate margins. $\delta^{18}O$ values of the minerals in the Patagonian xenoliths range between 4.9 and 6.5 per mil, with no difference in the range of values for either garnet or spinel peridotites (Fig. 7.9d). Sm–Nd and Rb–Sr isotope systematics of the peridotites indicate oceanic basalt-like source compositions, recent non-modal metasomatism, and the introduction of phlogopite veins that originated from subducted oceanic crust and sediments (Stern *et al.* 1986). Despite chemical, Nd–Sr isotopic, and mineralogical evidence for extensive metasomatism of the continental lithosphere under the Patagonian Plateau, only a few samples have olivines with aberrant $\delta^{18}O$ values (Fig. 7.9(d)). Furthermore, in contrast to the ^{18}O-rich hydrous phases in lherzolites with reversed pyroxene–olivine fractionations reported by Kempton *et al.* (1988) for the Eifel xenoliths, many of the $\delta^{18}O$ values of vein and disseminated phlogopite in the Patagonian xenoliths are less than 6.0 per mil and the olivine and orthopyroxene in many xenoliths appear unaffected by this metasomatism (Fig. 7.10).

Detailed analyses of the oxygen isotopic composition of the minerals in two metasomatized garnet harzburgites from the Patagonian Plateau (Fig. 7.11(a),(b)) exemplify the complexities associated with modal metasomatism: only occasionally are minerals in the environs of the 'metasomatic fluid' affected. The isotopic composition of the minerals in the xenolith represented in Fig. 7.11(a) demonstrates that the fluid can be ^{18}O-rich and produce phlogopite with high $\delta^{18}O$ values, but that only the $\delta^{18}O$ values of minerals next to

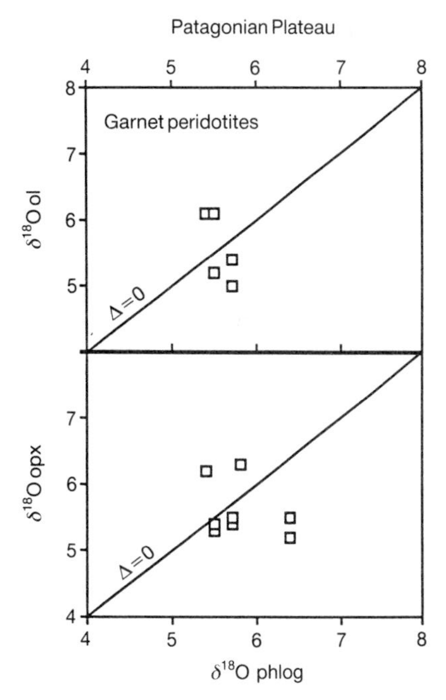

Fig. 7.10. Comparison of $\delta^{18}O$ values in olivine and orthopyroxene (opx) with phlogopite (phlog) in garnet peridotite xenoliths from alkali basalts of the Patagonian Plateau (Kyser and Stern 1988). All points should plot along $\Delta = 0$ at equilibrium in the lithosphere so that phlogopite, which occurs in veins and disseminated in the peridotite xenoliths, has not affected the isotope systematics of other minerals. Note that most phlogopites have $\delta^{18}O$ values near 5.5 per mil.

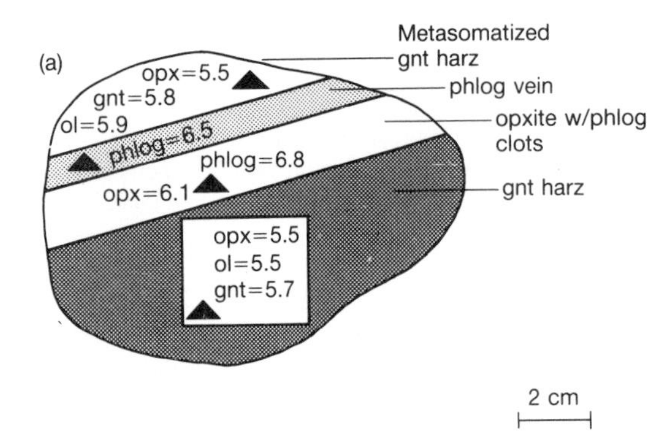

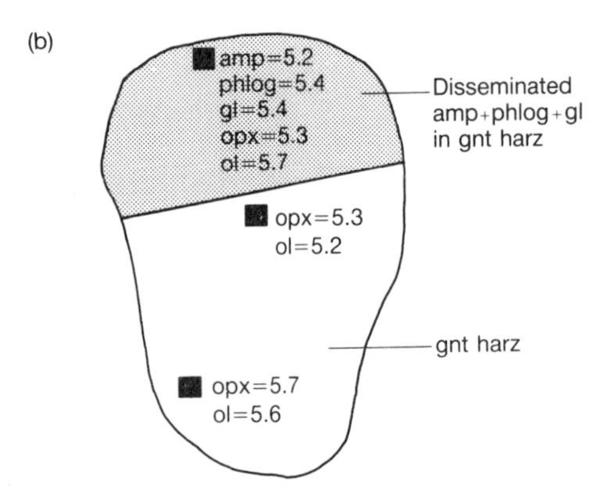

Fig. 7.11. Distribution of $\delta^{18}O$ values for orthopyroxene (opx), garnet (gnt), olivine (ol), phlogopite (phlog), amphibole (amp), and glass (gl) in two garnet harzburgite xenoliths from the Patagonian Plateau illustrating complexities associated with the effect of modal metasomatism on the lithosphere (data from Kyser and Stern, unpublished).

the vein are affected by metasomatism. The sample in Fig. 7.11(b) shows that olivine can be ^{18}O-rich relative to pyroxene in metasomatized peridotites, but the olivine in this sample was unaffected by the fluid that produced the amphibole and phlogopite, which have low $\delta^{18}O$ values. Relations among the $\delta^{18}O$ values of minerals in these two samples in conjunction with the wide range of the $\delta^{18}O$ values of hydrous phases in the Patagonian xenoliths collectively suggest that modal metasomatism, in some cases, may have little to do with the mechanism by which olivine becomes more ^{18}O-rich than pyroxene in the continental lithosphere.

Using the oxygen isotopic composition of peridotites from the continental lithosphere under the Patagonian Plateau (Figs 7.9(d) and 7.10) in conjunction with cation geothermometers and geobarometers (Stern *et al.* 1986), an idealized cross-section of the continental lithospheric mantle under Patagonia may be constructed (Fig. 7.12). Vein phlogopite has $\delta^{18}O$ values ranging from 5.4 to 6.5 per mil although most values are near 5.6 per mil (Fig. 7.10). If the fluids which affected the $\delta^{18}O$ values of the olivines are related to the most ^{18}O-rich phlogopites, the model of Gregory and Taylor (1986*b*) would require the unlikely scenario that the most ^{18}O-rich olivines have been totally exchanged but the pyroxenes were unaffected by this fluid. Furthermore, the existence of phlogopite with a range of $\delta^{18}O$ values requires devolatization of isotopically distinct portions of subducted crust unless the fluids which produced the phlogopites with low $\delta^{18}O$ values were originally ^{18}O-rich but were modified by exchange with the

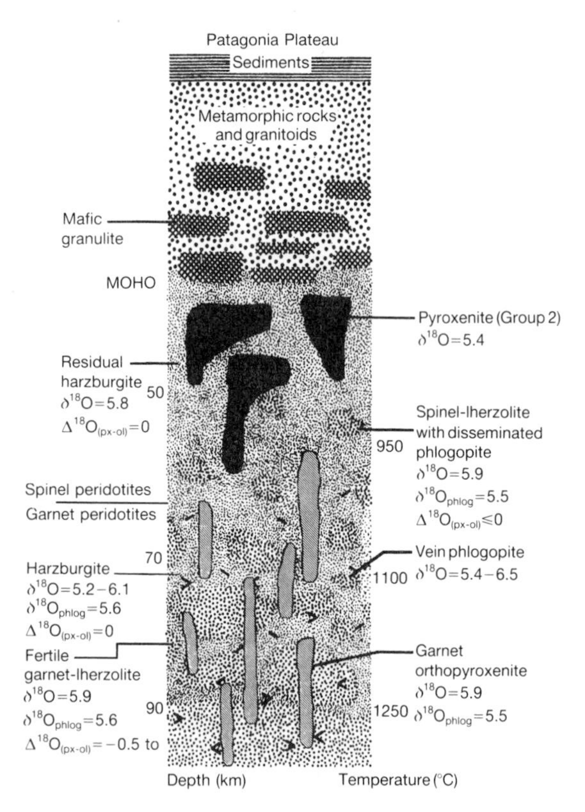

Fig. 7.12. Cross-section of the lithosphere under the Patagonian Plateau and possible distribution of rock types and their $\delta^{18}O$ values. Also indicated are $\delta^{18}O$ values of phlogopites and the range of $\Delta^{18}O$(px–ol) values. The depth of various peridotites were determined using cation geothermometers and geobarometers (modified from Stern *et al.* 1986).

peridotites (cf. Fig. 7.5). The cross-section for the Patagonian Plateau shown in Fig. 12 exemplifies the oxygen isotope heterogeneity in peridotites of the continental lithospheric mantle, and shows there is little relation between modal metasomatism and $\Delta^{18}O$(pyroxene–olivine) values as was suggested by Kempton *et al.* (1988) for the Eifel xenoliths.

Ultramafic inclusions from several localites can be classified into two groups (e.g. Frey and Prinz 1978; Irving 1980; Wilshire *et al.* 1980). Group 1, or Cr–diopside type xenoliths, are Mg- and olivine-rich peridotites which are residues of partial melts that have been slightly metasomatized to varying degrees by LILE-enriched melts. Group 2, or Al–augite type xenoliths, represent cumulates from mafic magmas or extensive interaction of metasomatic fluids with peridotites. Group 2 xenoliths from San Carlos, the Massif Central, and the Patagonian Plateau tend to have lower $\delta^{18}O$ values than Group 1 xenoliths from these areas (Fig. 7.13). Low $\delta^{18}O$ values of many Group 2 xenoliths are not consistent with their formation from SiO_2-undersaturated magmas because most alkalic lavas have $\delta^{18}O$ values in excess of 6.0 per mil (Kyser *et al.* 1982). Low $\delta^{18}O$ values imply formation from silicate liquids having low $\delta^{18}O$ values such as relatively primitive magmas from the asthenosphere (Kyser *et al.* 1982) and not from ^{18}O-rich fluids typically associated with metasomatism. Alternatively, higher $\delta^{18}O$ values of Group 1 relative to Group 2 could reflect the effect of the melt–mineral fractionations portrayed in Fig. 7.1 if Group 1 minerals consistently form at higher temperatures.

7.2.4. Oxygen isotopes in megacrysts

Megacrysts are large, single crystals that form at pressures in excess of 10 kbar and are found in many kimberlites and alkali basalts. The most common megacryst suite in kimberlite consists of ilmenite, clinopyroxene, orthopyroxene, garnet, and phlogopite, which have chemical compositions consistent with formation from disseminated nodules or from a differentiating alkalic magma (Jones 1987; Schulze 1987). Oxygen isotopic compositions have been measured only in phlogopite megacrysts from kimberlites (Fig. 7.14). $\delta^{18}O$ values vary from 5.1 to 7.2 per mil, similar to the range for minerals in peridotite xenoliths.

Alkali basalts contain megacryst suites consisting usually of clinopyroxene, amphibole, feldspar, and olivine. Irving (1984) classified these megacrysts into two groups, A and B, on the basis of their chemical compositions and mineralogy. Al–augite, olivine, and kaersutite amphibole are members of Group A and have chemical and Sr isotopic compositions that suggest they

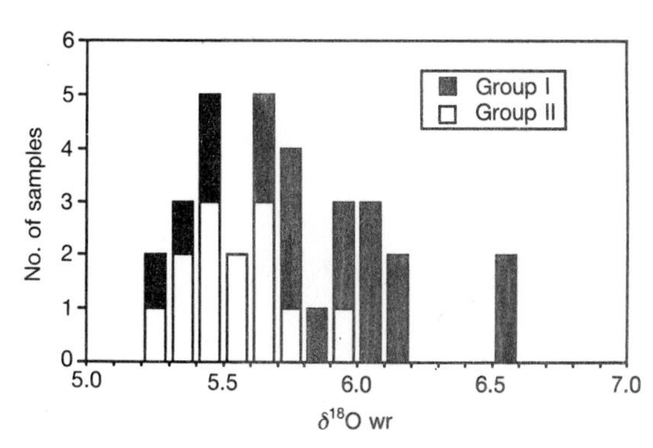

Fig. 7.13. Distribution of whole-rock (wr) $\delta^{18}O$ values for Group I (Cr-diopside) and Group II (Al-augite type) peridotite xenoliths from alkali basalts in North and South America. Group II xenoliths, which are genetically associated with mafic magmas or extreme metasomatism, have a lower and more restricted range of $\delta^{18}O$ values than Group I. Most alkali basalts have $\delta^{18}O$ values in excess of 6 per mil and may be related to Group II xenoliths only via disequilibrium processes (data from Kyser *et al.* 1982; Kyser and Stern 1988).

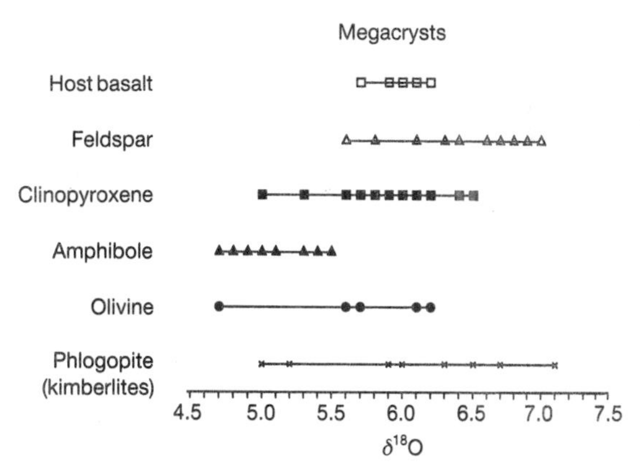

Fig. 7.14. $\delta^{18}O$ values of feldspar, pyroxene, amphibole, and olivine megacrysts and their host alkali basalts from the western USA and phlogopite megacrysts from kimberlites of South Africa. At any locality, $\delta^{18}O$ values of megacryst minerals are variable indicating most are not related to host lava or to each other. Phlogopite data from Sheppard and Epstein (1970), Sheppard and Dawson (1975), and Boettcher and O'Neil (1980); all other data from Kyser *et al.* (1981) and Kyser (unpublished).

formed at depth from alkalic basalts (Schulze 1987). However, in any individual suite, there usually is a range in compositions that signifies that many megacrysts must be xenocrysts unrelated to their hosts. This is especially true for Group B megacrysts in alkalic basalts, of which feldspar and mica are the most common.

$\delta^{18}O$ values of megacrysts from alkali basalts in the western USA are extremely variable, despite similar $\delta^{18}O$ values for the host basalts (Fig. 7.14). $^{18}O/^{16}O$ ratios generally increase in the order amphibole, olivine, pyroxene, feldspar although only the range in amphiboles is relatively restricted. At any locality, the range in the $\delta^{18}O$ values as well as chemical composition of a megacryst mineral is substantial, suggesting that the megacrysts are xenocrysts that may represent various trapped melts at depth (Wilshire and Pike 1975), in which case megacrysts are portions of alkalic sills or crystals plated on dyke walls at depth (Irving 1980), during which Soret effects may result in variable $\delta^{18}O$ values. Whatever the mechanism for the formation of megacrysts, the wide range in $\delta^{18}O$ values of megacrysts at any locality in the western USA indicates that: (1) many are xenocrysts; (2) many are unrelated, as supported by Sr isotopes (e.g. Kyser *et al.* 1981); and (3) there are silicate liquids in the continental lithospheric mantle with variable $\delta^{18}O$ values.

7.2.5. Oxygen isotopes and metasomatism

Gregory and Taylor (1986*b*) envisioned metasomatic fluids derived from: (1) exsolution from an ascending alkalic magma; or (2) devolatization or melting of subducted, hydrothermally altered oceanic crust. This fluid would be ^{18}O-rich and affect the equilibrium isotopic composition of nodules by metasomatism and transient, open-system exchange with olivine and spinel on a time-scale short enough so that more resistant pyroxenes are not affected. The high temperatures in the continental lithosphere (>900°C) would result in oxygen isotope equilibrium in a few hundred Ma so that nodules must be transported to the surface shortly after metasomatism to preserve disequilibrium. Re-equilibration will result in higher whole-rock $\delta^{18}O$ values than that of the original mantle of 5.7 per mil and $\Delta^{18}O$(olivine–pyroxene) values ≤ 0. Xenoliths with olivines having aberrantly high $\delta^{18}O$ values are not representative of most of the mantle in this model, which is supported by the data on nodules wherein only 20 per cent have positive $\Delta^{18}O$(olivine–pyroxene) values (Fig. 7.9(a)–(d)). Metasomatism and open-system exchange result in ^{18}O-enrichment of the mantle and are precursory to the production of ^{18}O-rich alkalic lavas. This ^{18}O-enrichment process from subducted altered oceanic crust must be world-wide: peridotite xenoliths from oceanic (Hawaii) areas devoid of subduction as well as continental areas have aberrant $\Delta^{18}O$(pyroxene–olivine) values (Kyser *et al.* 1981, 1982).

The range of $\delta^{18}O$ values of phlogopites and amphiboles of probable metasomatic origin in alkalic lavas, kimberlites, peridotites, and eclogites is substantial (Fig. 7.15) and comparable to the range of values in mafic lavas (Fig. 7.7) and minerals in peridotite xenoliths from the continental litospheric mantle (Fig. 7.9). However, the vast majority of hydrous minerals have $\delta^{18}O$ values of 5 to 6 per mil and no metasomatic minerals have values comparable to ^{18}O-depleted eclogitic xenoliths. Is this because metasomatic fluids or melts form from entire cross-sections of subducted oceanic crust, with an average $\delta^{18}O$ value of 5.7 (Gregory and Taylor 1981) or because most metasomatic fluids exchange with, and are buffered by, the $^{18}O/^{16}O$ ratio of the lithospheric mantle? Perhaps most metasomatic fluids originate in the asthenosphere, which may also have heterogeneous oxygen isotopic compositions because of the input of subducted components, and thus affect, but do not originate from, the lithospheric mantle. The high $\delta^{18}O$ values of megacrysts and aberrant $\delta^{18}O$ values of olivines in garnet perido-

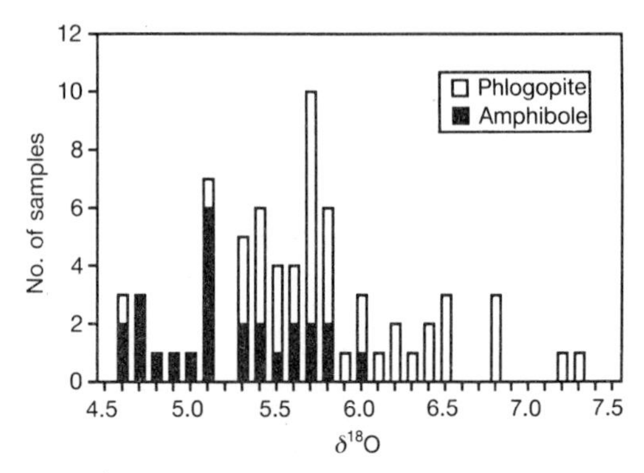

Fig. 7.15. Compilation of $\delta^{18}O$ values of phlogopite and amphibole veins and disseminations in xenoliths and megacrysts in alkali basalts and kimberlites. Most amphiboles are from alkali basalts and their spinel xenoliths, whereas most phlogopites are from kimberlites and their garnet peridotite xenoliths. If these hydrous minerals reflect metasomatic fluids, most fluids have $\delta^{18}O$ values of 5–6 per mil and most fluids associated with phlogopite are more ^{18}O-rich (data from Sheppard and Epstein 1970; Sheppard and Dawson 1975; Boettcher and O'Neil 1980; Kyser *et al.* 1982; Kyser and Stern 1988).

tites support this model in that both are related to melts or fluids that may have formed in the asthenosphere (see also Menzies, Chapter 4, this volume).

7.2.6. Oxygen isotopes in mafic lavas

Problems with using mafic lavas erupted on the continents to reflect the composition of the continental lithospheric mantle are threefold: (1) the components of mafic lavas are unstable near the surface and can be hydrothermally altered to lower $\delta^{18}O$ values or hydrate and weather at low temperatures to ^{18}O-rich products; (2) ascent of mafic lavas through the sialic crust may result in contamination of the magma and produce compositions similar to those of contamination by subducted material at the source of mafic melts; and (3) there is considerable debate about which chemical components in mafic melts originate in the lithosphere and which in the asthenosphere (see Menzies and Kyle, Chapter 8, this volume).

The effects of secondary alteration on the $^{18}O/^{16}O$ ratio of mafic lavas can be circumvented by analysing phenocrysts and calculating whole-rock $\delta^{18}O$ values assuming the fractionation for melt-plagioclase is zero, melt-pyroxene is 0.2 per mil, and melt-olivine is 0.5 per mil for the near-liquidus temperatures of mafic melts (Muehlenbachs and Kushiro 1974; cf. Fig. 7.1). Measured whole-rock $\delta^{18}O$ values can differ substantially from calculated whole-rock values as exemplified by data for boninites, andesites, tholeiites, and alkali basalts from several localities (Fig. 7.16(a)) and for Pliocene and Quaternary alkali basalts from the Patagonian Plateau in South America and cogenetic alkaline lavas from the Tuxtla Volcanic field, Mexico (Fig. 7.16(b)). There is no relation between differences in the measured and calculated $\delta^{18}O$ values and age of the basalt, chemical composition, or H_2O^+ contents for alkaline lavas from Mexico or for the Patagonian Plateau alkali basalts. This is not surprising because hydrothermal alteration can selectively affect only the oxygen isotopic composition of lavas (e.g. Forester and Taylor 1976) and low-temperature alteration can produce substantial ^{18}O-enrichments via hydration, formation of clay minerals, and non-hydrous authigenic phases. Studies that assume linear relations between H_2O^+ contents and $\delta^{18}O$ values to estimate the original $\delta^{18}O$ values prior to secondary alteration (e.g. Ferrara *et al.* 1985; Harmon *et al.* 1987) may not be valid for these reasons and also because initial H_2O^+ contents of the lavas must be assumed.

$\delta^{18}O$ values of mafic lavas estimated from analyses of phenocryst phases or from measurements of unaltered, pristine samples from oceanic and continental areas vary substantially and similarly (Fig. 7.17). The predominant type of lava on earth, mid-ocean ridge basalt (MORB), has a rather restricted range of $\delta^{18}O$ values, averaging +5.7 per mil, which compares favourably to lunar basalts (e.g. Clayton *et al.* 1971). Although the overlap in the $\delta^{18}O$ values of oceanic island basalts (OIB) and MORB is considerable, tholeiites from oceanic islands tend to have lower $\delta^{18}O$ values than MORB, averaging +5.4 per mil whereas oceanic alkalic lavas tend to be more ^{18}O-rich, averaging +6.1. Mafic lavas from continental areas have $\delta^{18}O$ values that range from +4.9 to +8.0 per mil, similar to that observed in OIB and peridotite nodules from the

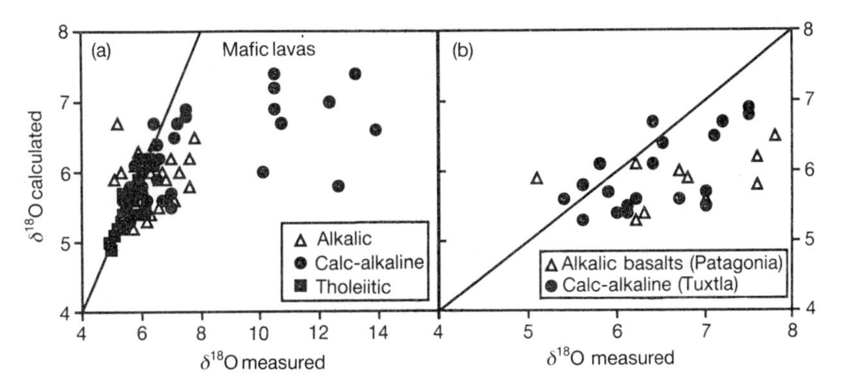

Fig. 7.16. Comparison of measured whole-rock $\delta^{18}O$ values with $\delta^{18}O$ values calculated using phenocrysts for: (a) alkalic lavas from the western USA, calc-alkaline basalts and andesites from Mexico, boninites from various areas (Kyser *et al.* 1986*a*), and tholeiites from oceanic basalts; and (b) alkali basalts from the Patagonian Plateau (Kyser and Stern 1988) and alkaline basalts and andesites from the Tuxtla Volcanic field, Mexico. Measured whole-rock $\delta^{18}O$ values do not reflect the original isotopic composition of the lava in many cases because of the effects of hydrothermal and low-temperature alteration of the groundmass.

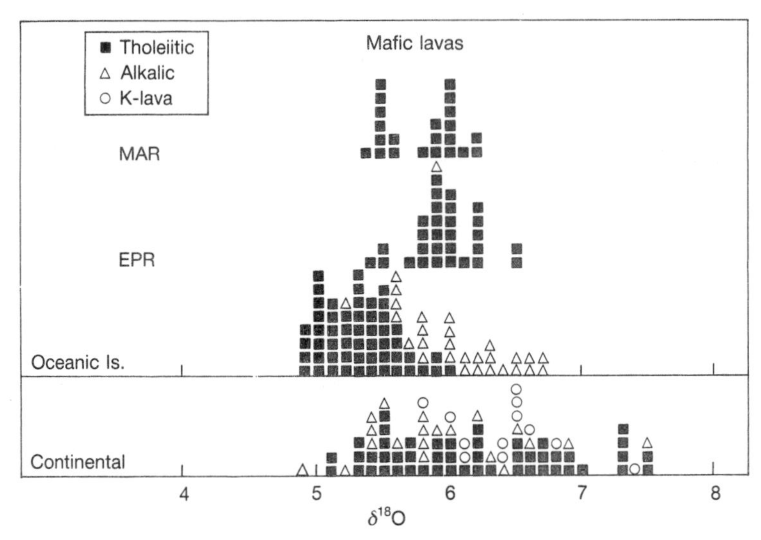

Fig. 7.17. Whole-rock $\delta^{18}O$ values of unaltered and uncontaminated tholeiites, alkalic basalts, and potassic lavas (K-lava) from oceanic and continental areas. Tholeiites from oceanic islands generally have lower values than many tholeiites from the Mid-Atlantic Ridge (MAR) and the East Pacific Rise (EPR) and alkalic lavas from oceanic islands. Tholeiites and alkalic lavas from continental areas span the entire range covered by oceanic basalts (after Kyser 1986).

continental mantle. Unlike the correlation between the oxygen isotopic composition and lava type that is evident in the oceanic environment, $\delta^{18}O$ values of both continental tholeiites and alkali basalts range from 5 to 7.5 per mil. Most potassic lavas from continental areas consistently have high $\delta^{18}O$ values of 6 to 8 per mil and andesites and boninites are also ^{18}O-rich with values between 6.5 and 8.0 per mil (e.g. Javoy 1970; Matsuhisa and Kurasawa 1983; Kyser *et al.* 1986*a*; Dobson and O'Neil 1987). The $\delta^{18}O$ value of the mantle beneath the continents and the oceans as sampled by mafic lavas is extremely heterogeneous (see also Haggerty, Chapter 5, this volume).

Convergent margin magmatism

Contamination of mafic lavas by continental crust has been suggested for calc-alkaline lavas on the basis of the relation between $\delta^{18}O$ values and Sr–Nd–Pb isotopes. James (1981) and Taylor and Sheppard (1986) have reviewed the expected relations between oxygen and strontium isotopic compositions for source contamination and crustal contamination of a magma (Fig. 7.18). Essentially, if the source (i.e. lithospheric or asthenospheric mantle) is contaminated by a fluid released from a subducted slab, the fluid will be enriched in incompatible elements including water and induce melting in the overlying mantle. The Sr and other incompatible elements will reflect the slab, but oxygen isotopes and other major elements will come primarily from the overlying mantle. The result on a $\delta^{18}O$ versus $^{87}Sr/^{86}Sr$, $^{144}Nd/^{143}Nd$, or $^{206}Pb/^{204}Pb$ plot will be a convex downward trend because the trace elements are affected most with the addition of the slab component (Fig. 7.18). The opposite is most likely with crustal contamination (i.e. convex upward patterns) because of the large reservoir of oxygen relative to trace elements in the crust and because the crust is usually distinct in its $\delta^{18}O$ value relative to the magma. Crystal fractionation, which must accompany crustal contamination, will produce variations in $\delta^{18}O$ values of less than 1 per mil (e.g. Taylor and Sheppard 1986) and more substantial variations in other elements depending on whether the phases crystallizing contain Sr, Nd, or Pb. Nevertheless, the convex upward pattern should persist.

Ito and Stern (1985/86) studied the oxygen and strontium isotope systematics in presumably unaltered andesitic rocks of the Mariana Arc where crustal contamination should be minimal. $\delta^{18}O$ values varied from 5.5 to 6.8 per mil with little variation in Sr isotopes. They attributed the variation in oxygen isotopes to a mixture of 1 per cent ^{18}O-rich sediment-derived fluids with OIB-type mantle. Woodhead *et al.* (1987) proposed a similar model in their study of volcanic rocks from the Northern Mariana Islands. Margaritz *et al.* (1978) reported more variable $\delta^{18}O$ values from 6.0 to 8.0 per mil for feldspars in andesites from the Banda Arc,

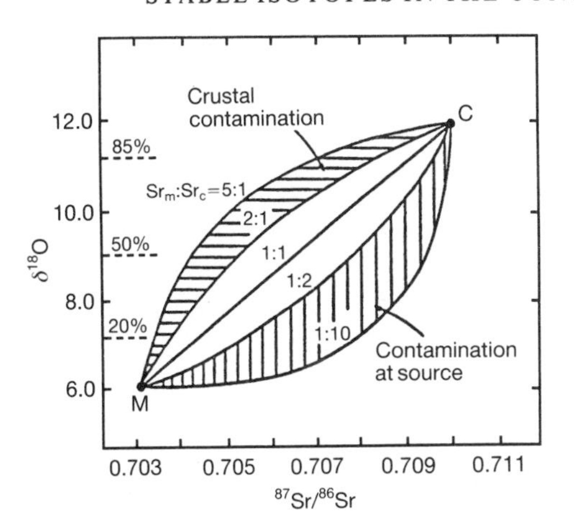

Fig. 7.18. Changes in the $\delta^{18}O$ value and $^{87}Sr/^{86}Sr$ ratio of a magma (M) during mixing with a crustal component (C). The ratio of oxygen in the magma and the crustal component are equal but the ratio of Sr in the magma (Sr_m) to Sr in the crustal component (Sr_c) varies from 5:1 to 1:10. Sr_m:Sr_c ratios in excess of 2:1 typify contamination of a magma as it ascends through continental crust, whereas ratios less than 1:2 are expected if a metasomatic fluid derived from subducted crust and rich in incompatible elements interacts with the source of the magma. Selected weight percentages of the crustal contamination are also indicated (after James 1981; Taylor and Sheppard 1986).

Indonesia, and ascribed the convex upward relation between oxygen and strontium isotopes to contamination of the source by up to 50 per cent of subducted continental crust.

Boninites are primitive magmas associated with subduction zones which have U-shaped REE patterns, OIB Nd isotopic compositions, and $\delta^{18}O$ values of 5.8 to 7.0 per mil (Kyser *et al.* 1986*a*; Dobson and O'Neil 1987). As with calc-alkaline rocks in island arcs, these relations have been interpreted to reflect mixing primarily between two components: a depleted portion of the lithosphere, which is the source of oxygen and other major elements, and a hydrous, LREE-enriched minor component (1–15 per cent) derived from subducted crust. Thus, most studies involving plate convergence at oceanic boundaries suggest that the source of calc-alkaline magmas is lithospheric mantle affected slightly by subducted components.

Interpretation of oxygen isotope and chemical data from calc-alkaline rocks at convergent margins involving continental crust is substantially different from interpretation of data involving island arcs. The determination of $^{18}O/^{16}O$ ratios in late Cainozoic Andean volcanic rocks have been part of several studies, most of which conclude that mafic magmas originate from a heterogeneous source in the asthenosphere or lithosphere and are variably contaminated by continental crust (e.g. James 1982, 1984; Longstaffe *et al.* 1983; Harmon and Hoefs 1984; Stern *et al.* 1984). This is especially true for basalts of the central volcanic zone of the Andes, where the crust is thickest and the $\delta^{18}O$ values of the basalts are all high, from 7.0 to 8.2 per mil. Contamination by continental crust of mafic magmas during ascent in the other volcanic zones of the Andes is based on assumed compositions of uncontaminated source regions, which are quite variable based on the isotopic compositions of xenoliths.

Detailed studies of the petrology of Cretaceous batholiths in California by Taylor and Silver (1978), Masi *et al.* (1981), and Hill *et al.* (1986) indicate that source regions for these batholiths consist of three major components: (1) MORB-like crust; (2) subducted oceanic basalts and sediments; and (3) continental lithospheric mantle (Fig. 7.19). The models are generally analogous to those involving the Andes and confirm

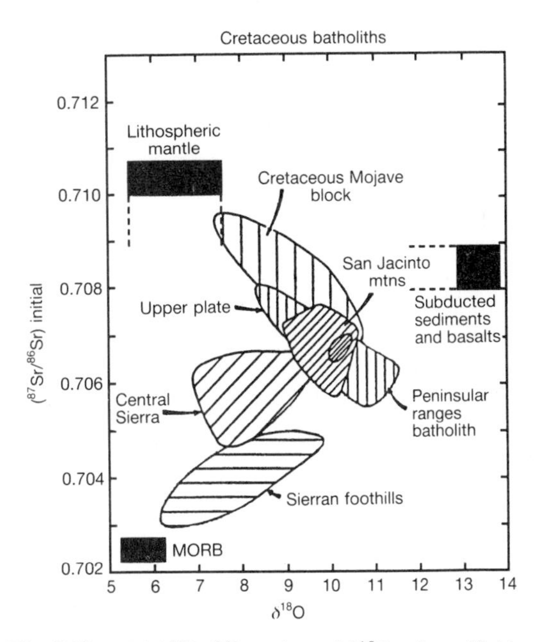

Fig. 7.19. Initial $^{87}Sr/^{86}Sr$ ratios and $\delta^{18}O$ values of fields of data for various groups of Cretaceous plutonic granites from California and possible source compositions. Values for mafic intrusive rocks from each area are similar to those of the granites (modified after Taylor and Sheppard 1986).

that subducted components and the continental lithosphere are instrumental in the production of calc-alkaline magmas. Note that both the lithospheric and MORB-like component of Fig. 7.19 could have variable $\delta^{18}O$ values of 5 to 7.5 per mil without affecting the model.

Potassic lavas

K-rich volcanic rocks from Italy have been studied in detail to determine the origin of these unique magmas. These magmas are alkali-rich, usually silica undersaturated, and devoid of xenoliths indicating slow ascent rates, thus making them prime candidates for studies on the effects of crustal contamination. $\delta^{18}O$ values in Italian K-rich lavas vary from 7.5 to 12 per mil and correlate roughly with $^{87}Sr/^{86}Sr$ ratios of 0.707 to 0.711.

One group of researchers (Taylor *et al.* 1979; Ferrara *et al.* 1985; Turi *et al.* 1986) believe that the K-rich volcanic rocks reflect mixing among four components: (1) continental lithospheric mantle, having $\delta^{18}O$ values of 5.7 per mil; (2) a metasomatic fluid, which enriches the lithosphere in ^{18}O to $\delta^{18}O$ values of 7.5 per mil; (3) ^{18}O-rich continental crust or crustal magmas which contaminate the K-rich magmas during ascent; and (4) hydration, further enriching some lavas in ^{18}O. They cite as evidence a maximum of 7.5 per mil for primary melts from the mantle, convex $\delta^{18}O$ versus $^{87}Sr/^{86}Sr$ patterns, and major and minor element variations with $\delta^{18}O$ values. Another interpretation (Holm and Munksgaard 1982, 1986; Holm *et al.* 1982) views most of the variation in oxygen and Sr isotopes as primarily source-related, with fluids released from ^{18}O- and ^{87}Sr-rich subducted sediments metasomatizing overlying continental lithospheric mantle up to $\delta^{18}O$ values of 9.4 per mil with limited ^{18}O-enrichment by crustal contamination or hydration. They cite as evidence the difficulty in maintaining primitive magma chemistry of some ^{18}O-rich samples during contamination, and the uncertainty of extrapolating $\delta^{18}O$ values from H_2O^+ contents as was done by the Taylor *et al.* group. Both groups share the same fundamental view involving ^{18}O-rich metasomatism of normal lithospheric mantle but differ about the possible degree of ^{18}O-enrichment with metasomatism.

Continental lithospheric mantle xenoliths and megacrysts all have $\delta^{18}O$ values less than 7.5 per mil as do hydrous minerals directly associated with metasomatism. Magmas such as boninites and island arc basalts, which have a component (albeit a small component) of subducted crust in their sources, also have $\delta^{18}O$ values less than 7.5 per mil so it seems unlikely that metasomatized lithosphere can yield silicate liquids with $\delta^{18}O$ values in excess of 7.5 per mil. However, it is also unlikely that there is a unique composition for either the metasomatic fluid or the lithosphere so that modelling on the basis of single end-members may be unrealistic.

Much of the controversy about the Italian K-rich lavas relies on oxygen isotope analyses of whole-rock samples or leucite phenocrysts, both of which are easily altered by secondary fluids (e.g. Taylor *et al.* 1984). Petrography is no criteria for unaltered $^{18}O/^{16}O$ whole-rock ratios as evidenced by the substantial differences between measured $\delta^{18}O$ values and those calculated from phenocryst $\delta^{18}O$ values for petrographically 'fresh' calc-alkaline lavas of the Tuxtla volcanic field or the alkali basals of the Patagonian Plateau shown in Fig. 7.16(b). Consequently, the primary K-rich lavas are ^{18}O-rich, but to what extent the source is heterogeneous or what effects crustal contamination and low-temperature processes have are yet to be convincingly resolved.

7.3. HYDROGEN ISOTOPES

Hydrogen is a major constituent of metasomatic fluid and, thus, its isotopic composition is a potential tracer of fluid flow in the continental lithospheric mantle. Kyser and O'Neil (1984) examined the hydrogen isotope systematics of oceanic basalts and concluded that their primary δD values were all -80 ± 5 per mil, despite substantial variations in their Sr, Nd, Pb, rare gas, and oxygen isotopic compositions. This δD value is similar to that of magmatic waters calculated for the interior of the alkaline Lilloise complex (Sheppard *et al.* 1977) and at the low end of the range from -80 to -40 per mil measured in primary micas from continental lithospheric xenoliths (Sheppard and Epstein 1970; Sheppard and Dawson 1975; Boettcher and O'Neil 1980). In contrast, subducted water has δD values of -80 to -30 as estimated from studies of subducted sediments (Magaritz and Taylor 1986) and hydrothermally altered rocks from ophiolite complexes (Heaton and Sheppard 1977; Gregory and Taylor 1981).

7.3.1. Hydrogen isotope effects at high temperatures

Hydrogen isotopes may fractionate among various phases in the continental lithosphere provided the hydrogen is bound to elements which have distinct chemical characteristics. For example, Kyser and O'Neil (1984) suggest that degassing of water vapour from a magma will result in lower δD values for the water remaining because water vapour is enriched in deuterium relative to water (H_2O or OH) dissolved in the magma. In contrast, loss of reduced hydrogenous gases such as H_2 or CH_4, both of which are extremely depleted in deuterium relative to water, will result in higher δD values for the hydrogen remaining in the magma (see also Haggerty, Chapter 5, this volume). These effects will only be pronounced in systems where a distinct vapour separates from a liquid phase.

Richet *et al.* (1986) suggested that the D/H fractionation between water vapour and albite melt should be in excess of 25 per mil at lithospheric temperatures based on experiments in the water–albite system. Suzuoki and Epstein (1976) measured the fractionation between water and muscovite, hornblende, and biotite at temperatures up to 850°C and found deuterium was enriched in the aqueous phase relative to the mineral phase. The fractionations were small, ranging from 0 per mil for H_2O–muscovite to +20 per mil for H_2O–biotite at 850°C. Extrapolation of their data to temperatures of 900–1300°C in the continental lithosphere suggest that the δD values of coexisting water, phlogopite, and amphibole should be similar. The effects of the chemical composition of amphiboles and micas on D/H fractionations appear to be minimal at such high temperatures (e.g. Sheppard *et al.* 1977).

7.3.2. Hydrogen isotopes in eclogite and peridotite xenoliths

The earliest work on D/H ratios in the mantle involved analyses of assumed xenocrystic and phenocrystic phlogopites in kimberlites and a few micas in peridotite xenoliths in South African kimberlites (Sheppard and Epstein 1970; Sheppard and Dawson 1975). δD values ranged from −70 to −40 per mil but, with the exception of the micas of metasomatic origin in the peridotite xenoliths, the origin of the other phlogopites was not known. These phlogopites could have been megacrysts, true phenocrysts, or disaggregated xenoliths of mantle and crustal material. Subsequent to these studies, Boettcher and O'Neil (1980) used petrographical and chemical criteria of megacrysts and xenoliths in South African kimberlites to select phlogopites for isotope analysis (Fig. 7.20(a)). The range in $\delta^{18}O$ values is 5.0 to 5.9 per mil and the range in δD values is only −70 to −63 per mil. These phlogopites have nearly ideal H_2O contents, with only 10 to 20 per cent halide substitution for OH.

Eclogite xenoliths in African kimberlites sometimes contain veins of secondary phlogopite. δD values for these metasomatic phlogopites are −80 to −70 per mil, despite variations of almost 2 per mil in their $\delta^{18}O$ values. These δD values are similar to those in oceanic basalts, which Kyser and O'Neil (1984) contend represent the primary hydrogen isotopic composition of the mantle. Because oceanic basalts represent melts originating at least in part from the asthenosphere, δD values of −80 to −70 per mil in phlogopites from eclogites and some peridotite xenoliths may also reflect fluids from the asthenosphere (see also Richardson, Chapter 3, this volume).

Garnet and spinel peridotite xenoliths from alkali basalts of the Patagonian Plateau also contain disseminated and vein phlogopite of metasomatic origin. The δD values of these phlogopites potentially reflect several

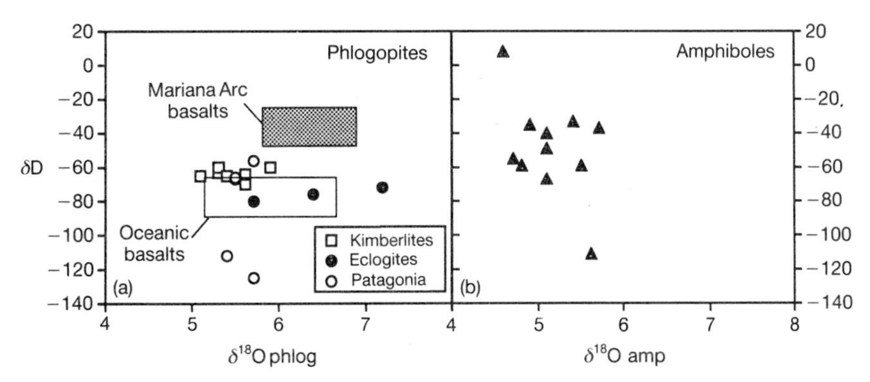

Fig. 7.20. Relation between δD and $\delta^{18}O$ values of: (a) vein and disseminated phlogopite (phlog) of presumed metasomatic origin from garnet peridotite and eclogite xenoliths in kimberlites from southern Africa and peridotite xenoliths in the alkali basalts of the Patagonian Plateau; and (b) amphibole (amp) in megacrysts and veins in xenoliths in alkali basalts from the USA and Australia. Also shown is the field for unaltered oceanic basalts (Kyser and O'Neil 1984), which reflects the composition of oceanic asthenosphere, and the field for basalts from the Mariana Arc (Poreda 1985), which reflects a component of subducted oceanic crust. Phlogopite and amphibole data are from Boettcher and O'Neil (1980), Kyser (1986), and Kyser and Stern (1988).

sources including: (1) metasomatic fluids from subducted oceanic crust of the Nazca Plate; (2) fluids from the continental asthenospheric or lithospheric mantle; and (3) alteration effects from local meteoric water, which has very low δD values.

The δD value of subducted oceanic crust that affects the lithospheric mantle has been estimated from the values in submarine basalts from the Mariana Arc (Poreda 1985) and from estimates of the original values in boninites from Bonin Island (Dobson and O'Neil 1987). δD values of basalts associated with subduction processes are high, near -40 per mil (Fig. 7.20(a)). The fluid from subducted material, which is hydrous and incompatible-element rich, is the primary reservoir of hydrogen in these environments because the lithospheric mantle is normally devoid of water. The highest δD value for the Patagonian phlogopites is -56 per mil, slightly D-enriched relative to many phlogopites associated with kimberlites and substantially D-enriched relative to oceanic basalts. If the Mariana Arc basalts do reflect the subducted component in their D/H ratios, the fluids that produced the Patagonian and African phlogopites may contain a substantial portion of this component, provided the other component has a δD value of -80 and is from the asthenosphere.

Two phlogopites from Patagonian xenoliths have low δD values but $\delta^{18}O$ values similar to other mantle phlogopites (Fig. 7.20(a)). The phlogopite in these samples has a much finer grain size than others from Patagonia. Low δD values have never been reported for phlogopite from African kimberlites or xenoliths, but Sheppard and Epstein (1970) measured δD values of -92 per mil for phlogopite phenocrysts in Proterozoic kimberlites from Bachelor Lake, Quebec. The Patagonian Plateau and Quebec both have meteoric waters with low δD values, whereas southern Africa does not, suggesting that phlogopites with low δD values may record alteration via hydrogen isotope exchange with meteoric waters. Low-temperature exchange of hydrogen in the phlogopites of the Patagonian xenoliths with meteoric water is probably enhanced by the fine grain size of the phlogopite. Because only the hydrogen and not the oxygen is affected by this alteration, the process must involve only hydrogen exchange and not hydrothermal alteration. Thus, only the Patagonian xenoliths with the highest δD values may reflect the composition of metasomatic fluids in the continental lithospheric mantle.

7.3.3. Hydrogen isotopes in megacrysts

Unlike the restricted range in δD values for phlogopites in the continental lithosphere, amphibole megacrysts in alkali basalts vary substantially from -113 to $+8$ per mil (Fig. 7.20(b)). However, most amphiboles have δD values between -70 and -35 per mil and many are more D-rich and ^{18}O-depleted relative to phlogopite. When phlogopite and amphibole occur together in a metasomatic vein, they are rarely in oxygen or hydrogen isotope equilibrium. The chemical composition of the amphiboles is also variable, with a wide range of water, F, Cl, and Fe^{3+} contents. Boettcher and O'Neil (1980) interpreted these chemical and isotopic variations in the amphiboles to result from complex processes involving discrete aqueous fluids. The relatively shallow depth of formation of the amphibole megacrysts, which would facilitate separation of a vapour phase, and their varied chemical compositions make this model appealing, but the origin of these discrete fluids remains an enigma (see also Watson *et al.*, Chapter 6, this volume).

7.4. SULPHUR ISOTOPES

Sulphur has four stable isotopes (Table 7.1). Isotope δ values are reported as the ratio $^{34}S/^{32}S$ in the sample relative to the same ratio in troilite from an iron-rich meteorite, Canyon Diablo (CDT for Canyon Diablo Troilite). The isotopic composition of most sulphides in other meteorites is similar to CDT and $\delta^{34}S$ values of meteorites are 0 ± 2 per mil (Thode *et al.* 1961; Nielson 1978). This value is also assumed for average sulphur in the bulk earth. $\delta^{34}S$ values of oceanic basalts are also 0 ± 2 per mil (Grinenko *et al.* 1975; Sakai *et al.* 1984) suggesting that the asthenosphere has a $\delta^{34}S$ value of about 0 per mil.

The $\delta^{34}S$ value of the continental lithospheric mantle, however, is much more variable than meteorites or oceanic basalts, with values between -5 and $+11$ per mil (e.g. Chaussidon *et al.* 1987, 1989). The $\delta^{34}S$ value of subducting oceanic crust will vary between 0 per mil, the value of sulphides in most massive sulphide deposits, and $+10$ per mil, the average value of sulphur in sediments. Sea water has a $\delta^{34}S$ value of about $+20$ per mil.

7.4.1. High-temperature fractionations and continental basalts

As with hydrogen, the fractionation of sulphur isotopes between oxidized forms, such as SO_2, and reduced forms, such as S^{2-}, are significant even at high temperatures. For example, at 1200°C, $\delta^{34}S(SO_2) - \delta^{34}S(S^{-2})$ is about 2 per mil. Consequently, loss of SO_2 will usually deplete a magma of ^{34}S. Fractionation of sulphur isotopes among sulphide minerals and reduced forms of sulphur are only a few tenths per mil at high temperatures so that precipitation of sulphides from a fluid will have little effect on $\delta^{34}S$ values.

MORB and submarine tholeiites from Hawaii have 500–1500 p.p.m. sulphur and $\delta^{34}S$ values of about 0 ± 2 per mil (Fig. 7.21). Sub-aerial Hawaiian tholeiites have only about 100 p.p.m. sulphur, because of degassing on eruption, and lower $\delta^{34}S$ values, down to -5 per mil (Sakai *et al.* 1982, 1984; Kyser 1986). Loss of about 80 per cent of the sulphur in the Hawaiian tholeiites during eruption results in preferential loss of ^{34}S and implies that the degassed sulphur species is oxidized, e.g. SO_2. Schneider (1970) reported that alkali basalts from Europe have higher $\delta^{34}S$ values than the associated tholeiites because of the loss of SO_2 from the latter during ascent. He estimates the $\delta^{34}S$ values of the continental mantle under Europe to be 1–1.5 per mil, more ^{34}S-rich than the oceanic asthenosphere.

Calc-alkaline magmas may also be enriched in ^{34}S relative to MORB and, hence, oceanic mantle. Basalts from the Japanese Island Arc (Ueda and Sakai 1984) and Mariana Arc (Woodhead *et al.* 1987) have $\delta^{34}S$ values up to 5 per mil that may be related to subduction of ^{34}S-rich components. However, all $\delta^{34}S$ values of sub-aerial lavas have been affected differentially by degassing and, potentially, all the $\delta^{34}S$ values of continental magmas can be affected by crustal sulphur contamination.

7.4.2. Sulphur isotopes in eclogite and peridotite xenoliths

Sulphides occur in a variety of forms in samples of the continental lithospheric mantle, including metasomatic veins in xenoliths, inclusions in silicate minerals of xenoliths and megacrysts, inclusions in diamonds, and as discrete minerals. The major sulphide phase is monosulphide solid solution between S, Fe, Ni, and Cu and pyrite, pentlandite, and cubanite are minor phases. The petrography and chemical compositions of sulphides in peridotite xenoliths are consonant with their origin as an immiscible sulphide formed during partial melting (e.g. Bishop *et al.* 1975; DeWaal and Calk 1975; Dromgoole and Pasteris 1987) although some favour an origin by infiltration of the lithospheric mantle by metasomatic fluids or the host magma (e.g. Irving 1980; Tsai *et al.* 1979).

Sulphur contents in whole-rock samples of xenoliths in alkali basalts from the western USA and from kimberlites in South Africa are usually lower than MORB (Fig. 7.21). $\delta^{34}S$ values are quite variable, from 0 to 11 per mil, and do not correlate with sulphur content. $\delta^{34}S$ values of spinel lherzolites from Dish Hill

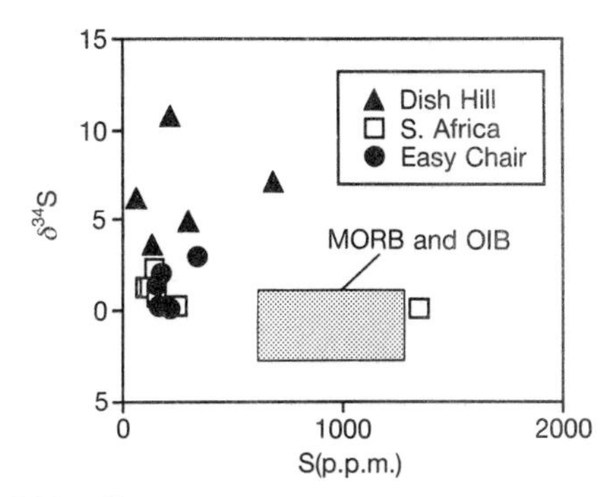

Fig. 7.21. $\delta^{34}S$ values and sulphur contents for whole-rock samples of spinel lherzolite xenoliths from Dish Hill, CA, garnet peridotite xenoliths from Africa, and amphibole-rich peridotite xenoliths from Easy Chair Crater, NV. Field of MORB compiled from Sakai *et al.* (1982, 1984) and data for xenoliths from unpublished data of Kyser.

are exceptionally ^{34}S-rich, even more so than the metasomatized pyroxenites from Easy Chair. Two samples of garnet peridotites from kimberlites shown in Fig. 7.21 have reversed $\Delta^{18}O$(pyroxene–olivine) values indicative of open-system exchange with metasomatic fluids. $\delta^{34}S$ values of these two peridotites are identical to those without aberrant olivine $\delta^{18}O$ values. These data suggest that both modal and trace metasomatism do not always result in elevated $^{34}S/^{32}S$ ratios in the continental lithosphere.

Whole-rock samples of eclogite inclusions in kimberlites yield $\delta^{34}S$ values from 0 to 3 per mil (Tsai *et al.* 1979). However, using an ion microprobe to measure the $^{34}S/^{32}S$ ratio of specific areas, Chaussidon *et al.* (1987, 1989) observed substantial total variations from -5 to 8 per mil in the $\delta^{34}S$ values of sulphides in eclogites, peridotites, and megacrysts (Fig. 7.22). They attribute variations in the $\delta^{34}S$ values of refractory peridotites primarily to isotopic fractionation between melt and Ni-rich sulphide globules, which are retained in the residue. In their model, Ni-rich sulphides are produced by 10–20 per cent partial melting of a source in the lithosphere containing 300 p.p.m. sulphur, with the residual sulphide being ^{34}S-enriched by 3 per mil relative to sulphur dissolved in the melt. Unfortunately, no such fractionations have been determined in laboratory experiments or observed in natural samples. For pyroxenites, which have more variable $\delta^{34}S$ values than refractory peridotites, they envision formation of low Ni sulphides via 'mantle heterogeneity' involving discrete fluids. This somewhat obscure model is similar to that proposed by Boettcher and O'Neil (1980) to explain the

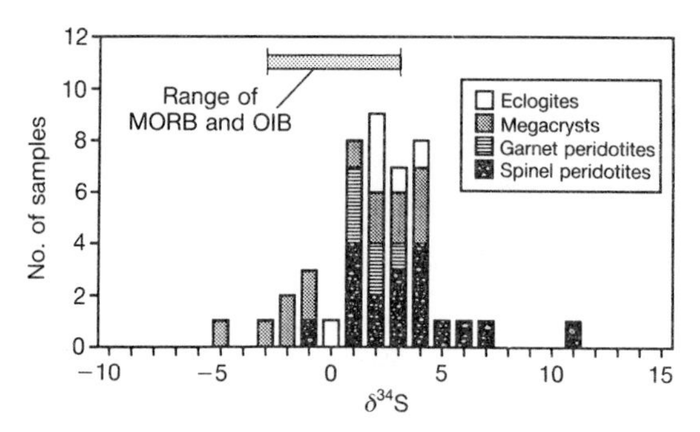

Fig. 7.22. Histogram of $\delta^{34}S$ values in megacrysts and eclogitic and peridotitic xenoliths showing the degree of heterogeneity in the sulphur isotopic composition of the continental lithospheric mantle relative to the range of $\delta^{34}S$ values in MORB and OIB. All of the megacryst and eclogite data and some of the peridotite data were determined on sulphides using the ion microprobe (Chaussidon *et al.* 1987, 1989), whereas others are whole-rock values.

extreme D/H ratios in amphibole megacrysts. Chaussidon *et al.* (1987) measured $\delta^{34}S$ values of up to +10 per mil in sulphide inclusions in diamonds so that, if the same discrete fluids are required for both hydrogen and sulphur, these fluids must exist at all depths in the continental lithospheric mantle.

7.5. CONCLUDING REMARKS

Isotope ratios of oxygen, hydrogen, and sulphur in minerals from alpine peridotites, mafic lavas, and mantle xenoliths indicate that both the oceanic and continental lithospheric mantles vary more substantially in isotope composition of these elements than expected from small fractionations at high temperatures. $\delta^{18}O$ values, which are particularly significant because oxygen is the major element in the lithosphere, vary from 5 to 7.5 per mil in the oceanic lithosphere whereas samples from the continental lithosphere reflect

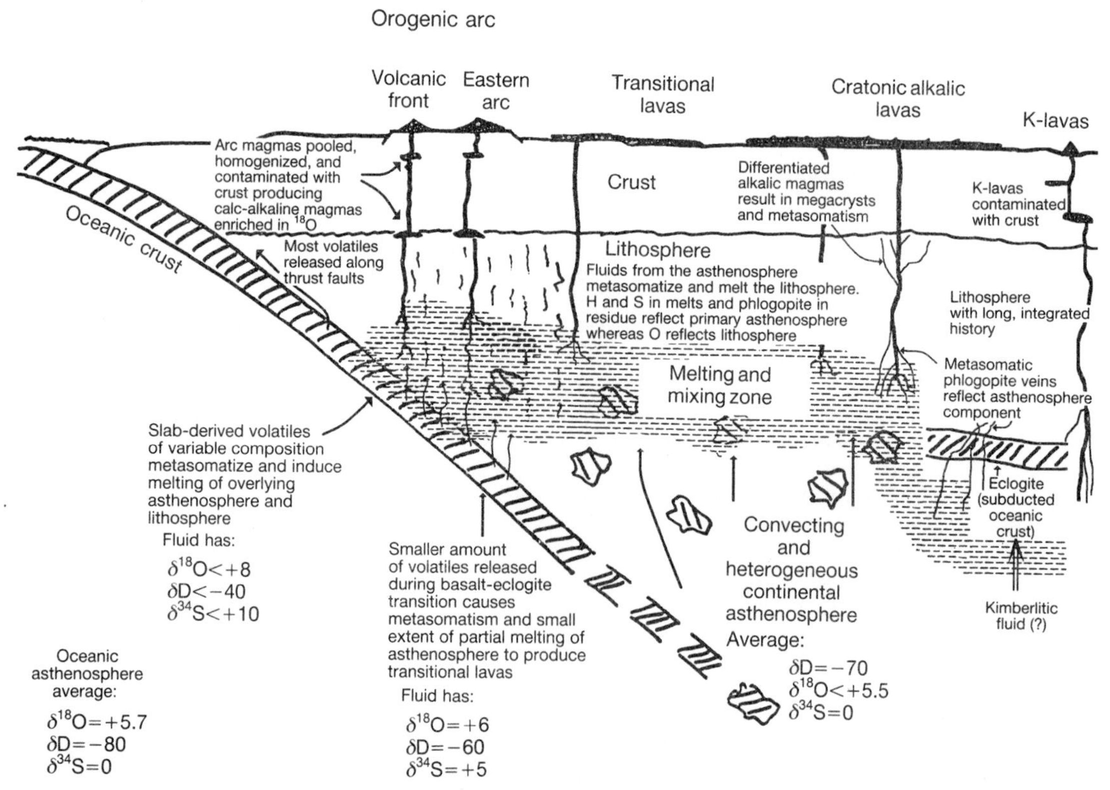

Fig. 7.23. Idealized diagram showing possible relations among the stable isotopic compositions of the continental lithospheric mantle, asthenosphere, crust, and subducted components (modified after Stern *et al.* 1989).

a much greater variation from 2 to 8 per mil. Variations in the stable isotope ratios of minor elements such as hydrogen and sulphur in the lithosphere are more pronounced than those of oxygen, with δD values ranging from about -120 to $+8$ per mil and $\delta^{34}S$ values varying from -5 to $+11$ per mil. As with oxygen, however, the continental lithospheric mantle appears to be more heterogeneous in its isotopic composition of hydrogen and sulphur than does the oceanic mantle, which has relatively constant δD and $\delta^{34}S$ values near -80 and -1 per mil, respectively.

There are many processes that can affect the stable isotopic composition of the continental lithosphere as depicted in Figure 7.23. The two most important appear to be: (1) subduction of material altered at the surface, which releases volatiles slightly enriched in ^{18}O, D, and ^{34}S into the lithosphere, inducing melting, and also mixes some oceanic crustal material (e.g. eclogites and peridotites) into the lithosphere and asthenosphere; (2) melts from the asthenosphere, which are heterogeneous in part because of subduction, interact with the continental lithosphere and differentiate, producing metasomatic fluids some of which induce melting of the lithosphere. These lithospheric-derived melts can also differentiate as they ascend and produce megacrysts, type 2 xenoliths, and other metasomatic fluids. Equilibrium melt–mineral fractionation, differential exchange rates, and Soret effects may be significant in producing the aberrant $\delta^{18}O$ values of pyroxene and olivine in peridotite xenoliths and in producing some of the variation in the δD and $\delta^{34}S$ values of the continental lithospheric mantle.

REFERENCES

Anderson, Don L. (1987). The depths of mantle reservoirs. In *Magmatic processes: physicochemical principles*, Geochemical Society Special Publication No. 1 (ed. B. O. Mysen), pp. 3–12. Geochemical Society, Pennsylvania.

Ater, P. C., Eggler, D. H., and McCallum, M. E. (1984). Petrology and geochemistry of mantle eclogite xenoliths from Colorado–Wyoming kimberlites: recycled oceanic crust? In *Kimberlites II: the mantle and crust–mantle relationships* (ed. J. Kornprobst), pp. 309–18. Elsevier, Amsterdam.

Bigeleisen, J. and Mayer, M. G. (1947). Calculation of equilibrium constants for isotopic exchange reactions. *Journal of Chemical Physics*, **15**, 261–7.

Bishop, F. C., Smith, J. V. and Dawson, J. B. (1975). Pentlandite–magnetite intergrowth in DeBeers spinel lherzolite. In Review of sulphides in nodules (ed. L. H. Ahrens *et al.*). *Physics and Chemistry of the Earth*, **9**, 323–37.

Boettcher, A. L. and O'Neil, J. R. (1980). Stable isotope, chemical, and petrographic studies of high-pressure amphiboles and micas: evidence for metasomatism in the mantle source regions of alkali basalts and kimberlites. *American Journal of Science*, **280A**, 594–621.

Bottinga, Y. and Javoy, M. (1973). Comments on oxygen isotope geothermometry. *Earth and Planetary Science Letters*, **20**, 250–63.

Bottinga, Y. and Javoy, M. (1975). Oxygen isotope partitioning among minerals in igneous and metamorphic rocks. *Reviews of Geophysics and Space Physics*, **13**, 410–18.

Boyd, F. R. and Mertzman, S. A. (1987). Composition and structure of the Kaapvaal lithosphere, southern Africa. In *Magmatic processes: physicochemical processes*, Geochemical Society Special Publication No. 1 (ed. B. O. Mysen), pp. 13–24. Geochemical Society, Pennsylvania.

Caporuscio, F. A., Kyser, T. K., and Smyth, J. R. (1987). Oxygen isotopes in mantle eclogites from South Africa. *Transactions of the American Geophysical Union*, **68**, 1551–2.

Carlson, R. W. (1983). Comment on "Implications of oxygen-isotope data and trace-element modeling for a large-scale mixing model for the Columbia River basalts", by D. O. Nelson. *Geology*, **11**, 735–6.

Carlson, R. W. (1984). Isotopic constraints on Columbia River flood basalt genesis and the nature of the subcontinental mantle. *Geochimica et Cosmochimica Acta*, **48**, 2357–72.

Chaussidon, M., Albarè, F., and Sheppard, S. M. F. (1987). Sulphur isotope heterogeneity in the mantle from ion microprobe measurements of sulphide inclusions in diamonds. *Nature*, **330**, 242–4.

Chaussidon, M., Albarè, F., and Sheppard, S. M. F. (1989). Sulphur isotope variations in the mantle from ion microprobe analyses of micro-sulphide inclusions. *Earth and Planetary Science Letters*, **92**, 144–56.

Chiba, H., Chacko, T., Clayton, R. N., and Goldsmith, J. R. (1988). Oxygen isotope fractionation in the diopside–forsterite–calcite system. *Transactions of the American Geophysical Union*, **69**, 511.

Clayton, R. N., Onuma, N., and Mayeda, T. K. (1971). Oxygen isotope fractionation in Apollo 12 rocks and soils. *Proceedings of the Second Lunar Science Conference*, pp. 1417–20. The MIT Press, Cambridge, Massachusetts.

Clayton, R. N., Goldsmith, J. R., and Mayeda, T. K. (1989). Oxygen isotope fractionation in quartz, albite, anorthite and calcite. *Geochimica et Cosmochimica Acta*, **53**, 725–33.

Cocker, J. D., Griffin, B. J., and Muehlenbachs, K. (1982). Oxygen and carbon isotope evidence for sea-water-hydrothermal alteration of the Macquarie Island ophiolite. *Earth and Planetary Science Letters*, **61**, 112–22.

Cohen, R. S., O'Nions, R. K., and Dawson, J. B. (1984). Isotope geochemistry of xenoliths from East Africa: implications for development of mantle reservoirs and their interaction. *Earth and Planetary Science Letters*, **68**, 209–20.

DeWaal, S. A. and Calk, L. C. (1975). The sulfides in the garnet pyroxenite xenoliths from Salt Lake Crater, Oahu. *Journal of Petrology*, **16**, 134–53.

Dobson, P. F. and O'Neil, J. R. (1987). Stable isotope compositions and water contents of boninite series volcanic rocks from Chichijima, Bonin Islands, Japan. *Earth and Planetary Science Letters*, **82**, 75–86.

Dromgoole, E. L. and Pasteris, J. D. (1987). Interpretation of the sulfide assemblages in a suite of xenoliths from Kilbourne Hole, New Mexico. In *Mantle metasomatism and alkaline magmatism*, Geological Society of America Special Paper No. 215 (ed. E. M. Morris and J. D. Pasteris), pp. 25–46. Geological Society of America, Boulder, Colorado.

Ferrara, G., Laurenzi, M. A., Taylor, H. P. Jr, Tonarini, S., and Turi, B. (1985). Oxygen and strontium isotope studies of K-rich volcanic rocks from the Alban Hills, Italy. *Earth and Planetary Science Letters*, **75**, 13–28.

Forester, R. W. and Taylor, H. P. Jr (1976). ^{18}O depleted igneous rocks from the Tertiary complex of the Isle of Mull, Scotland. *Earth and Planetary Science Letters*, **32**, 11–17.

Frey, F. A. and Prinz, M. (1978). Ultramafic inclusions from San Carlos, Arizona: petrologic and geochemical data bearing on their petrogenesis. *Earth and Planetary Science Letters*, **38**, 129–76.

Gregory, R. T. and Criss, R. E. (1986). Isotopic exchange in open and closed systems. In *Stable isotopes in high temperature geological processes*, Mineralogical Society of America Reviews of Mineralogy, Vol. 16 (ed. J. W. Valley, H. P. Taylor, Jr, and J. R. O'Neil), pp. 91–126. Mineralogical Society of America, Washington, DC.

Gregory, R. T. and Taylor, H. P. Jr (1981). An oxygen isotope profile in a section of Cretaceous oceanic crust, Samail Ophiolite, Oman: Evidence for ^{18}O buffering of the oceans by deep (5 km) seawater-hydrothermal circulation at mid-ocean ridges. *Journal of Geophysical Research*, **86**, 2737–55.

Gregory, R. T. and Taylor, H. P. Jr (1986*a*). Possible non-equilibrium oxygen isotope effects in mantle nodules, a criticism of the Kyser–O'Neil–Carmichael $^{18}O/^{16}O$ geothermometer. *Contributions to Mineralogy and Petrology*, **93**, 114–19.

Gregory, R. T. and Taylor, H. P. Jr (1986*b*). Non-equilibrium, metasomatic $^{18}O/^{16}O$ effects in uper mantle mineral assemblages. *Contributions to Mineralogy and Petrology*, **93**, 124–35.

Grinenko, V. A., Dimitriev, L. V., Migdisov, A., A., and Sharas'kin, A. Ya. (1975). Sulfur contents and isotopic compositions from igneous and metamorphic rocks from mid-ocean ridges. *Geochemistry International*, **1**, 132–7.

Harmon, R. S. and Hoefs, J. (1984). O-isotope relationships in Cenozoic volcanic rocks: evidence for heterogeneous mantle source and open-system magma genesis. In *Proceedings of the ISEM Field Conference on Open Magmatic Systems* (ed. M. A. Dungan, T. L. Grove, and W. Hildreth), pp. 66–8.

Harmon, R. S., Kempton, P. D., Stosch, H. G., Hoefs, J., Kovalenko, V. I., and Eonov, D. (1986/87). $^{18}O/^{16}O$ ratios in anhydrous spinel lherzolite xenoliths from the Shavaryn-Tsaram volcano, Mongolia. *Earth and Planetary Science Letters*, **81**, 193–202.

Harmon, R. S., Hoefs, J., and Wedepohl, K. H. (1987). Stable isotope (O, H, S) relationships in Tertiary basalts and their mantle xenoliths from the Northern Hessian Depression, W. Germany. *Contributions to Mineralogy and Petrology*, **95**, 350–69.

Hatton, C. J. and Gurney, J. J. (1979). A diamond–graphite eclogite from the Roberts Victor mine. In *The mantle sample: inclusions in kimberlites and other volcanics* (ed. F. R. Boyd and H. O. A. Meyer), pp. 29–36, American Geophysical Union, Washington, DC.

Hatton, C. J. and Gurney, J. J. (1987). Roberts Victor eclogites and their relation to the mantle. In *Mantle xenoliths* (ed. P. H. Nixon), pp. 453–63. John Wiley and Sons, Chichester.

Heaton, T. H. E. and Sheppard, S. M. F. (1977). Hydrogen and oxygen isotope evidence for seawater hydrothermal alteration and ore deposition, Troodos complex, Cyprus. *Volcanic processes in ore genesis*, Special Paper of the Geological Society, London No. 7, pp. 42–57. Blackwells, Oxford.

Helmstaedt, H. and Doig, R. (1975). Eclogite nodules from kimberlite pipes in the Colorado Plateau—samples of subducted Franciscan type oceanic lithosphere. *Physics and Chemistry of the Earth*, **9**, 95–111.

Hill, R. I., Silver, L. T., and Taylor, H. P. Jr (1986). Coupled Sr–O isotope variations as an indicator of source heterogeneity for the Northern Peninsular Ranges batholith. *Contributions to Mineralogy and Petrology*, **92,** 351–61.

Hoefs, J. (1980). *Stable isotope geochemistry*. Springer-Verlag, New York, 208 pp.

Holm, P. M. and Munksgaard, N. C. (1982). Evidence for mantle metasomatism: oxygen and strontium isotope study of the Vulsinian District, Central Italy. *Earth and Planetary Science Letters*, **60,** 376–88.

Holm, P. M. and Munksgaard, N. C. (1986). Reply to: a criticism of the Holm–Munksgaard oxygen and strontium isotope study of the Vilsinian district, Central Italy. *Earth and Planetary Science Letters*, **78,** 454–9.

Holm, P. M., Lou, S., and Nielson, Å. (1982). The geochemistry and petrogenesis of the lavas of the Vulsinian District, Roman Province, Central Italy. *Contributions to Mineralogy and Petrology*, **80,** 367–78.

Irving, A. J. (1980). Petrology and geochemistry of composite ultramafic xenoliths in alkalic basalts and implications for magmatic processes within the mantle. *American Journal of Science, Jackson Volume*, **280A,** 389–426.

Irving, A. J. (1984). Polybaric mixing and fractionation of alkalic magmas: Evidence from megacryst suites. *EOS Transactions of the American Geophysical Union*, **65,** 1153.

Ito, K. and Kennedy, G. C. (1970). *The fine structure of the basalt–eclogite transition*, Special Paper of the Mineralogical Society of America, No. 3, pp. 77–83. Mineralogical Society of America, Washington, DC.

Ito, K. and Stern, R. J. (1985/86). Oxygen- and strontium-isotopic investigations of subduction zone volcanism: the case of the Volcano Arc and the Marianas Island Arc. *Earth and Planetary Science Letters*, **76,** 312–20.

Jagoutz, E., Dawson, J. B., Hornes, S., Spettel, B., and Wanke, H. (1984). Anorthositic oceanic crust in the Archean. *Lunar and Planetary Science*, **15,** 395–6.

James, D. E. (1981). The combined use of oxygen and radiogenic isotopes as indicators of crustal contamination. *Annual Reviews of Earth and Planetary Science*, **9,** 311.

James, D. E. (1982). A combined O, Sr, Nd, and Pb isotopic and trace element study of crustal contamination in central Andean lavas. I. Local geochemical variations. *Earth and Planetary Science Letters*, **57,** 47–62.

James, D. E. (1984). Quantitative models of crustal contamination in andesitic volcanic rocks from the Northern and central Andes. In *Chemical and isotopic constraints on Andean magmatism* (ed. R. S. Harmon and B. Barreiro), pp. 124–38. Shiva Press, Nantwich.

Javoy, M. (1970). Utilization des isotopes de l'oxygène en magmatologie. D.Phil. Dissertation, University of Paris.

Javoy, M. (1980). $^{18}O/^{16}O$ and D/H ratios in high temperature peridotites. *Colloques International du CNRS*, **272,** 279–87.

Jones, R. A. (1987). Strontium and neodymium isotopic and rare earth element evidence for the genesis of megacrysts in kimberlites of southern Africa. In *Mantle xenoliths* (ed. P. H. Nixon), pp. 711–24. John Wiley and Sons, Chichester.

Kempton, P. D., Harmon, R. S., Stosch, H. G., Hoefs, J., and Hawkesworth, C. J. (1988). Open-system O-isotope behaviour and trace element enrichment in the sub-Eifel mantle. *Earth and Planetary Science Letters*, **89,** 273–87.

Kieffer, S. W. (1982). Thermodynamics and lattice vibrations of minerals: 5. Applicationss to phase equilibria, isotopic fractionation and high-temperature thermodynamic properties. *Reviews of Geophysics and Space Physics*, **20,** 827–49.

Kramers, J. D. (1979). Lead, uranium, strontium, potassium and rubidium in inclusion-bearing diamonds and mantle-derived xenoliths from southern Africa. *Earth and Planetary Science Letters*, **42,** 58–70.

Kramers, J., Roddick, J., and Dawson, J. (1983). Trace element and isotope studies on veined, metasomatic and 'MARID' xenoliths from Bultfontein, South Africa. *Earth and Planetary Science Letters*, **65,** 90–106.

Kyser, T. K. (1986). Stable isotope variation in the mantle. In *Stable isotopes in high temperature geological processes*, Mineralogical Society of America, Reviews of Mineralogy, Vol. 16 (ed. J. W. Valley, H. P. Taylor, Jr, and J. R. O'Neil), pp. 141–64. Mineralogical Society of America, Washington, DC.

Kyser, T. K. and O'Neil, J. R. (1984). Hydrogen isotope systematics of submarine basalts. *Geochimica et Cosmochimica Acta*, **48,** 2123–33.

Kyser, T. K. and Stern, C. R. (1988). Oxygen and hydrogen isotope systematics in alkali basalts and mantle xenoliths from the Patagonian Plateau [Abstr.]. *Geological Society of America Annual Meeting*. Geological Society of America, Boulder, Colorado.

Kyser, T. K., O'Neil, J. R., and Carmichael, I. S. E. (1981). Oxygen isotope thermometry of basic lavas and mantle nodules. *Contributions to Mineralogy and Petrology*, **77,** 11–23.

Kyser, T. K., O'Neil, J. R., and Carmichael, I. S. E. (1982). Genetic relations among basic lavas and ultramafic nodules: evidence from oxygen isotope compositions. *Contributions to Mineralogy and Petrology*, **81,** 88–102.

Kyser, T. K., Cameron, W. E., and Nisbet, E. G. (1986*a*). Boninite petrogenesis and alteration history: constraints from stable isotope compositions of boninites from Cape Vogel, New Caledonia and Cyprus. *Contributions to Mineralogy and Petrology*, **93,** 222–6.

Kyser, T. K., O'Neil, J. R., and Carmichael, I. S. E. (1986*b*). Reply to "Possible non-equilibrium oxygen isotope effects in mantle nodules, an alternative to the Kyser–O'Neil–Carmichael $^{18}O/^{16}O$ geothermometer". *Contributions to Mineralogy and Petrology*, **93,** 120–3.

Longstaffe, F. J., Clark, A H., McNutt, R. H., and Zentilli, M. (1983). Oxygen isotopic compositions of Central Andean plutonic and volcanic rocks, latitudes 26°–29° south. *Earth and Planetary Science Letters*, **64,** 9–18.

MacGregor, I. D. and Carter, J. L. (1970). The chemistry of clinopyroxenes and garnets of eclogite and peridotite xenoliths from the Roberts Victor mine, South Africa. *Physics of the Earth and Planetary Interiors*, **3,** 391–7.

MacGregor, I. D. and Manton, W. J. (1986). Roberts Victor eclogites: ancient oceanic crust. *Journal of Geophysical Research*, **91,** 14 063–79.

Magaritz, M. and Taylor, H. P. Jr (1986). Oxygen 18/oxygen 16 and D/H studies of plutonic granitic and metamorphic rocks across the Cordilleran batholiths of southern British Columbia. *Journal of Geophysical Research*, **91,** 2193–217.

Magaritz, M., Whitford, D. J., and James, D. E. (1978). Oxygen isotopes and the origin of high-$^{87}Sr/^{86}Sr$ andesites. *Contributions to Mineralogy and Petrology*, **40,** 220–30.

Manton, W. I. and Tatsumoto, M. (1971). Some Pb and Sr isotopic measurements on eclogites from the Roberts Victor mine, South Africa. *Earth and Planetary Science Letters*, **10,** 217–26.

Massi, U., O'Neil, J. R., and Kistler, R. W. (1981). Stable isotope systematics in Mesozoic granites of central and northern California and Southwestern Oregon. *Contributions to Mineralogy and Petrology*, **76,** 116–26.

Matsuhisa, Y. and Kurasawa, H. (1983). Oxygen and strontium isotopic characteristics of calc-alkaline volcanic rocks from the central and western Japan arcs: evaluation of contribution of crustal component to the magmas. *Journal of Volcanology and Geothermal Research*, **18,** 483–510.

McCulloch, M. T. (1986). Sm–Nd systematics in eclogite and garnet peridotite nodules from kimberlites: implications for the early differentiation of the Earth [abstr.]. *Fourth International Kimberlite Conference*. Geological Society of Australia.

Menzies, M. (1987). Alkaline rocks and their inclusions: a window on the Earth's interior, Geological Society of London, Special Publication, No. 30. Blackwells, Oxford. In *Alkaline Rocks* (ed. J. G. Fitton and B. G. J. Upton), pp. 15–27.

Menzies, M. A. and Wass, S. Y. (1983). CO_2 and LREE-rich mantle below eastern Australia: A REE and isotopic study of alkaline magmas and apatite-rich mantle xenoliths from the Southern Highlands Province, Australia. *Earth and Planetary Science Letters*, **65,** 287–302.

Muehlenbachs, K. and Clayton, R. N. (1976). Oxygen isotope composition of the oceanic crust and its bearing on seawater. *Journal of Geophysical Research*, **81,** 4365–9.

Muehlenbachs, K. and Kushiro, I. (1974). Oxygen isotope exchange and equilibrium of silicates with CO_2 or O_2. *Carnegie Institution of Washington, Yearbook*, **73,** 232–6.

Nelson, D. O. (1983*a*). Implications of oxygen-isotope data and trace-element modeling for a large-scale mixing model for the Columbia River Basalt. *Geology*, **11,** 248–51.

Nelson, D. O. (1983*b*). Implications of oxygen-isotope data and trace-element modeling for large-scale mixing model for the Columbia River basalt, reply to a comment by R. W. Carlson. *Geology*, **11,** 735–6.

Nielson, H. (1978). Sulfur isotopes in nature. In *Handbook of geochemistry* (ed. K. H. Wedepohl), Chapter 16-B. Springer-Verlag, Berlin.

Ongley, J. S., Basu, A. R., and Kyser, T. K. (1987). Oxygen isotopes in coexisting garnets, clinopyroxenes and phlogopites of Roberts Victor eclogites: implications for petrogenesis and mantle metasomatism. *Earth and Planetary Science Letters*, **83,** 80–4.

Poreda, R. (1985). Helium-3 and deuterium in back-arc basalts: Lau Basin and the Mariana Trough. *Earth and Planetary Science Letters*, **73,** 244–54.

Richet, P., Roux, J., and Pineau, F. (1986). Hydrogen isotope fractionation in the system H_2O–liquid $NaAlSi_3O_8$: new data and comments on D/H fractionation in hydrothermal experiments. *Earth and Planetary Science Letters*, **78,** 115–20.

Sakai, H., Casadevall, T. J., and Moore, J. G. (1982). Chemistry and isotope ratios of sulfur in basalts and volcanic gases at Kilauea volcano, Hawaii. *Geochimica et Cosmochimica Acta*, **46,** 729–38.

Sakai, H., Des Marais, D. J., Ueda, A., and Moore, J. G. (1984). Concentrations of isotope ratios of carbon, nitrogen and sulfur in ocean floor basalts. *Geochimica et Cosmochimica Acta*, **48**, 2433–41.

Schiffman, P., Williams, A. E., and Evarts, R. C. (1984). Oxygen isotope evidence for submarine hydrothermal alteration of the Del Puerto ophiolite, California. *Earth and Planetary Science Letters*, **70**, 207–20.

Schneider, A. (1970). The sulfur isotope composition of basaltic rocks. *Contributions to Mineralogy and Petrology*, **25**, 95–124.

Schulze, D. J. (1987). Megacrysts from alkalic volcanic rocks. In *Mantle xenoliths* (ed. P. H. Nixon), pp. 433–51. John Wiley and Sons, Chichester.

Sheppard, S. M. F. (1980). Isotopic evidence for the origins of water during metamorphic processes in oceanic crust and ophiolite complexes. *Associations Mafiques Ultra-Mafiques dans les Orogénes*, pp. 135–47. C.N.R.S., Paris, France.

Sheppard, S. M. F. and Dawson, J. B. (1975). Hydrogen, carbon and oxygen isotope studies of megacryst and matrix minerals from Lesothan and South African kimberlites. *Physics and Chemistry of the Earth*, **9**, 747–63.

Sheppard, S. M. F. and Epstein, S. (1970). D/H and $^{18}O/^{16}O$ ratios of minerals of possible mantle or lower crustal origin. *Earth and Planetary Science Letters*, **9**, 232–9.

Sheppard, S. M. F., Brown, P. E., and Chambers, A. D. (1977). The Lilloise Intrusion, East Greenland: hydrogen isotope evidence for the efflux of magmatic water into the contact metamorphic aureole. *Contributions to Mineralogy and Petrology*, **63**, 129–47.

Spooner, E. T. C., Beckinsale, R. D., Fyfe, W. S., and Smewing, J. D. (1974). ^{18}O-enriched ophiolitic metabasic rocks from E. Liguria (Italy), Pindos (Greece), and Troodos (Cyprus). *Contributions to Mineralogy and Petrology*, **47**, 41–62.

Stern, C., de Wit, M. J., and Lawrence, J. R. (1976). Igneous and metamorphic processes associated with the formation of Chilean ophiolites and their implication for ocean floor metamorphism, seismic layering, and magnetism. *Journal of Geophysical Research*, **81**, 4370–80.

Stern, C. R., Futa, K., and Muehlenbachs, K. (1984). Isotope and trace element data for orogenic andesites from the Austral Andes. In *Chemical and isotopic constraints on Andean magmatism* (ed. R. S. Harmon and B. Barreiro), pp. 31–46. Shiva Press, Nantwich.

Stern, C. R., Futa, K., Saul, S., and Milka, A. S. (1986). Nature and evolution of the subcontinental mantle lithosphere below southern South America and implications for Andean magma genesis. *Revista Geologica de Chile* **27**, 41–53.

Stern, C. R., Frey, F. A., Futa, K., Zartman, R. E., Peng, Z., and Kyser, T. K. (1990). Trace-element and Sr, Nd, Pb, and O isotopic composition of Pliocene and Quaternary alkali basalts of the Patagonian Plateau lavas of southernmost South America. *Contributions to Mineralogy and Petrology* (in press).

Suzuoki, T. and Epstein, S. (1976). Hydrogen isotope fractionation between OH-bearing silicate minerals and water. *Geochimica et Cosmochimica Acta*, **40**, 1229–40.

Taylor, H. P. Jr and Epstein, S. (1970). Oxygen and silicon isotope ratios of lunar rock 12013. *Earth and Planetary Science Letters*, **9**, 208–10.

Taylor, H. P. Jr and Sheppard, S. M. F. (1986). Igneous rocks: I. Processes of isotopic fractionation and isotope systematics. In *Stable isotopes in high temperature geological processes*, Mineralogical Society of America, Reviews of Mineralogy, Vol. 16 (ed. J. W. Valley, H. P. Taylor, Jr, and J. R. O'Neil), pp. 227–71. Mineralogical Society of America, Washington, DC.

Taylor, H. P. Jr and Silver, L. T. (1978). Oxygen isotope relationships in plutonic igneous rocks of the Peninsular Ranges batholith, southern and Baja California. *US Geological Survey Open File Report*, **78–701**, 423–7.

Taylor, H. P. Jr, Giannetti, B., and Turi, B. (1979). Oxygen isotope geochemistry of the potassic igneous rocks from the Roccamonfina Volcano, Roman Comagmatic Region, Italy. *Earth and Planetary Science Letters*, **46**, 81–106.

Taylor, H. P. Jr, Turi, B., and Cundari, A. (1984). $^{18}O/^{18}O$ and chemical relationships in K-rich volcanic rocks from Australia, East Africa, Antarctica and San Venanzo-Cupaello, Italy. *Earth and Planetary Science Letters*, **69**, 263–75.

Thode, H. G., Monster, J., and Dunford, H. B. (1961). Sulfur isotope geochemistry. *Geochimica et Cosmochimica Acta*, **25**, 159–74.

Tsai, H., Shieh, Y., and Meyer, H. O. A. (1979). Mineralogy and $^{34}S/^{32}S$ ratios of sulfides associated with kimberlite, xenoliths, and diamonds. In *The mantle sample* (ed. F. R. Boyd and H. O. A. Meyer), pp. 87–103. American Geophysical Union, Washington, DC.

Turi, B., Taylor, H. P. Jr, and Ferrara, G. (1986). A criticism of the Holm–Munksgaard oxygen and strontium isotope study of the Vulsinian District, Central Italy. *Earth and Planetary Science Letters*, **78,** 447–53.

Ueda, A. and Sakai, H. (1984). Sulfur isotope study of Quaternary volcanic rocks from the Japanese Islands Arc. *Geochimica et Cosmochimica Acta*, **48,** 1837–48.

Urey, H. C. (1947). The thermodynamic properties of isotopic substances. *Journal of the Chemical Society*, **6,** 562–81.

Vogel, D. E. and Garlick, G. D. (1970). Oxygen isotope ratios in metamorphic eclogites. *Contributions to Mineralogy and Petrology*, **28,** 183–91.

Walker, D. and DeLong, S. E. (1982). Soret separation of mid-ocean ridge basalt magma. *Contributions to Mineralogy and Petrology*, **79,** 231–40.

Walker, D., Lesher, D. E., and Hays, J. F. (1981). Soret separation of Lunar liquid. *Proc. Lunar Planet. Geochimica et Cosmochimica Acta*, **12B,** 991–9.

Wilshire, H. G. and Pike, J. E. N. (1975). Upper-mantle diapirism: evidence from analogous features in alpine peridotite and ultramafic inclusions in basalt. *Geology*, **3,** 467–70.

Wilshire, H. G., Pike, J. E. N., Meyer, C. E., and Schwarzmann, E. C. (1980). Amphibole-rich veins in lherzolite xenoliths, Dish Hill and Deadman Lake, California. *American Journal of Science*, **280A,** 576–93.

Woodhead, J. D., Harmon, R. S., and Fraser, D. G. (1987). O, S, Sr, and Pb isotope variations in volcanic rocks from the Northern Mariana Islands: implications for crustal recycling in intra-oceanic arcs. *Earth and Planetary Science Letters*, **83,** 39–52.

8

Continental volcanism: a crust–mantle probe

Martin A. Menzies and Philip R. Kyle

8.1. INTRODUCTION

Areas of continental volcanism and oceanic ridge systems differ in one fundamental aspect, the existence of a continental lithospheric plate of variable thickness and age (Boyd *et al.* 1985; Anderson, Chapter 1, this volume; Richardson, Chapter 3, this volume). As a result of variable amounts of post-stabilization interaction with small volume melts (Watson *et al.*, Chapter 6, this volume), chemically different lithospheres (Haggerty, Chapter 5, this volume; Kyser, Chapter 7, this volume; Menzies, Chapter 4, this volume) underplate the continental crust. Lithosphere is, however, cold and stable and for it to contribute to surface volcanism melting must occur by: (1) lowering the peridotite solidus by the introduction of volatiles; (2) increasing the potential temperature of the asthenosphere; or (3) thinning the lithosphere thus encouraging upwelling of the asthenosphere (McKenzie and Bickle 1988; Latin *et al.* 1989). The recent demonstration of a link between hot spots and continental volcanism indicates that such volcanism may be the surface expression of the collision of lithospheric plates (crust and mantle) with upwelling deep mantle structures (Duncan 1978; Storey *et al.* 1988; Watts and Cox 1989; White and McKenzie 1989). The world-wide distribution of mantle plumes coincides with geoid highs (Chase 1979) and regions of low seismic velocity in the mantle (Silver *et al.* 1988), which have been interpreted as areas of upward counterflow in a global recycling system. In particular, the majority of hot spots are centred on the southern Pacific, southern and eastern Atlantic Oceans, and west Africa. This upward chemical flux is believed to be 'driven' by the downward flow of cold subducted slabs (high seismic velocity) occurring around the periphery of the Pacific Ocean. Consequently, correlations between plate-tectonic features, seismic tomography, the geoid, hot spots, and continental flood basalt provinces (cfb) may indicate that cfb have a significant *deep mantle* component (>650 km) whereas normal oceanic ridge volcanism may contain a dominant *shallow mantle* contribution due to the absence of low velocity characteristics at depths ≥ 350 km beneath mid-ocean ridges (see Anderson, Chapter 1, this volume).

With regard to continental intraplate volcanism recent theoretical considerations have demonstrated that the volume and composition of melt generated during continental lithospheric extension depends on the amount of extension (β = final/initial surface area), the temperature, and the amount of melting (McKenzie and Bickle 1988). During asthenosphere–lithosphere interaction or continental extension the degree of melting increases and a switch may occur from eruption of volcanic rocks that tap mainly asthenospheric reservoirs to volcanic rocks that tap mainly lithospheric reservoirs (mantle and crust). This may result in a temporal change from a brief episode of potassic/alkaline volcanism to the eruption of volumetrically more significant tholeiites and their fractionated and 'contaminated' equivalents (e.g. rhyolites, ignimbrites).

Petrogenetic models involving asthenosphere–lithosphere or plume–lithosphere interaction seem to be the most appropriate in any consideration of the genesis of continental volcanic rocks. Evidence for such petrogenetic processes within the continental lithosphere is forthcoming from studies of high-presssure inclusions in diamonds and kimberlite-borne and basalt-borne xenoliths. Metasomes in high-pressure xenolith suites reveal that most melts that have stagnated in the lithosphere (100–150 km) have isotopic similarities to ocean island basalts (OIB) and not to mid-ocean ridge basalts (MORB). This lends support to the idea that continental lithosphere may have been progressively eroded by deep mantle OIB plumes with the eventual eruption of volcanic material at the surface. The presence of potassic, volatile-rich (>50 per cent) melts (180–200 km) with OIB isotopes as micro-inclusions in megacrystic diamonds (Navon *et al.* 1988 and references therein) and the existence of kimberlitic derivatives with OIB isotopes as the source of the metasomatism in garnet peridotites (100–150 km) (Kramers *et al.* 1983) attest to the existence of an OIB component in 'cratonic' lithosphere of Archaean age. This is also the case for lamproites and kimberlites that crystallized as mica–

amphibole–rutile–ilmenite–diopside (MARID) xenoliths at a depth of 100–120 km within the lithosphere (Waters 1987). Similarly, thinner 'oceanic' lithosphere that underplates post-Archaean crust has been infiltrated with OIB-like magmas that crystallized as pyroxenite and amphibolite vein networks in spinel peridotites at depths of around 50 km (Irving 1980). While we can formulate a case for different lithospheres beneath continents of different ages (Richter 1988; Anderson, Chapter 1, this volume; Menzies, Chapter 4, this volume), it is apparent that most of the secondary processes in these lithospheres have a common origin in the sublithospheric mantle (McKenzie 1989). Essentially the melt or fluid enrichment and metasomatism observed in Archaean cratonic and post-Archaean oceanic lithosphere is an expression of plume–lithosphere or asthenosphere–lithosphere interaction.

Potentially, any continental volcanic rock suite can contain contributions from the lower mantle, asthenosphere (MORB reservoir and recycled materials), lithospheric mantle (thermal and mechanical boundary layers), and the crust (lower and upper). It is worth noting that the debate to date has concentrated on continental volcanic rocks being derived from: (1) the asthenosphere with a dominant crustal contamination overprint; or (2) the asthenosphere and the mantle section of the lithosphere with a minor crustal overprint. Any attempt to resolve this issue has been partly confused by the overwhelming and uncritical embrace of the evidence for enriched mantle.

From Fig. 8.1 it is apparent that chemically distinct lithospheres exist beneath Archaean and post-Archaean crust and that there is no conclusive evidence for enriched reservoirs beneath post-Archaean terrains. Archaean 'cratonic' lithosphere is undoubtedly heterogeneous for radiogenic and stable isotopes and is isotopically similar in many instances to average 'lower continental crust' (Richardson *et al.* 1984; Richardson, Chapter 3, this volume) but very different from Proterozoic and Phanerozoic 'oceanic' lithosphere. Since most continental flood basalt provinces were erupted through post-Archaean crust the plumbing was established in relatively depleted post-Archaean lithospheric mantle and did not necessarily encounter enriched Archaean domains. Therefore to invoke the participation of enriched mantle as an explanation for the isotopic characteristics of cfb belies the character of the post-Archaean lithosphere. Ironically, while post-Archaean lithosphere may not be isotopically heterogeneous enough to account for the isotopic heterogeneity observed in cfb, it is more fertile than Archaean lithospheric mantle (Boyd 1989) and, as such, is a more viable source for tholeiitic melts (Falloon and Green 1988).

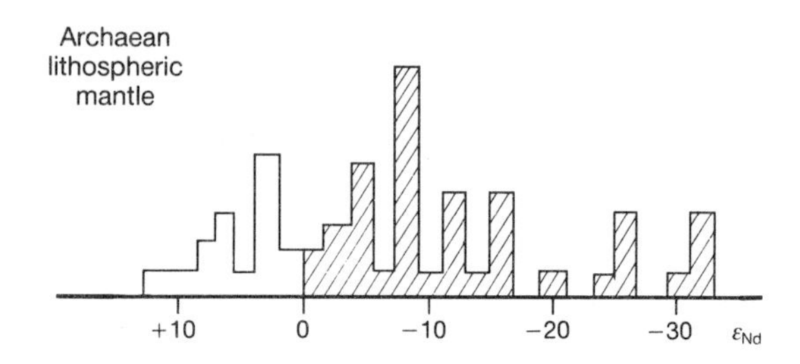

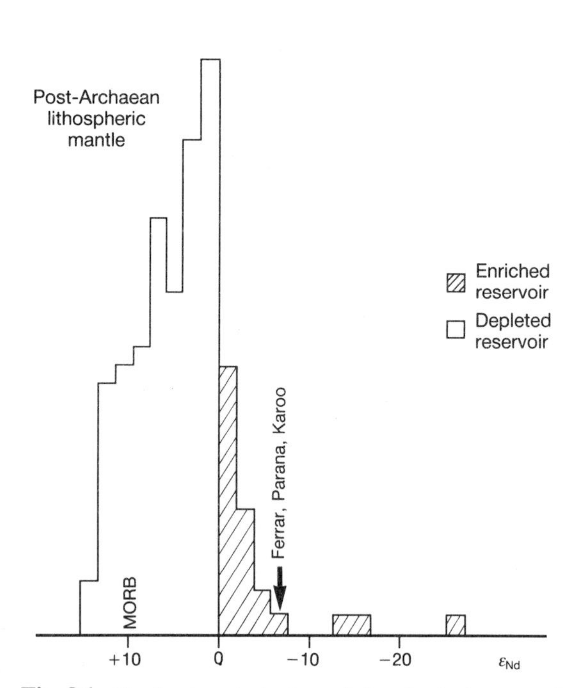

Fig. 8.1. Neodymium isotopic variation in Archaean and post-Archaean lithospheric mantle. Note that the bulk of the post-Archaean reservoir encountered by continental flood basalts (cfb) and entrained as basalt-borne xenoliths and megacrysts is 'oceanic' in character (i.e. depleted in isotopes) whereas the Archaean reservoir sampled by lamproites and micaceous kimberlites and entrained as kimberlite-borne xenoliths and megacrysts is 'cratonic' in character (i.e. enriched in isotopes). The approximate location of three flood basalt provinces is shown as is the approximate position of the MORB (i.e. DMM) reservoir. The Ferrar, Parana, and Karoo flood basalt provinces are all located in the southern hemisphere and petrogenetic models may require an enriched mantle component in their genesis. It is apparent from this histogram that the post-Archaean lithospheric mantle is an unlikely location for that enriched component.

8.2. VOLCANIC CASE STUDIES

Recent studies of continental volcanic rocks (MacDougall 1988; Leeman and Fitton 1989) have alluded to the participation of two components in their genesis: (1) a dominant 'depleted' asthenospheric component similar in composition to ocean island basalts (OIB) or mid-ocean ridge basalts (MORB); and (2) a minor 'enriched' lithospheric component derived from Proterozoic or Phanerozoic lithospheric mantle (e.g. Carlson and Hart 1988; Dickin 1988; Mahoney 1988; Zhou *et al.* 1988).

With regard to the first component, it should be stressed that the sources of OIB and MORB may be very different. To what extent a MORB source could act as a contributory reservoir for cfb is contingent upon the ability of a MORB source to be activated at the depths necessary to erode the base of the continental lithosphere. Moreover, we have already outlined the lack of evidence, in peridotites of unequivocal subcontinental origin, for the presence of MORB melts or the derivatives thereof. It would appear that an OIB component is a more appropriate contributor to the source of flood basalts than a MORB component.

With regard to the second component and the involvement of 'enriched' post-Archaean lithosphere it is worth once again stressing the similarity between post-Archaean lithosphere and asthenosphere (Fig. 8.1). Indeed, they are so alike that post-Archaean lithosphere is frequently termed 'oceanic' (e.g. Boyd 1989). Consequently, it may be very difficult, or indeed impossible, to tell the difference between contributions from the lithosphere and the asthenosphere unless the lithosphere is accreted and carries a 'subduction signature' (Cox 1988). An identifiable subduction signature in the lithospheric mantle may conveniently account for some of the trace element characteristics (i.e. Nb and Ta depletions) in flood basalts and their overall andesitic character. *Caveat*—even this is not without its problems in that the asthenosphere may contain a deep recycled component with a subduction signature.

The problems that plague the investigation of cfb are to some degree not so apparent in recent studies of continental intraplate volcanic rocks that tend to be volumetrically less important than cfb. Most recent studies have concentrated on defining the geochemical characteristics of mantle domains (i.e. DMM, EM1, EM2, and HIMU) in the asthenosphere or the lithosphere (Cameron *et al.* 1989; Luhr *et al.* 1989; Hegner and Pallister 1989; Farmer *et al.* 1989; Lum *et al.* 1989; Pier *et al.* 1989). Moreover, careful selection of data has allowed some investigators to use continental volcanic rocks as a deep mantle probe enabling the geometry of individual mantle domains to be 'mapped' (Menzies and Halliday 1988; Hart *et al.* 1989; Menzies 1989).

To help illustrate some of the problems with the interpretation of the petrology and geochemistry of continental volcanic rocks the remainder of this chapter will deal with two case studies—a small continental intraplate volcanic field, the Zuni-Bandera volcanic field New Mexico, USA, and an extremely large flood basalt province, the Ferrar Supergroup, Antarctica. In both case studies we will attempt to identify the composition of crustal and/or mantle components involved in their genesis.

8.2.1. Continental intraplate volcanism—Zuni Bandera, New Mexico

The Zuni-Bandera volcanic field (ZBVF) is located to the west of the Rio Grande Rift in west-central New Mexico in the transition zone between the Colorado Plateau (crust >40 km thick) and the Basin and Range (crust <25 km thick). The ZBVF was erupted in the Quaternary (1.38 Ma to AD 400) through Proterozoic and Phanerozoic crust and consists of several individual flows, viz. McCartys, Bandera, Hoya de Cibola, Paxton, Zuni, Oso, Twin Crater, and El Calderon (Maxwell 1986). The ZBVF is situated adjacent to the Rio Grande rift where it is apparent that an asthenospheric plume has either helped 'drive' extension or has upwelled in response to lithospheric extension (Olsen *et al.* 1987).

Phanerozoic or Proterozoic crust

Initial studies of the ZBVF ascribed the genesis of the alkaline and tholeiitic magmas to contamination (Laughlin *et al.* 1971; 1972*a,b*; Carden and Laughlin 1974) of upper mantle magmas with Proterozoic crust mainly because of the presence of radiogenic strontium isotopic ratios in the ZBVF—isotope ratios that were different from mid-ocean ridge basalts. To what extent assimilation of post-Archaean crust accompanied evolution of the ZBVF can be better evaluated by a broader consideration of the petrology and geochemistry of the volcanic rocks.

Temporal changes occurred within the ZBVF from concurrent eruption of tholeiitic and alkaline basalts at the beginning of the Pleistocene (0.78–1.38 Ma), eruption of alkaline basalts in the mid-Pleistocene, to contemporaneous eruption of alkaline and tholeiitic basalts at the end of the Holocene. Most ZBVF samples are MgO-rich and fractionation may have been limited to olivine and pyroxene. In detail the tholeiitic magmas tend to be richer in SiO_2 and poorer in K_2O, Na_2O, Rb, and Sr relative to the alkaline magmas. More importantly, the tholeiitic magmas have the highest $^{87}Sr/^{86}Sr$ ratios and, in a general sense, the $^{87}Sr/^{86}Srr$ decreases with increasing K_2O+Na_2O, Rb, and Sr and decreasing SiO_2.

MORB-normalized trace element abundance patterns (Fig. 8.2) indicate a range in the concentration of trace elements from an alkaline basalt with the highest concentration of incompatible and high field strength elements to a tholeiitic basalt with the lowest concentration of these elements. In general, the volcanic rocks have rather smooth patterns that reveal an overall enrichment in incompatible elements and a depletion in compatible elements relative to MORB. The relatively unfractionated state of the ZBVF rocks is consistent with the lack of trace element anomalies that one would normally associate with fractional crystallization (e.g. apatite: P anomaly; plagioclase: Sr anomaly; pyroxene: Ti anomaly). Overall the continental volcanic rocks from the ZBVF have many of the attributes of *oceanic* alkaline volcanic rocks (cf. Fitton and Dunlop 1985).

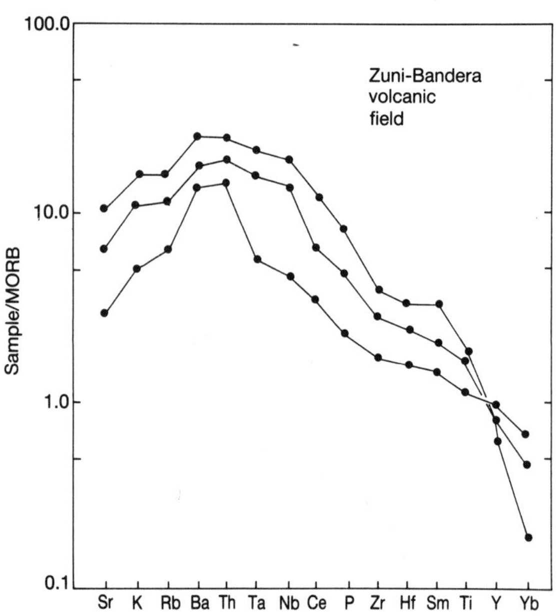

Fig. 8.2. MORB-normalized trace element concentrations in alkaline and tholeiitic volcanic rocks from the Zuni-Bandera volcanic field, New Mexico. Note the overall smoothness of the profiles and the lack of any negative or positive anomalies normally associated with fractionation and or assimilation of crust. The tholeiitic basalt does appear, however, to have an anomalous Th/Ta ratio perhaps the result of crustal contamination or involvement of accreted lithosphere? Normalization values after Pearce (1983).

If crustal rocks are assimilated by a magma during fractionation, correlations exist between indices of fractionation (MgO, SiO_2) and indices of contamination ($^{87}Sr/^{86}Sr$). In Fig. 8.3 the variation in trace elements for the ZBVF can be compared with variations

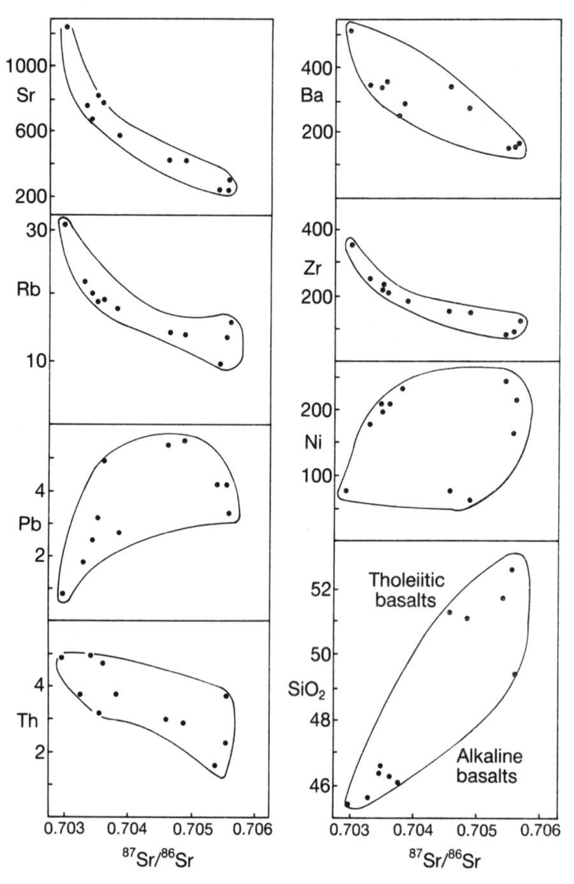

Fig. 8.3. Element–isotope variation diagram for the Zuni-Bandera volcanic field, New Mexico. The elements shown were chosen for several reasons: (1) Sr is normally depleted in the crust; (2) Rb, Pb, Th, and Ba are normally enriched in the crust; (3) Zr is concentrated in small volume melts; and (4) Ni and SiO_2 are rough indices of fractionation. Note that Sr, Rb, Th, Ba, and Zr decrease with increasing $^{87}Sr/^{86}Sr$ ratio and, whilst the decrease in the concentration of Sr and the increase in the concentration of Pb and SiO_2 with increasing $^{87}Sr/^{86}Sr$ ratio is compatible with contamination with crustal material, the increase in silica from alkaline to tholeiitic magmas could also indicate changing degrees of partial melting from small degrees (high Zr) to low degrees (low Zr). If crustal contamination is invoked, the decrease in the concentration of Th and Rb is problematical.

in $^{87}Sr/^{86}Sr$ ratio. With an increase in $^{87}Sr/^{86}Sr$ ratio we observe a decrease in the content of Sr, Rb, Th, Ba, Zr, and an increase in Pb and SiO_2. While the gradual increase in strontium isotopic ratio with increasing silica is compatible with assimilation of crust rich in ^{87}Sr and fractional crystallization changes to the SiO_2 content, the decreasing content of Sr, Rb, Th, Ba (and TiO_2, Na_2O, K_20) cannot be easily explained by contamination with average crust unless the crustal rock had a basic composition as well as a *high* $^{87}Sr/^{86}Sr$ *ratio*. Essentially, the highest concentrations of those elements normally found in the crust (e.g. Na_2O, K_2O, Ba, and Rb) and the lowest concentration of SiO_2, Ni, and Cr occur in the volcanic rock with the lowest $^{87}Sr/^{86}Sr$ ratio.

This is unusual in that the concentration of those elements and isotopes found in the crust should increase with contamination. The addition of ^{87}Sr from crustal rocks with low strontium contents leads to a negative correlation between $^{87}Sr/^{86}Sr$ and the relative abundance of Sr. In Fig. 8.4 the ZBVF data are plotted on a $^{87}Sr/^{86}Sr$ versus Sr diagram and the array defined by the Zuni-Bandera volcanic rocks is exactly that expected from crustal contamination processes (cf. British Tertiary Volcanic province). Moreover, with increasing $^{87}Sr/^{86}Sr$ ratio there is a decrease in the Na_2O content of the ZBVF basalts, a feature tht is compatible with experimental observations where a loss of 'basaltic' sodium to a crustal melt was noted (Watson and Jurewicz 1985). A problem still exists, however, with the ZBVF in that the highest concentrations of large ion lithophile (LIL), light rare earth (LRE), and high field strength (HFS) elements occur in the basaltic rock with the lowest $^{87}Sr/^{86}Sr$ ratio and the highest MgO content. This problem is accentuated by the fact that concentrations of these elements *decrease* systematically from alkaline to tholeiitic compositions with a concomitant *increase* in $^{87}Sr/^{86}Sr$. If we invoke crustal contamination as an explanation for the concentration of 'crustal' elements like rubidium and barium in the basalts, then the crust in question must have had a very *low* $^{87}Sr/^{86}Sr$ *ratio*. If, in contrast, we wish to invoke crustal contamination to explain the high $^{87}Sr/^{86}Sr$ ratios in the basalts, then the crust in question must have been depleted in those elements normally enriched in the crust (e.g. Rb, Th, etc.).

Since we now appear to have reached an impasse, perhaps we can utilize other geochemical parameters to resolve this issue. The involvement of crustal material in the genesis of volcanic rocks can be monitored with the

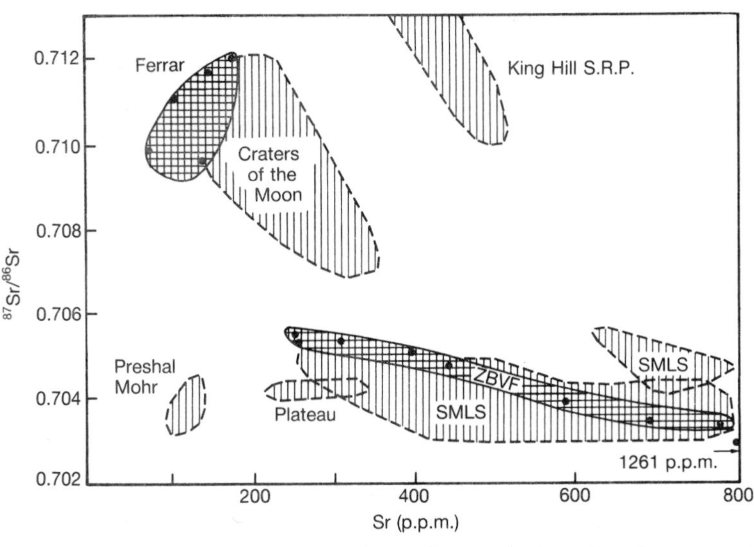

Fig. 8.4. Element–isotope variation diagram as a measure of crustal contamination. Crustal contamination of a magma (low $^{87}Sr/^{86}Sr$ ratios and high Sr content) with crustal rocks (high $^{87}Sr/^{86}Sr$ ratios and low Sr content) produces a negative correlation on this diagram. Note the *negative* correlation of the ZBVF; King Hill, Snake River Plain (SRP) and the Skye Main Lava Series (SMLS) data and the *positive* correlation of the Ferrar Supergroup flood basalts (FS). Data sources are Moorbath and Thompson (1980), Mensing *et al.* (1984), Kyle *et al.* (1990), Leeman and Menzies (unpublished data).

use of the Th/Ta ratio since Th is concentrated in crustal rocks (Fig. 8.5) and Ta is generally depleted (Pearce 1983). This diagram does not, however, distinguish between crustal material added to a source region during accretion of lithospheric mantle or addition of crustal material during stoping of wall rock. Nevertheless, the ZBVF data have a range in Th/Ta ratio similar to MORB and intraplate volcanic rocks with a slight tendency toward higher Th/Ta ratios in the tholeiitic magmas. From these data and the MORB-normalized diagrams (Fig. 8.2) one can infer that the alkaline

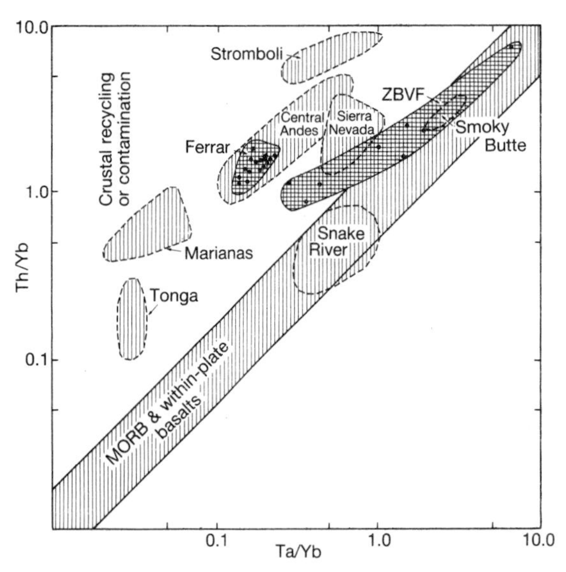

Fig. 8.5. Th/Yb versus Ta/Yb diagram used as a measure of the involvement of crustal rocks in magmagenesis either as a recycled component or as a high-level addition through crustal contamination processes. Crustal rocks have a high Th/Ta ratio and plot in the top left-hand quadrant of the diagram. Mid-ocean ridge and within-plate volcanic rocks plot along the diagonal band of constant Th/Ta ratio indicating no preferential enrichment of Th over Ta. The ZBVF alkaline rocks plot within this band and the tholeiitic members plot off it. The basaltic andesites of the Ferrar Supergroup plot off this band due to high Th/Ta ratios in the same field as volcanic rocks from the Andes, Tonga, and the Marianas. From this diagram it is apparent that a 'crustal component' played a significant role in the genesis of the FS and a possible minor role in the genesis of the tholeiites of the ZBVF. In contrast, the alkaline volcanic rocks of the ZBVF have strong intraplate characteristics like ocean island basalts. Data sources are Ellam (1986), Fraser *et al.* (1985), Van Kooten (1980), Kyle *et al.* (1990), and Pearce (1983 and references therein).

volcanic rocks were produced by intraplate processes with little or no discernible involvement of crustal materials. In contrast, the more siliceous tholeiitic volcanic rocks with the most radiogenic $^{87}Sr/^{86}Sr$ ratio have the highest Th/Ta ratio (Figs 8.2 and 8.5), a feature that may point to crustal involvement in the genesis of the more tholeiitic rocks. If assimilation of Proterozoic and Phanerozoic crust, or recycling of a crustal component, played a role in the genesis of the ZBVF rocks, then its effects are most apparent in the tholeiitic magmas ($SiO_2 = 49$–52 per cent). Several other aspects of the ZBVF data, however, are at variance with such a model. First and foremost the majority of the ZBVF rocks are relatively unfractionated alkaline to tholeiitic rocks ($MgO > 6.0$ per cent), a feature that is inconsistent with any hypothesis that invokes fractionation as the main 'energy' source for assimilation of crustal wall rocks. Second, the generally smooth MORB-normalized patterns of the bulk of alkaline and tholeiitic basalts contrast with the 'spiky' MORB-normalized patterns of contaminated volcanic rocks from the western USA (Menzies *et al.* 1984; McMillan and Dungan 1988). Why would consumption of crustal rocks, or partial melts thereof, maintain smooth trace element profiles? Third, the highest concentrations of crustal elements (e.g. potassium, rubidium, barium) are to be found in the alkaline basalt with the lowest $^{87}Sr/^{86}Sr$ ratio. Since K_2O has been shown to be markedly increased by crustal contamination, a new form of this process must be invoked—one that increases the K_2O content (and most other incompatible elements) and lowers the $^{87}Sr/^{86}Sr$ ratio. Stoping of considerable quantities of MORB-like material underplated on the crust–mantle boundary may be necessary to explain such features. The only aspect of the ZBVF that is compatible with an assimilation and fractional crystallization (AFC) model is the presence of tholeiites with higher $^{87}Sr/^{86}Sr$ ratios, Th/Ta ratios, and SiO_2 content than contemporaneous alkaline volcanic rocks.

Proterozoic accreted lithosphere

A considerable amount of evidence exists for growth of the Proterozoic lithosphere in the western USA by arc accretion (e.g. Ernst 1988) and, consequently, the subcrustal lithosphere may also have the characteristics of accreted lithosphere produced above active subduction zones. One could monitor the participation of such accreted lithosphere in the genesis of the ZBVF by consideration of those elements that define a subduction signature. For example, trace elements that are normally enriched in arc basalts include Sr, Ba, and Th and those normally depleted in arc basalts include Nb and Ta. We have already noted a lack of any preferential enrichment (or depletion) of Th in most of the ZBVF rocks as part of our consideration of the evidence for crustal contamination. The lack of any marked enrichment in Th relative to Ta (Fig. 8.5), except perhaps in the most tholeiitic material, made it rather unlikely that high-level crustal contamination had been a dominant process. One could use the same data, however, to argue against supra-subduction enrichment processes as a major contributor to alkaline basaltic volcanism in the ZBVF. Furthermore, the high Zr/Ba (or high Ta/Sr, low La/Nb) ratios of the ZBVF rocks are not compatible with supra-subduction enrichment processes (cf. Ormerod *et al.* 1988), and are more akin to the trace element enrichment processes associated with the genesis of ocean island basalts. Similarly the Ba/Ta ratio has frequently been used as a diagnostic feature of volcanism involving supra-subduction processes (i.e. Ba/Ta >450) because of the high concentration of barium in recycled material and the depletion in Ta (Gill 1981). It is noteworthy that the volcanic rocks from the ZBVF have low Ba/Ta ratios (i.e. 107–250) outside the field occupied by arc basalts. The only evidence of possible supra-subduction processes (or crustal contamiantion) is the observation that the highest La/Nb, Ba/Ta, and Ta/Th ratios are in the rocks with the highest $^{87}Sr/^{86}Sr$ ratio—the tholeiites. This could be explained by addition of a minor amount of crustal material during eruption or involvement of subduction modified lithosphere. It is important to stress, however, that the dominant control on the geochemistry of the ZBVF rocks does not appear to be processes associated with contamination by crustal rocks or melting of accreted subcrustal lithosphere.

Asthenosphere

In the previous two sections we have concluded that the volcanic rocks of the ZBVF are magnesian and that they have undergone limited fractionation. Moreover, there is no unequivocal evidence in the chemistry of the ZBVF rocks for processes involving high-level assimilation of post-Archaean crust or deeper involvement of accreted Proterozoic–Phanerozoic lithosphere. If further analysis supports this viewpoint, one must conclude that much of the chemical variation results from processes either within the asthenosphere or lithosphere of different composition to the accreted lithosphere considered above (e.g. MORB). To properly evaluate such a claim we must consider the chemistry of the ZBVF rocks in relation to the data base that exists for oceanic rocks.

The range in strontium isotopic ratio in the ZBVF rocks (Fig. 8.3) covers most of the variation known to exist in ocean island (Frey and Roden 1987) and mid-ocean ridge basalts (Le Roex 1987). Moreover, the relationship between $^{87}Sr/^{86}Sr$ ratio and Rb/K ratio for the ZBVF rocks indicates a systematic increase in Sr isotopic ratio with increasing Rb/K ratio revealing that source heterogeneity or source mixing (Galer and O'Nions 1986) best explains the trace element and isotopic characteristic of the ZBVF data and that crustal influences are minimal. This results from a high potassium content (by a factor of two) in the basalt with the lowest $^{87}Sr/^{86}Sr$ ratio. Plots of K_2O/Th versus La/Ta which apparently distinguish contaminated from uncontaminated basaltic rocks (Thompson and Morrison 1988) indicate that virtually *all* the ZBVF rocks fall within the extremely *small* field occupied by oceanic and continental volcanic rocks as exemplified by the Miocene–Recent rift-related volcanism in Colorado (Leat *et al.* 1988), the Quaternary volcanism of the Basin and Range (Menzies *et al.* 1983), MORB of the southern oceans (LeRoex 1987), and the Honolulu alkaline volcanic rocks (Frey and Roden 1987).

A partial melting diagram can be used to help better understand these intraplate processes and to define the composition of possible reservoirs. On such a diagram (Fig. 8.6) small degrees of melt have high Zr/Y ratios and low Zr/Nb ratios whilst large degrees of melt have high Zr/Nb and low Zr/Y ratios. In Fig. 8.6 the variation in Zr/Y versus Zr/Nb is plotted for the ZBVF and the variation compared with plume-type, transitional-type, and normal-type MORB (LeRoex 1987). The oceanic samples define a mixing hyperbola between a large degree melt of MORB mantle ($^{87}Sr/^{86}Sr <0.703$) and a small degree melt of an enriched mantle plume ($^{87}Sr/^{86}Sr >0.703$). In contrast, the continental samples from the ZBVF define a more extensive mixing hyperbola between a small degree melt of MORB mantle and a large degree melt of an enriched mantle source ($^{87}Sr/^{86}Sr >0.705$). The hyperbola defined by the ZBVF overlaps with that part of the oceanic array occupied by plume and transitional MORB (i.e. P-type). In contrast to the oceanic data set the alkaline ZBVF rocks have very high Zr/Y ratios and can as such be treated as a small degree melt. Thompson *et al.* (1983) calculated that small degrees of melting of an OIB source would produce a Ta/Yb ratio of 8–10, a figure that compares favourably with a Ta/Yb ratio of 6.0 for the ZBVF rocks. Moreover, some of the alkaline basalts from the ZBVF have higher Th/Ta ratios than the Smoky Butte lamproites (Fig. 8.5) (small degree melt of *Archaean lithosphere*; Fraser *et al.* 1985) and a $(La/Yb)_N$ ratio twice that of the Honolulu nephelinites (small degree melt of *Phanerozoic lithosphere*; cf. Clague and Frey 1982). The alkali basalt from the ZBVF can clearly be interpreted as a small degree melt of a mantle source that has an isotopic composition similar to the source of

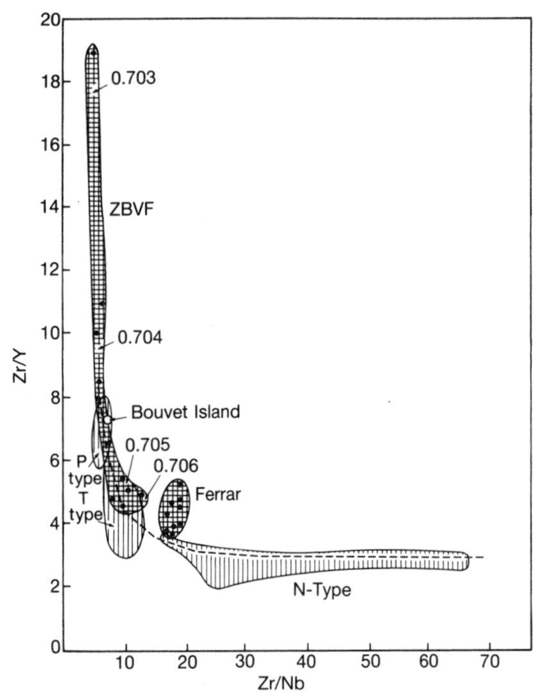

Fig. 8.6. Zr/Y versus Zr/Nb as a measure of the amount of partial melting involved in the genesis of volcanic rocks. Low degree melts have very high Zr/Y ratios and low Zr/Nb ratios and high degree melts have low Zr/Y ratios and high Zr/Nb ratios. The fields of N, T, and P type MORBs from the southern Ocean are shown as a classic demonstration of mantle source mixing between a plume (e.g. Bouvet Island) and MORB mantle. The ZBVF data define a 'mixing hyperbola' that extends from very high Zr/Y to low Zr/Y ratios consistent with the mixing of small and large degree melts. Note that within the ZBVF the $^{87}Sr/^{86}Sr$ ratio increases as the Zr/Y ratio decreases indicating that the origin of these volcanic rocks involves at least two discrete components. The low $^{87}Sr/^{86}Sr$ alkaline basalts are small volume melts with within-plate characteristics (Fig. 8.5) and the high $^{87}Sr/^{86}Sr$ tholeiitic basalts represent larger degrees of melting and may have a crustal component in their source. The basaltic andesites of the Ferrar Supergroup have slightly higher Zr/Y ratios than the range for N-type MORB. From this diagram one would conclude that the ZBVF contains magmas that were formed by very different degrees of partial melting. Data sources are LeRoex (1987) and Kyle *et al.* (1990).

MORB and, on the basis of low $(La/Yb)_N$, low Zr/Y, and high Zr/Nb ratios (Fig. 8.6), the tholeiitic basalts can be interpreted as a much larger degree melt of a more isotopically enriched reservoir. If one accepts that crustal processes have not dominated the geochemistry of these volcanic rocks, we must decide on a source in the mantle for: (1) a 'tholeiitic' magma enriched in the LIL and LRE elements and with a radiogenic $^{87}Sr/^{86}Sr$ ratio; and (2) an 'alkaline' magma markedly enriched in the LIL and LRE elements and with a non-radiogenic $^{87}Sr/^{86}Sr$ ratio.

Several combinations of mantle sources can be envisaged:

1. An asthenospheric source for the alkaline magmas (i.e. DMM) and derivation of the tholeiitic magmas from Proterozoic lithospheric mantle (i.e. EM 2).
2. Derivation of the alkaline and tholeiitic magmas from heterogeneous asthenosphere as has been proposed for the Cameroon Line (Fitton and Dunlop 1985).
3. Derivation of the alkaline magmas from the asthenosphere ($^{87}Sr/^{86}Sr < 0.703$) or the thermal boundary layer at the base of the continental lithosphere (i.e. DMM) and a lower mantle or plume origin for the tholeiitic magmas ($^{87}Sr/^{86}Sr > 0.706$). This model is similar to that invoked for Hawaiian magmatism (Frey and Roden 1987).

In order to evaluate these possibilities we need to look at the character of the lithosphere and the asthenosphere beneath the ZBVF. We have already dismissed any involvement of accreted lithosphere in the genesis of the ZBVF as we found no evidence for a dominant subduction or crustal signature in the volcanic rocks. This makes hypothesis 1 rather untenable. With regard to hypothesis 2, whilst the ZBVF data may support the idea of isotopic heterogeneities in the asthenosphere we do not accept that isotopic heterogeneities covering the range of mid-ocean ridge basalts and most ocean island basalts can survive within a single convecting reservoir. We therefore find hypothesis 2 unacceptable. With regard to hypothesis 3 one can point to the existence of small volume melts in the MORB asthenosphere beneath the western USA due to the widespread occurrence of xenolith-bearing basanites erupted through regions of thin crust and high heat flow (i.e. the southern Basin and Range). Moreover the protolith that constitutes most of the sub-continental lithosphere beneath the south-western USA has an isotopic composition very similar to the asthenosphere. Therefore alkaline magmas similar to those erupted at the ZBVF may have had a source in the asthenosphere or the thermal boundary layer at the base of the lithosphere. The larger degrees of melt represented by the enriched tholeiites may have a source (as in Hawaii) in a deep or lower mantle enriched plume (i.e. $^{87}Sr/^{86}Sr > 0.706$). It is of interest to note that certain compositional similarities exist between the ZBVF and Hawaiian magmatism. For example the overall range in $^{87}Sr/^{86}Sr$ composition for the ZBVF is similar to the Waianae tholeiitic and alkaline volcanics, the Honolulu alkaline volcanics, and the Koolau tholeiitic volcanic rocks (Frey and Roden 1987). As is the case in the ZBVF, the alkaline volcanic rocks in Hawaii have the lowest $^{87}Sr/^{86}Sr$ ratio and the highest concentration of incompatible elements and the tholeiites have the highest $^{87}Sr/^{86}Sr$ ratio and the lowest concentration of incompatible elements. Whilst the Hawaiian alkaline rocks (silica undersaturated) are believed to represent small degree melts of a shallow lithospheric source, the tholeiitic magmas (silica saturated) are believed to constitute deep plume magmatism. This may also apply to the ZBVF.

The geochemical data for the ZBVF are compatible with a model of interaction between the asthenosphere or the continental lithospheric mantle and a lower mantle plume. The basanites (small degree alkaline melts) are believed to be derived from the asthenosphere ($^{87}Sr/^{86}Sr < 0.703$) and the tholeiites (large degree melts) from deeper enriched sources related to upwelling plume material ($^{87}Sr/^{86}Sr > 0.706$). Over the last 1.4 Ma the interplay between asthenospheric and lithospheric reservoirs has been expressed on the surface as the concurrent eruption of alkaline and tholeiitic magmas.

ZBVF as a crust–mantle probe

Consideration of the genesis of individual Cainozoic volcanic fields throughout the western USA has provided us with an indication of the regional distribution of particular lithospheric and asthenospheric reservoirs (Leeman 1982; Menzies *et al.* 1983; Perry *et al.* 1988; Menzies 1989). These data can be correlated with physical parameters like crustal thickness (Fig. 8.7), age, and heat flow. For example, the volcanic rocks erupted through the Wyoming craton, an area of low heat flow and thick Archaean crust, have mainly tapped a lithospheric reservoir of enriched mantle (i.e. EMI). To the south within a region of Proterozoic crust of variable crustal thickness (25 and 50 km) and variable heat flow the volcanic rocks appear to tap a variety of reservoirs that include mixtures of depleted (i.e. DMM) and enriched domains (i.e. EM2). Evidence from lithospheric xenoliths indicates that Proterozoic mantle beneath the western USA can be thought of as a mixture of DMM, EM2, and HIMU domains (Menzies 1989; Menzies, Chapter 4, this volume). On the basis of Sr, Nd, and C isotopes the oldest domain appears to be depleted MORB mantle (DMM), which has been subsequently modified by trace element enrichments that have resulted in Sr, Nd, and Pb isotopic characteristics not unlike EM2 and HIMU. More recently, this has been overprinted by asthenospheric domains similar to DMM. In contrast in the southern Basin and Range, an area of the highest heat flow and thinnest Proterozoic crust, the majority of Cainozoic volcanic rocks are alkali basalts that tap an asthenospheric reservoir similar to depleted mid-ocean ridge basalt mantle (i.e. DMM) (see

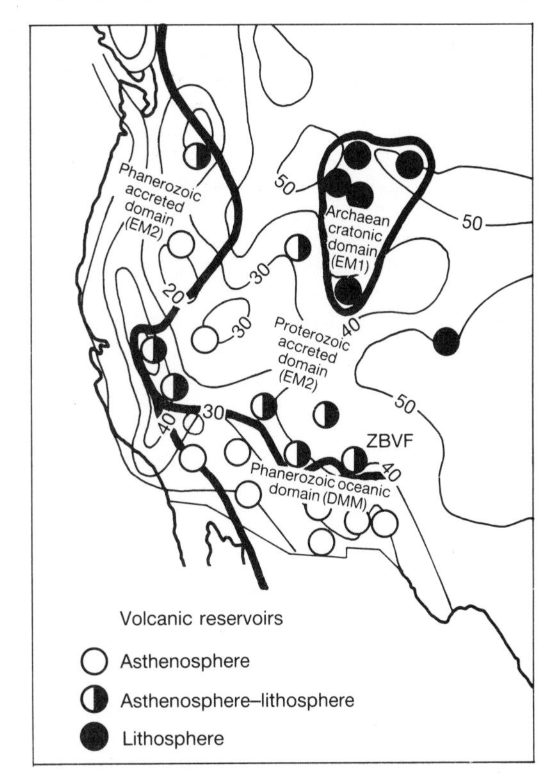

Fig. 8.7. Distribution of mantle domains in the western USA in relation to crustal thickness (contours with thickness in kilometers) based on volcanic rock data (modified after Menzies 1989). The presence of most of the Phanerozoic oceanic domains (DMM) coincide with the area of thin crust (high heat flow > 2HFU) whereas the Archaean cratonic domains (EM1) coincide with the area of thickest crust (low heat flow < 2HFU). The ZBVF is located within the transition zone from the Basin and Range (where basalt sources are predominantly in the asthenosphere) to the Colorado Plateau (where Proterozoic accreted lithosphere plays a significant role in the genesis of volcanic rocks). In the case of the ZBVF, magmagenesis involved two discrete reservoirs. The first, in the asthenosphere, permitted eruption of small volume melts with intraplate characteristics and low $^{87}Sr/^{86}Sr$ ratios. The second allowed for eruption of larger degree melts with high $^{87}Sr/^{86}Sr$ ratios.

Anderson, Chapter 1, this volume). The ZBVF data, however, indicate a more complex situation than the presence of cratonic (EM1) and oceanic (DMM and EM2) lithospheres beneath the western USA. The possible presence of an Archaean domain (EM1), a Proterozoic circum-cratonic domain (DMM), a Proterozoic–Phanerozoic domain (EM2), and a Phanerozoic domain (DMM) does not accommodate the isotopic variability seen at the ZBVF. The ZBVF data may point to the existence of greater sub-lithospheric heterogeneity in the form of enriched lower mantle plumes (e.g. Yellowstone).

8.2.2. Continental flood volcanism—Ferrar Supergroup

In the Jurassic, the Ferrar Supergroup (FS) was erupted into Proterozoic crust and comprises several distinct groups including the Kirkpatrick Basalt Group (volcanic–lava flows), Ferrar Dolerite Group (hypabyssal—dolerite sills and dykes), and the Dufek massif (plutonic—layered gabbro). The FS represents one of the world's largest continental tholeiitic basalt provinces (3000 km in length) with eruption or emplacement of approximately 10^6 km^3 of basaltic material (Faure *et al.* 1972, 1974; Kyle 1980; Kyle *et al.* 1981, 1983, 1990). On the basis of their silica saturation the volcanic rocks are quartz-tholeiites or basaltic andesites/andesites.

In the case of the FS we are dealing with one of the major continental flood basalt provinces (cfb), whereas in the case of the ZBVF we dealt with a small continental volcanic field. Despite this, cfb provide a valuable insight into crustal and mantle processes. The Ferrar Supergroup has been at the heart of a controversy that has raged for decades concerning the genesis of continental flood basalts from an isotopically depleted asthenospheric source (i.e. MORB or OIB) with extensive high-level crustal contamination or from an isotopically enriched lithospheric source with minimal high-level crustal contamiantion (Compston *et al.* 1968, Faure *et al.* 1972, 1974; Kyle 1980; Kyle *et al.* 1983, 1990; Mensing *et al.* 1984; Hoefs *et al.* 1980). In the case of the FS we will first examine the evidence for involvement of post-Archaean crust and then Proterozoic subcrustal lithosphere and asthenosphere.

Post-Archaean crust

Major elements show systematic vertical variations in the stratigraphic section at Gorgon Peak, Antarctica (Kyle *et al.* 1990). At the base of the Gorgon Peak section (Fig. 8.8) the volcanic rocks are rich in SiO_2, K_2O, Na_2O, Rb, and Sr and towards the top there is an overall decrease in these elements. Similarly, $^{87}Sr/^{86}Sr$ ratios vary throughout the section from high $^{87}Sr/^{86}Sr$ ratios at the base to lower $^{87}Sr/^{86}Sr$ ratios at the top. It should be noted, however, that the strontium isotopic ratios are much higher throughout the Gorgon Peak section than any MORB or OIB. Overall, the range in $^{87}Sr/^{86}Sr$ is more akin to that of upper crustal rocks or

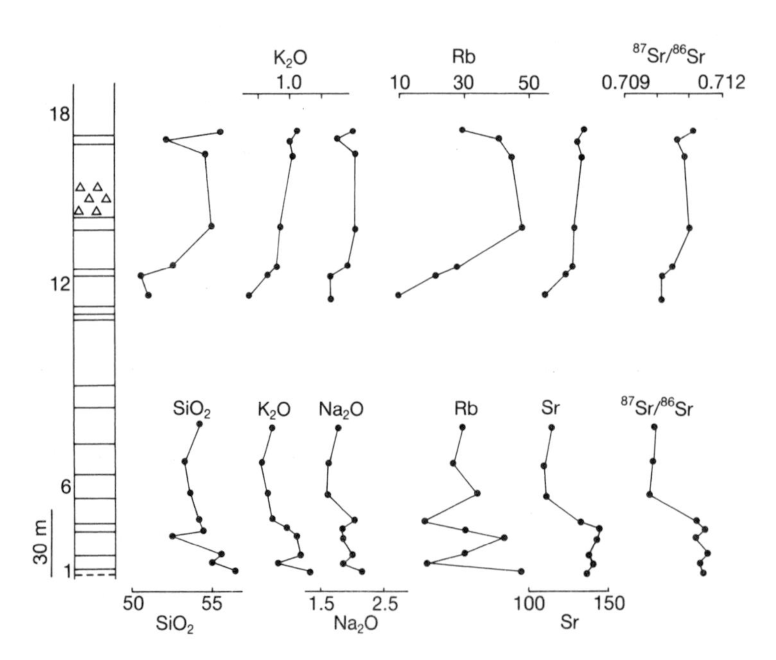

Fig. 8.8. Vertical variation in major, minor, and trace elements and strontium isotopes in the Gorgon Peak section of the Ferrar Supergroup, Antarctica (Kyle *et al.* 1990). At the base of the section the basaltic andesites are richer in SiO_2, K_2O+Na_2O, Rb, and Sr have higher $^{87}Sr/^{86}Sr$ ratios than the basaltic andesites higher in the section.

basalt-borne and kimberlite-borne xenoliths (Fig. 8.1) from Proterozoic to Phanerozoic lithospheric mantle (Menzies, Chapter 4, this volume).

Variation diagrams (a possible index of fractionation) reveal a good correlation between MgO and several major, minor, and trace elements consistent with fractional crystallization of plagioclase, pyroxene, and olivine (Fig. 8.9). Kyle *et al.* (1990) used observed phenocryst phase compositions and produced fractionation solutions of 7.7 per cent olivine, 3.1 per cent clinopyroxene, and 14.2 per cent plagioclase. The only exception was K_2O, which he concluded may have been enhanced by post-emplacement processes.

While on an AFM diagram the Gorgon Peak basalts show the typical iron enrichment trend of tholeiitic rocks, a MORB-normalized diagram (Fig. 8.10) better illustrates the elemental consequences of fractionation processes. The FS samples show distinct negative anomalies at Ba, Ta, Nb, P, and Ti. Fractionation of various mineral phases can be proposed as an explanation for the P (apatite) and Ti (pyroxene or magnetite) anomalies but not the Ba, Ta, and Nb anomalies. The depletion in Ta and Nb is believed to be related to the

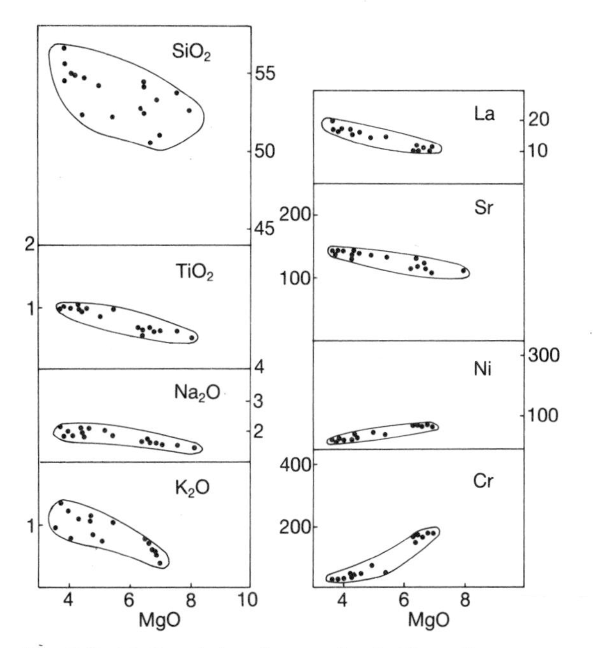

Fig. 8.9. MgO variation diagrams for the Ferrar Supergroup as a measure of fractional crystallization. The systematic behaviour of most elements reveals fractionation of olivine, clinopyroxene, and plagioclase. Data sources are Kyle *et al.* (1990) and Mensing *et al.* (1984).

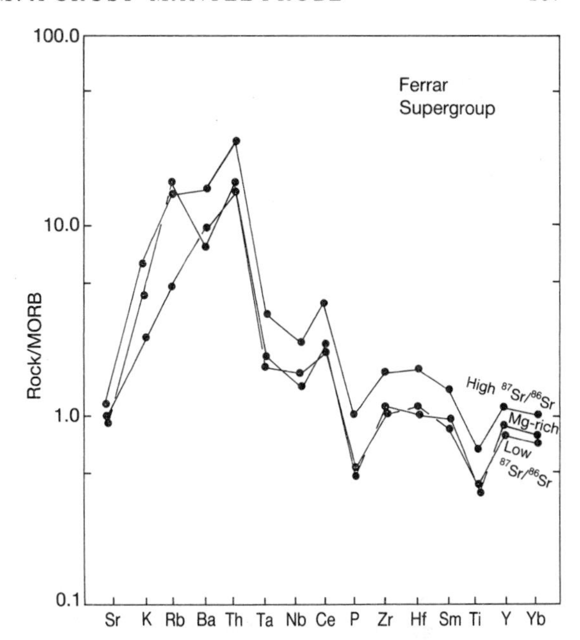

Fig. 8.10. MORB-normalized trace element concentrations in Ferrar Supergroup basaltic rocks (Kyle *et al.* 1990). The samples shown include the sample with the highest MgO content (least fractionated) and two samples that cover the greatest range in $^{87}Sr/^{86}Sr$ ratios (least to most contaminated?). The negative anomalies at P and Ti may be explained by fractionation of apatite (P) and pyroxene/magnetite (Ti) whereas the anomalies at Ta and Nb are similar to that observed in arc basalts and may indicate a 'subduction signature' in the source of the FS. Note that the FS data have a more spiky profile than the ZBVF (Fig. 8.2) data due to fractionation and the involvement of 'crust' either at high level in crustal magma chambers or in the mantle source. The change in $^{87}Sr/^{86}Sr$ with fractionation (parallelism of MORB-normalized data) is indicative of high-level crustal input.

source of the basalts, rather than to fractionation, a feature that will be dealt with later. The Ba anomaly may be an artefact of post-consolidation processes. In Fig. 8.10 the elemental variation in the most magnesian basalt from the Gorgon Peak section and those basalts from the section with the highest and lowest $^{87}Sr/^{86}Sr$ ratios are shown for comparison. The most magnesian basalt is very similar to the basalt with the lowest $^{87}Sr/^{86}Sr$ ratio and the basalt with the highest $^{87}Sr/^{86}Sr$ ratio has the highest concentrations of all elements except Rb. The overall shape of the normalized profile is, however, roughly parallel to that of the volcanic rock with the lowest $^{87}Sr/^{86}Sr$ ratio, a feature that can be adequately explained by fractionation. Fractionation cannot, however, alone explain the change in strontium isotopic composition in the more evolved FS rocks. The most fractionated sample in the MORB-normalized diagram has the highest $^{87}Sr/^{86}Sr$ ratio and the least fractionated sample has the lowest $^{87}Sr/^{86}Sr$ ratio and, therefore, a relationship may exist between fractionation and changing isotopic ratio. Kyle *et al.* (1990) reported that in the Ferrar Supergroup (i.e. Gorgon Peak) the $^{87}Sr/^{86}Sr$ ratio varied from 0.7098 in the most basic Mg-rich rocks to 0.7115 in the most evolved Mg-poor rocks and that there were good correlations between $(^{87}Sr/^{86}Sr)_O$ and some major and trace elements indicative of assimilation and fractional crystallization processes (i.e. AFC).

To further illustrate AFC processes in the evolution of the FS (Fig. 8.11) we have chosen: (1) several elements that would tend to be sensitive to addition of crustal rocks (e.g. Sr, Rb, Pb, Th); and (2) SiO_2 as a classic index of fractionation. Sr, Pb, Th, and SiO_2 increase with increasing $^{87}Sr/^{86}Sr$ ratio and Rb shows considerable scatter. If we compare the FS data with the ZBVF rocks it is apparent that only Pb and SiO_2 display a systematic increase with $^{87}Sr/^{86}Sr$ ratio, whereas the other elements decrease (e.g. Sr, Rb, Th). If the basaltic rocks underwent closed-system fractional crystallization, all fractionated derivatives (high silica) should have $^{87}Sr/^{86}Sr$ ratios identical to the most magnesian least fractionated rock type. This is not the case in the FS. The element–isotope correlation suggests that whilst fractionation proceeded it was associated with assimilation of crustal rocks rich in Pb and Th. Similar correlations between isotopes and elemental abundance data have been reported for continental flood basalts from around the world (Menzies *et al.* 1984; Hawkesworth *et al.* 1988; McMillan and Dungan 1988) and have been interpreted as evidence for crustal contamination.

The increase in Sr content with increasing $^{87}Sr/^{86}Sr$ ratio is the opposite of what is normally associated with crustal contamination. The positive correlation (Figs 8.4 and 8.11) indicates that the Sr content of the residual magma was increasing and had not been depleted by plagioclase fractionation. If plagioclase had already fractionated, there would be a negative correlation between $^{87}Sr/^{86}Sr$ ratio and Sr content but the observed correlation indicates that contamination had already started prior to plagioclase fractionation. The most likely contaminant for the FS was the Palaeozoic granitoids that constitute much of the crust in this region—the Granite Harbour Intrusive Group (Borg *et al.* 1987). This form of contamination involving a magma with a $(^{87}Sr/^{86}Sr)_O=0.709$ contrasts with the

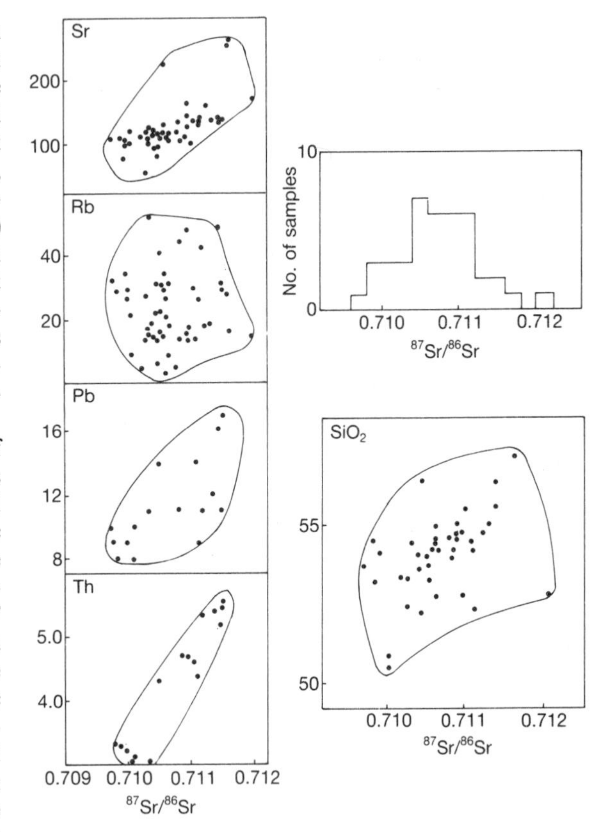

Fig. 8.11. Element–isotope variation diagrams utilizing elements commonly depleted in the crust (Sr), enriched in the crust (Rb, Pb, Th), and an index of fractionation (SiO_2). Note that in the FS the concentrations of Sr, Pb, Th, and SiO_2 increase with increasing $^{87}Sr/^{86}Sr$ ratio and the most incompatible element enriched magma has the highest $^{87}Sr/^{86}Sr$ ratio—the opposite was the case for the ZBVF rocks. Moreover, in both the ZBVF rocks and the FS volcanic rocks silica and lead concentrations vary systematically with $^{87}Sr/^{86}Sr$ ratio but the reasons for that variation may be very different. Data sources are Kyle *et al.* (1980) and Mensing *et al.* (1984).

situation in the Dufek massif, a large stratiform intrusion which is often thought of as an analogue to the high-level magma chambers from which the FS may have been extracted. Studies of the Dufek massif (Ford *et al.* 1986) report $(^{87}Sr/^{86}Sr)_O=0.70763$ and 0.70918 for pyroxene and plagioclase, respectively, the latter overlapping with the $(^{87}Sr/^{86}Sr)_O$ of other Ferrar Supergroup rocks. In the Dufek massif a change in $^{87}Sr/^{86}Sr$ ratio appears to have been synchronous with plagioclose fractionation because the magma from which the pyroxenes crystallized had a lower $^{87}Sr/^{86}Sr$ ratio than that which produced the plagioclase. The Dufek data may constrain an episode of crustal contamination which transformed an initial magma $((^{87}Sr/^{86}Sr)_O = 0.707)$ to a series of fractionated and contaminated derivatives.

Further evidence for crustal contamination in the genesis of the FS may be found in the following features:

1. Selective diffusion of K into a magma during AFC has been demonstrated in experimental runs (Watson 1982; Watson and Jurewicz 1985), a factor that may explain the apparent poor fractionation mass balance solutions for K_2O in the FS (Kyle *et al.* 1990).
2. Leaching of pyroxene removes about 50 per cent of the inventory of Rb and Sr but $(^{87}Sr/^{86}Sr)_O$ remains unchanged. In contrast, leaching of plagioclase removes the majority of the Sr and a small amount of Rb and $(^{87}Sr/^{86}Sr)_O$ increases dramatically from 0.71176 to 0.71354.
3. Late-stage exchange of oxygen isotopes with upper crustal material is clearly indicated by the elevated oxygen isotopic values (viz. $\delta^{18}O = 8.3$) in the groundmass of certain samples (Kyle *et al.* 1990). Furthermore, whole-rock oxygen isotope analyses show evidence of meteoric water isotopic exchange in the dolerites, but not the basalts.
4. The high Th/Ta ratio (Fig. 8.3) in the FS indicates either addition of Th during recycling processes in the mantle source or high-level crustal contamination. The FS data plot in a position similar to that of volcanic rocks from the Andes, a feature that is consistent with the overall andesitic nature of the FS rocks and the depletion in Nb and Ta observed in the MORB-normalized diagrams (Fig. 8.11).

We have already outlined the evidence for low-pressure fractionation of silicate phases and concomitant assimilation of crust (AFC) during the evolution of the FS. However, whether the magma that underwent assimilation and fractional crystallization came from the lithosphere or the asthenosphere remains a contentious issue. It is apparent from the isotope–element plots that the lowest $^{87}Sr/^{86}Sr$ end of any plot is characterized by an $^{87}Sr/^{86}Sr \leq 0.709$, much higher than mantle-derived magmas in oceanic basins. The issue at hand is whether or not that value represents that of: (1) a magma derived from the asthenosphere that was contaminated in the crust giving rise to a uniform high $^{87}Sr/^{86}Sr$ prior to being involved in the AFC outlined above; (2) a magma derived from accreted lithospheric mantle with a supra-subduction signature; or (3) a magma derived from a deeper source where recycling of crustal components and mixing with depleted mantle produced an enriched domain (i.e. EM2).

Asthenosphere

An asthenospheric source ($^{87}Sr/^{86}Sr = 0.702$–0.703) for the source magma to the FS is a simple solution and requires a large homogeneous reservoir in the upper mantle that can produce large volumes of tholeiitic magma at low pressure similar to that which occurs beneath mid-ocean ridges. Large magma chambers within the crust, similar to the Dufek massif, would then provide a platform to convert the chemistry of these magmas during fractionation and assimilation of crust. Note that in the case of the FS several cycles of contamination are required to produce the volcanic sequences observed at the surface. The erupted rocks could represent samples of an assimilation fractional crystallization continuum that formed in several magma chambers tapped at different stages. Part of the FS, the tholeiitic andesites at Gorgon Peak in South Victoria Land, Antarctica, can be shown to have evolved by an assimilation and fractional crystallization process. Contamination was by selective assimilation involving the addition of K, radiogenic Sr, non-radiogenic Nd, and slightly enriched oxygen. Low-temperature hydrothermal processes have also affected these basalts. With such unequivocal evidence for low- and high-temperaturee interaction with crust it would seem imprudent to assume that the lowest $^{87}Sr/^{86}Sr$ ratios in the FS or even the Dufek Massif are necessarily the most 'primary'. The vertical variations in elemental abundances and isotopic ratios in other cfb provinces (e.g. Deccan, India) reveal that the chemical heterogeneity observed in the Gorgon Peak volcanic rocks may represent a sample 'window' of the possible contaminated products generated within a large crustal magma chamber undergoing AFC.

Hergt *et al.* (1989) provide elegant reasons why crustal contamination of MORB parental magmas, or even magmas similar in composition to ocean island basalt (OIB) and island arc tholeiite (IAT), cannot adequately account for the geochemistry of the Tasmania dolerites. Similar reasoning applies to the FS. Any model involving contamination of a MORB magma with upper or lower crust needs to explain the following characteristics of the FS.

1. The andesitic nature of the FS and their similarity to magmas formed in arc environments (i.e. depletions in Ta and Nb).
2. The uniform oxygen isotope data for pyroxene

separates ($\delta^{18}O = +5.15$) is generally accepted as that of mantle-derived magmas which have not experienced any interaction with the crust (Hoefs *et al.* 1980; Mensing *et al.* 1984). Therefore, the oxygen isotope compositions of the pyroxenes leave little room for assimilation of *typical* continental crust.

3. The uniform initial strontium, neodymium, and lead isotopic data, indicative of an initial magma from a source enriched in Rb, U, and the light rare earth elements.
4. The existence of trace element and isotopic systematics in the Gorgon Peak magmas that are consistent with assimilation and fractional crystallization but the lack of such systematics in all other magmas with $^{87}Sr/^{86}Sr \leq 0.709$.
5. The lack of any volcanic rock with an $^{87}Sr/^{86}Sr$ similar to MORB or OIB within the Ferrar Supergroup which stretches across Antarctica for over 3000 km (Mensing *et al.* 1984; Kyle *et al.* 1990). All samples of the FS have $^{87}Sr/^{86}Sr \geq 0.709$.

Proterozoic subcrustal lithosphere

The parental magma to the FS is not believed to be derived from depleted asthenosphere and must therefore originate in the lithosphere or deeper mantle. The flood basalt volcanism associated with the break-up of Gondwanaland covers an area in excess of 1 000 000 km^2 and therefore required an enriched reservoir of substantial size (e.g. EM2). If this reservoir was located in the lithosphere, one would expect to find evidence for its presence in the prolific xenolith suites in South Africa, Antarctica, and Australia. A very important argument against a lithospheric mantle source for the parental magma to the FS is the lack of unequivocal evidence for a continuous, enriched reservoir in the lithospheric mantle beneath the Proterozoic of Antarctica or, for that matter, any other continent (i.e. $(^{87}Sr/^{86}Sr)_O > 0.7075$; isotopic ratio $\varepsilon_{Nd} = -4.1$ to -7.5) (Fig. 8.1). Any xenoliths with enriched compositions are limited in distribution and are frequently restricted to individual vents (McGibbon *et al.* 1987; McGibbon 1990; Gamble *et al.* 1988). If we consider the composition of the Proterozoic lithosphere on a world-wide scale (Menzies, Chapter 4, this volume), it would appear that the Gondwanaland flood basalts were derived from a source with a low $^{143}Nd/^{144}Nd$ ratio and a high $^{87}Sr/^{86}Sr$ ratio, features that are not a common characteristic of the post-Archaean lithosphere (Fig. 8.1).

Lower mantle plume

Finally, if we find the lack of evidence for an extensive, enriched, lithospheric reservoir beneath the Proterozoic of Antarctica disturbing and the asthenospheric melt-crustal contamination model incompatible with the available data, then we are left to consider one final aspect of global geodynamics. Many features of the geochemistry of the Ferrar Supergroup and other volcanic rocks associated with the break-up of Gondwanaland can be interpreted in terms of a recycled component similar to terrestrial sediments. The FS andesites are characterized by a depletion in Ta and Nb (Fig. 8.10), elements normally found to be depleted in arc volcanic rocks, and an enrichment in Th (Fig. 8.5) an element normally concentrated in crustal rocks. Moreover, the Sr, Nd, and Pb isotopic data for the FS coincide with the field of global terrigenous sediments as do the mantle-normalized petterns of basalts and dolerites from the Trans-Antarctic Mountains and Tasmania (Hergt 1987; Hergt *et al.* 1989). In particular, Sr and Nd isotope data (Fig. 8.12) for the Ferrar Supergroup are very similar to that recently reported for Pacific pelagic sediments (Woodhead 1989) (i.e. $^{87}Sr/^{86}Sr = 0.7080$–0.7090; $^{143}Nd/^{144}Nd = 0.51235$–$0.51245$). Finally, on all elemental and isotopic plots the low $^{87}Sr/^{86}Sr$ end of linear relationships consistently has a value of $(^{87}Sr/^{86}Sr)_O = 0.7095$–$0.710$ and Sr contents of 100 and 108 p.p.m. (Fig. 8.11). These chemical characteristics are similar to melts produced in source regions (e.g. EM2) that comprise a mix of depleted mantle (DMM) and recycled components. Whilst it may be convenient to argue that such a recycled component is most likely located in accreted lithosphere, we have already noted a lack of evidence for widespread enriched lithosphere.

An alternative source for the FS may involve significant input from deep mantle sources similar to those that produce the Dupal basalts of the southern hemisphere which have a strong subduction signature and $^{87}Sr/^{86}Sr = 0.704$–0.707 (Dupre and Allegre 1983). The Dupal signature has also been reported in volcanic rocks from Brazil (Hawkesworth *et al.* 1986), South Africa (LeRoex 1986), and Australia (Ewart *et al.* 1988) and is also apparent in lithospheric xenoliths (Stolz and Davies 1988; Menzies, Chapter 4, this volume). The age of the Dupal anomaly is somewhat controversial and estimates range from the *Archaean* (Hart 1984), to the *Proterozoic* (Palacz and Saunders 1986), and even the *Phanerozoic* (Zindler and Hart 1986). The latter is valid if one accepts that the strong subduction zone signature in the Dupal basalts may be due to Pangaean subduction. Similarly, the location of the Dupal 'reservoir' is as controversial as its age. Hawkesworth *et al.* (1986) and LeRoex (1986) believe in a lithospheric location for Dupal as it has been reported in kimberlites and continental flood basalts. This is in marked contrast to a

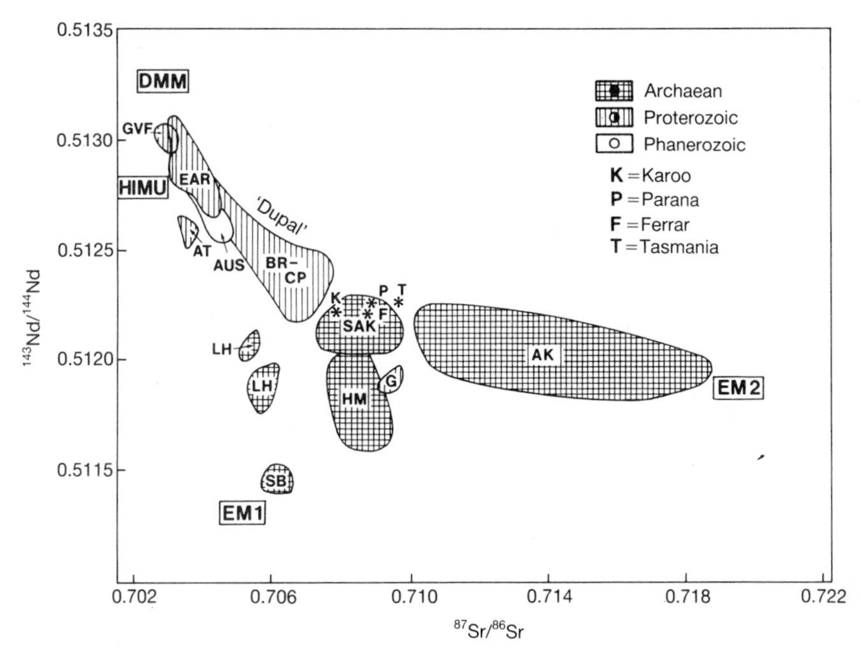

Fig. 8.12. Sr and Nd isotopic composition of volcanic rocks from the Ferrar, Parana, and Karoo flood basalt suites and the Tasmanian dolerites. Shown for comparison is the range in Sr and Nd isotopes for volcanic rocks erupted through Archaean, Proterozoic, and Phanerozoic crust. In a very general sense the volcanic rocks erupted through Archaean lithosphere have 'cratonic' isotopic ratios while those erupted through post-Archaean (Proterozoic and Phanerozoic) lithosphere have 'oceanic' isotopic ratios. Note that the lithospheric xenolith data summarized in Fig. 8.1 show a similar broad subdivision. The similarity in isotopic composition of the Ferrar, Parana, and Karoo flood basalt suites and the Tasmanian dolerites is believed to be due to the possible involvement of a recycled component in their genesis (e.g. Dupal), a deep mantle component that would account for the presence in these rocks of enriched isotopic ratios (^{87}Sr/^{86}Sr = 0.707) and a subduction signature similar to global terrigenous sediments or Pacific pelagic sediments (Woodhead 1989). Key to abbreviations and data sources: AT, Ataq, S. Yemen alkali basalts (Menzies and Murthy (1980); AK, Australian kimberlites (Fraser *et al.* 1985); AUS, Australian alkaline volcanic rocks (Menzies and Wass 1983); BR-CP, Basin and Range and Colorado Plateau of the western USA alkaline volcanic rocks (Menzies 1990 and references therein); EAR, East African Rift valley alkaline volcanic rocks (Vollmer and Norry 1978; Norry *et al.* 1980); F, Ferrar (Kyle *et al.* 1990); G, Gaussberg potassic volcanic rocks (Collerson and McCulloch 1983); GVF, Geronimo, Arizona basanites (Menzies *et al.* 1985); HM, Highwood Mountains, Montana (O'Brien *et al.* 1988); K, Karoo, South Africa (Erlank *et al.* 1984); LH, Leucite Hills, Wyoming (Vollmer *et al.* 1984); P, Parana, Brazil (Hawkesworth *et al.* 1988); SAK, South African kimberlites (Smith 1983); SB, Smoky Butte, Montana (Fraser *et al.* 1985); T, Tasmanian dolerite (Hergt *et al.* 1989).

lower mantle location as suggested by Castillo (1988). Perhaps both are correct in that the Dupal anomaly may originate from a deep mantle source and, during transfer to the surface the plumes or hot spots, impart a Dupal signature on the lithospheric mantle by processes of plume–lithosphere contamination. The geochemistry of the Dupal volcanic rocks may involve recycling of a crustal component thus explaining their position on an Sr–Nd isotope diagram between DMM and EM2 (Fig. 8.12). A similar model involving mixing of depleted mantle and a component similar to post-Archaean terrestrial shale ($\varepsilon_{Nd} = -16.5$ to -20; ^{87}Sr/^{86}Sr = 0.723–0.725) best accounts for the geochemistry of the Tasmanian dolerites (Hergt *et al.* 1989). It remains to be seen, however, whether this mixing occurred in the shallow mantle during Phanerozoic orogenesis prior to Jurassic volcanism or at depth in the lower mantle as part of a global recycling system (i.e. Dupal).

If we accept that the southern hemisphere Dupal anomaly is not lithospheric in origin and that it may have been a significant contributor to cfb volcanism, then its location and origin is perhaps better constrained by consideration of global geophysics. The recent demonstration (Hager and Gurnis 1987; Silver *et al.* 1988) of correlations between the geophysical and geochemical properties of the Earth and models involving subduction of oceanic lithosphere into the lower mantle (Creager and Jordan 1984) indicate that global mechanisms may be responsible for the development of the Dupal anomaly. The majority of Dupal hot spots exist within a region of major upwelling within the Earth, a region that covers the southern Pacific, Africa, and the eastern Atlantic. This upwelling region is characterized by low seismic velocities and is perhaps fuelled by areas of downwelling of cold slab material around the margins of the Pacific Ocean. Since geoid highs and lows mimic the late Phanerozoic plate tectonic pattern of the Earth for the last 200 Ma (the period within which the cfb and Dupal OIB were erupted), one can speculate that the existence of the Dupal or EM2 chemical anomaly beneath both the oceanic (Dupré and Allegré 1983) and continental plates (Ewart *et al.* 1988; Stolz and Davies 1988) in the southern hemisphere may be due to Phanerozoic recycling. In accordance with the ideas of Silver *et al.* (1988), global recycling of lithosphere would provide a subduction signature, with the characteristics of global terrigenous materials in areas of low seismic velocity (geoid highs).

We suggest that Dupal hot spots were parental ($^{87}Sr/^{86}Sr = 0.704$–0.707) to the southern hemisphere flood basalts, in particular the Ferrar Supergroup volcanism. A Dupal plume–lithosphere model for the Ferrar Supergroup not only provides a regionally acceptable source ($^{87}Sr/^{86}Sr = 0.705$–0.707), it also provides a subduction signature within that enriched source whose ultimate origin may be linked to global models of downwelling (cold slab) and upwelling (hot plume) in the Earth (Silver *et al.* 1988). The Dupal anomaly not only coincides with regions of low seismic velocity but also underlies areas of continental flood volcanism (e.g. Parana, Karoo). The source similarity between these provinces (Fig. 8.12) may argue for a regional sublithospheric reservoir that also feeds the ocean basins rather than a shallow lithospheric reservoir that provides for selected continental basalts and has to be delaminated to account for its occurrence in the ocean basalts. The volcanism of Ferrar–Parana–Karoo may indeed have resulted from Dupal–lithosphere interaction.

8.3. CONCLUDING STATEMENT

Crustal contamination undoubtedly occurs in the evolution of continental volcanic rock suites but it is scientifically imprudent to assume that all continental volcanic rocks are a 'lost cause' with regard to unravelling the complexities of the Earth's upper mantle. Continental intraplate magmas can be very effectively used in conjunction with entrained crustal and mantle xenoliths as a first approximation at chemical tomography providing us with information about crustal domains at < 50 km, lithospheric mantle domains at 50–150 km, and the nature of the asthenosphere at > 150 km. Domain systematics correlate with physical parameters like crustal thickness, age, and heat flow. Continental flood basalt provinces are much more complicated systems that can retain evidence of polybaric fractional crystallization and crustal contamination in the lithosphere and information regarding the nature of the sublithospheric plume that may have triggered that particular episode of continental volcanism. Careful consideration of the evidence for the presence of enriched lithospheric mantle (Menzies, Chapter 4, this volume) and its lateral and vertical continuity makes it rather unlikely that extensive enough reservoirs existed in the post-Archaean lithospheric mantle to produce flood basalts. We must exercise caution in assigning flood basalt sources to the lithosphere merely because there is no apparent evidence for crustal contamination. Global recycling and plume–lithosphere interaction appears to be a more appropriate 'driving' mechanism for cfb volcanism, particularly when one considers the close correlation between geophysical and geochemical features in the Earth.

REFERENCES

Borg, S. G., Stump, E., Chappell, B. W., McCulloch, M. T., Wyborn, D., Armstrong, R. L., and Holloway, J. R. (1987). Granitoids of northern Victoria Land; implications of chemical and isotopic variations to regional crustal structure and tectonics. *American Journal of Science*, **287**, 127–69.

Boyd, F. R. (1989). Compositional distinction between oceanic and cratonic lithosphere. *Earth and Planetary Science Letters*, **96**, 15–26.

Boyd, F. R., Gurney, J. J., and Richardson, S. H. (1985). Evidence for a 150–200 km thick Archaean lithosphere from diamond inclusion thermobarometry. *Nature*, **315**, 387–9.

Cameron, K. L., Nimz, G. J., Kuentz, D., Niemeyer, S., and Gunn, S. (1989). Southern Cordilleran basaltic andesite suite, southern Chihuahua, Mexico: a link between Tertiary continental arc and flood basalt magmatism in north America. *Journal of Geophysical Research*, **94**, 7817–40.

Carden, J. R. and Laughlin, A. W. (1974). Petrochemical variations within the McCarty's basalt flow, Valencia, County, New Mexico. *Geological Society of America, Bulletin*, **85**, 1479–84.

Carlson, R. W. and Hart, W. K. (1988). Flood basalt volcanism in the northwestern United States. In *Continental flood basalts* (ed. J. D. MacDougall), pp. 35–61. Kluwer Academic Publishers, Boston.

Castillo, P. (1988). The Dupal anomaly as a trace of the upwelling lower mantle. *Nature*, **336**, 667–70.

Chase, C. G. (1979). Subduction, the geoid and lower mantle convection. *Nature*, **282**, 464–8.

Clague, D. A. and Frey, F. A. (1982). Petrology and trace element geochemistry of the Honolulu Volcanics Oahu: implications for the oceanic mantle below Hawaii. *Journal of Petrology*, **23**, 447–504.

Collerson, K. D. and McCulloch, M. T. (1983). Nd and Sr isotope geochemistry of leucite-bearing lavas from Gaussberg, East Antarctica. *Proceedings of the Fourth International Symposium on Antarctic Earth Sciences* (ed. R. L. Oliver, P. R. James, and J. B. Jago), pp. 676–80. Cambridge University Press and Australian Academy of Science.

Compston, W., McDougall, I., and Heier, K. S. (1968). Geochemical comparison of the Mesozoic basaltic rocks of Antarctica, South Africa, South America and Tasmania. *Geochimica et Cosmochimica Acta*, **32**, 129–49.

Cox, K. G. (1988). The Karoo Province. In *Continental flood basalts* (ed. J. D. MacDougall), pp. 239–71. Kluwer Academic Publishers, Boston.

Creager, K. C. and Jordan, T. H. (1984). Slab penetration in the lower mantle. *Journal of Geophysical Research*, **89**, 3031–49.

Dickin, A. P. (1988). The North Atlantic Tertiary Province. In *Continental flood basalts* (ed. J. D. MacDougall), pp. 111–49. Kluwer Academic Publishers, Boston.

Duncan, R. A. (1978). Geochronology of basalts from the Ninetyeast Ridge and continental dispersion in the eastern Indian Ocean. *Journal of Volcanology and Geothermal Research*, **4**, 283–305.

Dupré, B. and Allegré, C. J. (1983). Pb–Sr isotope variation in Indian Ocean basalts and mixing phenomena. *Nature*, **303**, 142–6.

Ellam, R. M. (1986). The transition from calc-alkaline to potassic volcanism in the Aeolian Islands, southern Italy. Unpublished D.Phil. thesis, The Open University.

Erlank, A. J., Marsh, J. S., Duncan, A. R., Miller, R.McG., Hawkesworth, C. J., Betton, P. J., and Rex, D. C. (1984). Geochemistry and petrogenesis of the Etendeka volcanic rocks from south west Africa/Namibia. *Special Publication of the Geological Society*, **13**, 195–245.

Ernst, W. G. (1988). Metamorphic terranes, isotopic provinces, and implications for crustal growth of the western United States. *Journal of Geophysical Research*, **93**, 7634–42.

Ewart, A., Chappell, B. W., and Menzies, M. A. (1988). An overview of the geochemical and isotopic characteristics of the eastern Australian Cainozoic volcanic provinces. In *Oceanic and continental lithosphere: similarities and differences* (ed. M. A. Menzies and K. G. Cox), pp. 225–74. Oxford University Press, Oxford.

Falloon, T. J. and Green, D. H. (1988). Anhydrous partial melting and peridotite from 8 to 35 kb and the petrogenesis of MORB. In *Oceanic and continental lithosphere: similarities and differences* (ed. M. A. Menzies and K. G. Cox), pp. 379–414. Oxford University Press, Oxford.

Farmer, G. L., Perry, F. V., Semken, S., Crowe, B., Curtis, D., and DePaolo, D. J. (1989). Isotopic evidence on the structure and origin of subcontinental lithospheric mantle in southern Nevada. *Journal of Geophysical Research*, **94**, 7885–98.

Faure, G., Bowman, J. R., Elliot, D. H., and Jones, L. H. (1974). Strontium isotope composition and petrogenesis of the Kirkpatrick Basalt, Queen Alexandra Range, Antarctica. *Contributions to Mineralogy and Petrology*, **48**, 153–69.

Faure, G., Hill, R. L., Jones, L. M., and Elliot, D. H. (1972). Isotope composition of strontium and silica content of Mesozoic basalt and dolerite from Antarctica. In *Antarctica geology and Geophysics* (ed. R. H. Adiie), pp. 617–24. Oslo Universitetsforlaget, Oslo.

Fitton, J. G. and Dunlop, H. (1985). The Cameroon Line, West Africa and its bearing on the origin of oceanic and continental alkali basalt. *Earth and Planetary Science Letters*, **72**, 23–38.

Ford, A. B., Kistler, R. W., and White, L. D. (1986). Strontium and oxygen-isotope study of the Dufek intrusion. *Antarctic Journal of the United States*, **21**, 63–6.

Fraser, K. J., Hawkesworth, C. J., Erlank, A. J., Mitchell, R. H., and Scott-Smith, B. H. (1985). Sr-, Nd-, and Pb-isotope and minor element geochemistry of lamproites and kimberlites. *Earth and Planetary Science Letters*, **76,** 57–70.

Frey, F. A. and Roden, M. F. (1987). The mantle source for the Hawaiian Islands: constraints from the lavas and ultramafic inclusions. In *Mantle metasomatism* (ed. M. A. Menzies and C. J. Hawkesworth), pp. 423–60. Academic Press, London.

Galer, S. J. G. and O'Nions, R. K. (1986). Magmagenesis and the mapping of chemical and isotopic variations in the mantle. *Chemical Geology*, **56,** 45–61.

Galer, S. J. G. and O'Nions, R. K. (1989). Chemical and isotopic studies of ultramafic inclusions from the San Carlos volcanic field, Arizona. Bearing on their petrogenesis. *Journal of Petrology*, **30,** 1033–64.

Gamble, J. A., McGibbon, F., Kyle, P. R., Menzies, M. A., and Kirsch, I. (1988). Metasomatised xenoliths from Foster Crater, Antarctica: Implications for lithospheric structure and processes beneath the Transantarctic Montain Front. In *Oceanic and continental lithosphere: similarities and differences* (ed. M. A. Menzies and K. G. Cox), pp. 109–38. Oxford University Press, Oxford.

Gill, J. B. (1981). *Orogenic andesites and plate tectonics*. Springer-Verlag, Berlin.

Hager, B. H. and Gurnis, M. (1987). Mantle convection and the state of the earth's interior. *Reviews of Geophysics*, **25,** 1277–85.

Hart, S. R. (1984). A large-scale isotope anomaly in the Southern hemisphere mantle. *Nature*, **309,** 753–7.

Hawkesworth, C. J., Mantovani, M., Taylor, P. N., and Palacz, Z. (1986). Evidence from the Parana of south Brazil for a continental contribution to Dupal basalts. *Nature*, **322,** 356–9.

Hawkesworth, C. J., Mantovani, M., and Peate, D. (1988). Lithosphere remobilization during Parana CFB magmatism. In *Oceanic and continental lithosphere: similarities and differences* (ed. M. A. Menzies and K. G. Cox), pp. 205–24. Oxford University Press, Oxford.

Hegner, E. and Pallister, J. S. (1989). Pb, Sr and Nd isotopic characteristics of Tertiary Red Sea rift Volcanics from the Central Saudi Arabian Coastal Plain. *Journal of Geophysical Research*, **94,** 7749–56.

Hergt, J. M. (1987). The origin and evolution of the Tasmanian dolerites. Unpublished D.Phil. thesis, Australian National University.

Hergt, J. M., Chappell, B. W., McCulloch, M. T., McDougall, I., and Chivas, A. R. (1989). Geochemical and isotopic constraints on the origin of the Jurassic dolerites of Tasmania. *Journal of Petrology*, **30,** 841–84.

Hoefs, J., Faure, G., and Elliot, D. H. (1980). Correlation of $\delta^{18}O$-initial $^{87}Sr/^{86}Sr$ ratios in Kirkpatrick basalts on Mt. Falla, Transantarctic Mountain. *Contributions to Mineralogy and Petrology*, **75,** 199–204.

Irving, A. J. (1980). Petrology and geochemistry of composite xenoliths in alkalic basalts and implications for magmatic processes within the mantle. *American Journal of Science*, **280A,** 389–426.

Kramers, J., Roddick, J., and Dawson, J. B. (1983). Trace element and isotope studies on veined, metasomatic and MARID xenoliths from Bultfontein South Africa. *Earth and Planetary Science Letters*, **65,** 90–106.

Kyle, P. R. (1980). Development of heterogeneities in the sub-continental mantle: evidence from the Ferrar Group Antarctica. *Contributions to Mineralogy and Petrology*, **73,** 89–104.

Kyle, P. R., Elliot, D. H., and Sutter, J. F. (1981). Jurassic Ferrar Supergroup tholeiites from the Transantarctic Mountains, Antarctica and their relationship to the initial fragmentation of Gondwana. In *Gondwana five* (ed. M. M. Cresswell and P. Vella), pp. 283–7. AA. A. Balkema Rotterdam.

Kyle, P. R., Pankhurst, R. J., and Bowman, J. R. (1983). Isotopic and chemical variations in Kirkpatrick Basalt group rocks from southern Victoria Land. In *Antarctic earth science* (ed. R. L. Olivier, P. R. James, and I. B. Jago), pp. 234–7. Australian Academy of Sciences.

Kyle, P. R., Pankhurst, R., Bowman, R. J., and Menzies, M. A. (1990). Petrogenesis of the Ferrar volcanic rocks Antarctica. Preprint available from Department of Geosciences, N.M.I.M.T., Socorro, New Mexico.

Latin, D. M., Dixon, J. E., Fitton, J. G., and White, N. (1990). Mesozoic magmatic activity in the North Sea Basin: implications for stretching history. In *Tectonic movements responsible for Britain's oil and gas reserves* (ed. R. F. P. Hardman and J. R. V. Brooks). Geological Society of London (in press).

Laughhlin, A. W., Brookins, D. G., Kudo, A. M., and Causey, J. D. (1971). Chemical and strontium isotopic investigations of ultramafic inclusions and basalt, Bandera Crater, New Mexico. *Geochimica et Cosmochimica Acta*, **35,** 107–13.

Laughlin, A. W., Brookins, D. G., and Carden, J. R. (1972*a*). Variations in the initial strontium ratios of a single basalt flow. *Earth and Planetary Science Letters*, **14,** 79–82.

Laughlin, A. W., Brookins, D. G., and Causey, J. D. (1972*b*). Late Cenozoic basalts from the Bandera lava field, Valencia County, New Mexico. *Geological Society of America, Bulletin*, **83,** 1443–552.

Leat, P. T., Thompson, R. N., Morrison, M. A., Hendry, G. L., and Dickin, A. P. (1988). Compositionally diverse Miocene to Recent rift-related magmatism in N.W. Colorado: partial melting and mixing of mafic magmas from three different asthenospheric and lithospheric mantle sources. In *Oceanic and continental lithosphere: similarities and differences* (ed. M. A. Menzies and K. G. Cox), pp. 351–78. Oxford University Press, Oxford.

Leeman, W. P. (1982). Tectonic and magmatic significance of strontium isotopic variations in Cenozoic volcanic rocks from the western United States. *Geological Society of America, Bulletin*, **93,** 487–503.

Leeman, W. P. and Fitton, J. G. (eds) (1989). Magmatism and lithospheric extension. *Journal of Geophysical Research*, **94,** 7682–986.

LeRoex, A. P. (1986). Geochemical correlation between southern African kimberlites and South Atlantic hotspots. *Nature*, **324,** 243–5.

LeRoex, A. P. (1987). Source regions of mid-ocean ridge basalts: evidence for enrichment processes. In *Mantle metasomatism* (ed. M. A. Menzies and C. J. Hawkesworth), pp. 389–422. Academic Press, London.

Luhr, J. F., Aranda-Gomez, J. J., and Pier, J. G. (1989). Spinel lherzolite bearing Quaternary volcanic centers in San Luis Potosi Mexico. 1, Geology, Mineralogy and Petrology. *Journal of Geophysical Research*, **94,** 7916–40.

Lum, C. C. L., Leeman, W. P., Foland, K. A., Kargel, J. A., and Fitton, J. G. (1989). Isotopic variations in continental basaltic lavas as indicators of mantle heterogeneity: examples from the western U.S. Cordillera. *Journal of Geophysical Research*, **94,** 7871–84.

MacDougall, J. D. (ed.) (1988). *Continental flood basalts*. Kluwer Academic Publishers, Boston.

Mahoney, J. J. (1988). Deccan traps. In *Continental flood basalts* (ed. J. D. MacDougall), pp. 151–94. Kluwer Academic Publishers, Boston.

Maxwell, C. H. (1986). Geologic map of El Malpais lava field and surrounding areas, Cibola County, New Mexico. *U.S. Geological Survey, miscellaneous investigation map*, I-1595.

McGibbon, F. G., Hawkesworth, C. J., Menzies, M. A., and Kyle, P. (1987*a*). Geochemistry of ultramafic xenoliths from McMurdo Sound, Antarctica: implications for Jurassic Ferrar magmatism. *Terra Cognita*, **7,** 396.

McGibbon, F. G. (1990). Crust–mantle evolution across the eastern margin of the Ross Sea. *Proceedings of the Fifth International Symposium on Antarctic Earth Sciences* (ed. M. R. A. Thomson, J. A. Crame, and J. W. Thomson. Cambridge University Press, Cambridge (in press).

McKenzie, D. (1989). Some remarks on the movement of small melt fractions in the mantle. *Earth and Planetary Science Letters*, **95,** 53–72.

McKenzie, D. and Bickle, M. J. (1988). The volume and composition of melt generated by extension of the lithosphere. *Journal of Petrology*, **29,** 625–79.

McMillan, N. J. and Dungan, M. A. (1988). Open system magmatic evolution of the Taos Plateau Volcanic Field, northern New Mexico: 3 petrology and geochemistry of andesite and dacite. *Journal of Petrology*, **29,** 527–58.

Mensing, T. M., Faure, G., Jones, L. M., Bowman, J. R., and Hoefs, J. (1984). Petrogenesis of the Kirkpatrick basalt, Solo Nunitak, Northern Victoria Land, Antarctica, based on isotopic compositions of strontium, oxygen and sulfur. *Contributions to Mineralogy and Petrology*, **87,** 101–8.

Menzies, M. A. (1989). Cratonic, circum-cratonic and oceanic mantle domains beneath the western U.S.A. *Journal of Geophysical Research*, **94,** 7899–915.

Menzies, M. A. and Wass, S. Y. (1983). CO_2 and LREE-rich mantle below eastern Australia: a REE and isotopic study of Australian magmas and apatite-rich mantle xenoliths from the Southern Highlands Province, Australia. *Earth and Planetary Science Letters*, **65,** 287–302.

Menzies, M. A., Leeman, W. P., and Hawkesworth, C. J. (1983). Isotope geochemistry of Cenozoic volcanic rocks reveals mantle heterogeneity below western U.S.A. *Nature*, **303,** 205–9.

Menzies, M. A., Leeman, W. P., and Hawkesworth, C. J. (1984). Geochemical and isotopic evidence for the origin of continental flood basalts with particular reference to the Snake River Plain and Columbia River. *Philosophical Transactions of the Royal Society, London*, **A310,** 643–60.

Menzies, M. A., Kempton, P. D., and Dungan, M. (1985). Interaction of continental lithosphere and asthenospheric melts below the Geronimo volcanic field Arizona, U.S.A. *Journal of Petrology*, **26,** 663–93.

Moorbath, S. and Thompson, R. N. (1980). Strontium isotope geochemistry and petrogenesis of the early Tertiary lava pile of the Isle of Skye, Scotland, and other basic rocks of the British Tertiary Province: an example of magma–crust interaction. *Journal of Petrology*, **21,** 295–321.

Navon, O., Hutcheon, I. D., Rossman, G. R., and Wasserburg, G. J. (1988). Mantle-derived fluids in diamond micro-inclusions. *Nature*, **335**, 784–9.

Norry, M. J., Trackle, R. H., Lippard, S. J., Hawkesworth, C. J., Weaver, S. D., and Marriner, G. F. (1980). Isotopic and trace element evidence from lavas bearing on mantle heterogeneity beneath Kenya. *Philosophical Transactions of the Royal Society, London*, **A297**, 259–71.

O'Brien, H. E., Irving, A. J., McCallum, I. S., and Thirlwall, M. F. T. (1988). Characterization of source components of potassic mafic magmas of the Highwood Mountains Province, Montana. *Transactions of the American Geophysical Union*, **9**, 519.

Olsen, K. H., Baldridge, W. S., and Callender, J. F. (1987). Rio Grande rift: an overview. *Tectonophysics*, **143**, 110–39.

Ormerod, D. S., Hawkesworth, C. J., Leeman, W. P., and Menzies, M. A. (1988). The identification of subduction-related and within-plate components in basalts from the western U.S.A. *Nature*, **333**, 349–53.

Palacz, Z. A. and Saunders, A. D. (1986). Coupled trace element and isotope enrichment in the Cook–Austral–Samoa islands, southwest Pacific. *Earth and Planetary Science Letters*, **79**, 270–80.

Pearce, J. A. (1983). Role of sub-continental lithosphere in magma genesis at active continental margins. In *Continental basalts and mantle xenoliths* (ed. C. J. Hawkesworth and M. J. Norry), pp. 230–49. Shiva Publishers, England.

Perry, F. V., Baldridge, W. S., and DePaolo, D. P. (1987). Role of asthenosphere and lithosphere in the genesis of late Cenozoic basaltic rocks from the Rio Grande rift and adjacent region sof the southwestern United States. *Journal of Geophysical Research*, **92**, 9193–213.

Pier, J. G., Podosek, F. A., Luhr, J. F., Brannon, J. C., and Aranda-Gomez, J. J. (1989). Spinel lherzolite bearing Quaternary volcanic centres in San Luis Potosi, Mexico. 2, Sr and Nd isotopic systematics. *Journal of Geophysical Research*, **94**, 7941–51.

Richardson, S. H., Gurney, J. J., Erlank, A. J., and Harris, J. W. (1984). Origin of diamonds in old enriched mantle. *Nature*, **310**, 198–202.

Richter, F. M. (1988). A major change in the thermal state of the earth at the Archean–Proterozoic boundary: consequences for the nature and preservation of continental lithosphere. In *Oceanic and continental lithosphere: similarities and differences* (ed. M. A. Menzies and K. G. Cox), pp. 39–52. Oxford University Press, Oxford.

Silver, P. G., Carlson, R. W., and Olson, P. (1988). Deep slabs, geochemical heterogeneity and the large-scale structure of mantle convection. *Annual Reviews of Earth and Planetary Science*, 477–541.

Smith, C. B. (1983). Pb, Sr and Nd isotopic evidence for sources of southern African Cretaceous kimberlites. *Nature*, **304**, 51–4.

Stolz, A. J. and Davies, G. R. (1988). Chemical and isotopic evidence from spinel lherzolite xenoliths for episodic metasomatism of the upper mantle beneath southeastern Australia. In *Oceanic and continental lithosphere: similarities and differences* (ed. M. A. Menzies and K. G. Cox), pp. 303–12. Oxford University Press, Oxford.

Storey, M., Saunders, A. D., Tarney, J., Leat, P., Thirwall, M. F., Thompson, R. N., Menzies, M. A., and Marriner, G. F. (1988). Geochemical evidence for plume–mantle interactions beneath Kerguelen and Heard Islands, Indian Ocean. *Nature*, **336**, 371–4.

Thompson, R. N. and Morrison, M. A. (1988). Asthenospheric and lower-lithospheric mantle contributions to continental extensional magmatism: an example from the British Tertiary Province . *Chemical Geology*, **68**, 1–15.

Thompson, R. N., Morrison, M. A., Dickin, A. P., and Hendry, G. L. (1983). Continental flood basalts . . . arachnids rule O.K.? In *Continental basalts and mantle xenoliths* (ed. C. J. Hawkesworth and M. J. Norry), pp. 158–85. Shiva Publishers, England.

VanKooten, G. K. (1980). Mineralogy, petrology and geochemistry of an ultrapotassic basaltic suite, central Sierra Nevada, California. *Journal of Petrology*, **21**, 651–84.

Vollmer, R. and Norry, M. J. (1978). Possible origin of K-rich volcanic rocks from Virunga, East Africa by metasomatism of continental crustal material: Pb, Nd and Sr isotopic evidence. *Earth and Planetary Science Letters*, **64**, 374–86.

Vollmer, R., Ogden, P., Schilling, J. G., Kingsley, R. H., and Waggoner, D. G. (1984). Nd and Sr isotopes in ultrapotassic volcanic rocks from the Leucite Hills, Wyoming. *Contributions to Mineralogy and Petrology*, **87**, 359–68.

Waters, F. G. (1987). A suggested origin of MARID xenoliths in kimberlites by high pressure crystallisation of an ultrapotassic rock such as lamproite. *Contributions to Mineralogy and Petrology*, **95**, 523–33.

Watson, E. B. (1982). Basalt contamination by continental crust: some experiments and models. *Contributions to Mineralogy and Petrology*, **80,** 73–87.
Watson, E. B. and Jurewicz, S. R. (1985). Behaviour of alkalies during diffusive interaction of granitic xenoliths with basaltic magma. *Journal of Geology*, **92,** 121–31.
Watts, A. B. and Cox, K. G. (1989). The Deccan Traps: an interpretation in terms of progressive lithospheric flexure in responses to a migrating load. *Earth and Planetary Science Letters*, **93,** 85–97.
White, R. and MacKenzie, D. (1989). Magmatism at Rift Zones: the generation of volcanic continental margins and flood basalts. *Journal of Geophysical Research*, **94,** 7685–730.
Woodhead, J. (1989). Geochemistry of the Mariana Arc (West Pacific): source composition and processes. *Chemical Geology*, **76,** 1–24.
Zhou, X. H., Zhu, B. Q., Liu, R. X., and Chen, W. J. (1988). Cenozoic basaltic rocks in Eastern China. In *Continental flood basalts* (ed. J. D. MacDougall), pp. 311–30. Kluwer Academic Publishers, Boston.
Zindler, A. and Hart, S. R. (1986). Chemical geodynamics. *Annual Reviews of Earth and Planetary Science*, 493–571.

Index